EMERGING DEVICES FOR LOW-POWER AND HIGH-PERFORMANCE NANOSYSTEMS

Pan Stanford Series on Intelligent Nanosystems

Series Editor

Simon Deleonibus

Titles in the Series

Published

Vol. 1
Intelligent Integrated Systems: Devices, Technologies, and Architectures
Simon Deleonibus, ed.
2014
978-981-4411-42-4 (Hardcover)
978-981-4411-43-1 (eBook)

Vol. 2
Integrated Nanodevice and Nanosystem Fabrication: Materials, Techniques, and New Opportunities
Simon Deleonibus, ed.
2017
978-981-4774-22-2 (Hardcover)
978-1-315-18125-7 (eBook)

Vol. 3
Emerging Devices for Low-Power and High-Performance Nanosystems: Physics, Novel Functions, and Data Processing
Simon Deleonibus, ed.
2018
978-981-4800-11-2 (Hardcover)
978-0-429-45873-6 (eBook)

Forthcoming

Vol. 4
Convergence of More Moore, More Than Moore, and Beyond Moore: Materials, Devices, and Nanosystems
Simon Deleonibus, ed.

Pan Stanford Series on Intelligent Nanosystems Volume 3

EMERGING DEVICES FOR LOW-POWER AND HIGH-PERFORMANCE NANOSYSTEMS

Physics, Novel Functions, and Data Processing

edited by
Simon Deleonibus

PAN STANFORD PUBLISHING

Published by

Pan Stanford Publishing Pte. Ltd.
Penthouse Level, Suntec Tower 3
8 Temasek Boulevard
Singapore 038988

Email: editorial@panstanford.com
Web: www.panstanford.com

British Library Cataloguing-in-Publication Data
A catalogue record for this book is available from the British Library.

Emerging Devices for Low-Power and High-Performance Nanosystems: Physics, Novel Functions, and Data Processing

ISBN 978-981-4800-11-2 (Hardcover)
ISBN 978-0-429-45873-6 (eBook)

Contents

Part II
Revised Actuation, Sensing, and Data Storage Modes in Emerging Devices for Alternative Computing

Preface

A few years ago, Pan Stanford Publishing (Singapore) invited me to edit the Pan Stanford Series on Intelligent Nanosystems. The series is now quite successful and attracts many internationally well-known researchers as authors on today's hottest topics of nanoelectronics and nanosystems. The notion of system has been evolving in the past and will continue to evolve throughout the future history of technologies. In the field of nanoelectronics and nanosystems, the down-scaling of components has made possible the integration of more and more complex functions which can today be handheld. As an example, the telephone was a complex technology which was possessed by few people until the middle of the 20th century and, in most cases, a fixed home or office appliance. Nowadays, one can do so many things other than just placing a call with a mobile phone! Moreover, every human being on our planet is susceptible to own a mobile phone which delivers access to information, services, and leisure applications with multimodal communication. Challenges are ahead of us to meet the growing demand for energy-efficient and sustainable technologies not necessarily available with today's current schemes. In such a context, alternative devices and approaches are benchmarked for sensing, actuating, data storage, and processing. This evolution already requests new technology integration paradigms, which leave the floor to integrated intelligent nanosystems to match low power and high performance by either hybrid or heterogeneous CMOS in 2D/3D and give opportunities to emerging devices. This is what the series is aimed at highlighting.

The first volume, published in April 2014, contains 13 chapters and is dedicated to *Intelligent Integrated Systems: Devices, Technologies, and Architectures.* The second volume, published in November 2017, contains 8 chapters and is dedicated to *Integrated*

Nanodevice and Nanosystem Fabrication: Materials, Techniques, and New Opportunities.

The current volume, the third, contains 10 chapters and is titled *Emerging Devices for Low-Power and High-Performance Nanosystems: Physics, Novel Functions, and Data Processing*. Following the introduction, the book is divided into two parts. Part 1 (Chapters 1–4), titled Hybrid and Heterogeneous CMOS for Ultra-Low-Power Data Processing, reports on junctionless transistors, III–V/Si heterostructure–based tunnel field-effect transistors, NEMS switches for logic and memory, and NEMS switches for ultra-low-power adiabatic architecture design. Part 2 (Chapters 5–9), titled Revised Actuation, Sensing, and Data Storage Modes in Emerging Devices for Alternative Computing, reviews single-spin quantum bit manipulation in pMOSFET or electron-based coupled quantum dots both on silicon, metal/insulator/metal, or metal/phase change material/metal-based memristor potential to dual store or process data, bio-inspired neuromorphic 1 transistor-1 resistor circuit architectures, and nanomagnetic logic architectures based on multilayered nanomagnets exchanging bias vertically and horizontally.

This book and the series are intended for graduate students, educators, researchers, and engineers. I hope to give the readers the opportunity to enjoy the highest-level tutorial reports on the most advanced research results, written by worldwide famous authors. I am very happy to present this book, which, I am sure, will be useful to boost their own research or simply their curiosity and knowledge.

Simon Deleonibus
Former Chief Scientist, CEA–LETI
Fellow, IEEE
Fellow, The Electrochemical Society
CEA Research Director

simonedit2016@gmail.com

Acknowledgments

I wish to congratulate all co-authors of the nine chapters of *Emerging Devices for Low-Power and High-Performance Nanosystems: Physics, Novel Functions, and Data Processing*, the third volume published of the Pan Stanford Series on Intelligent Nanosystems, for the results of the invaluable quality of their research, as well as for their excellent contributions, including a review of the state of the art in their respective fields. They are world-renowned researchers and professionals belonging to top-level universities, institutes, or companies. Their determination and energy are of utmost importance to the success of this ambitious ongoing project.

It is a great honor to receive support from Prof. Leon Chua, University of California at Berkeley (CA, USA), Prof. Michiharu Tabe, Shizuoka University (Japan), and Prof. Mark Lundstrom, Purdue University (IN, USA), in promoting this book. Their influence in the Nanoelectronics, Nanotechnology and Nanoscience world underlines the high scientific level of the different contributions.

I have special thanks to address to Mr. Stanford Chong, Ms. Jenny Rompas, Mr. Sarabjeet Garcha, and the staff of Pan Stanford Publishing for their availability throughout the whole process from defining the book's content to publishing and promoting its final edition. Their trust and enthusiasm throughout the long process of the publication of the series is also essential to its success.

Finally, the constant support and patience of my wife Geneviève and my son Tristan, as well as encouragement from my friends, while I take care of the series, are certainly of prime importance to me. I dedicate this work to them.

Simon Deleonibus

Introduction

Cramming More Functions into Integrated Systems toward a Sustainable Information Technology World

Simon Deleonibus
Past Chief Scientist, Université Grenoble Alpes, CEA, LETI, MINATEC Campus, Grenoble, France
simonedit2016@gmail.com

The history of information and communication technologies (ICTs) has been paved by both evolutive paths and challenging alternatives, so-called emerging devices and architectures. Their introduction demands sole consideration or consideration in combination of new state variables, information processing paradigms, and process integration schemes in 2D, 3D, or above integrated circuit (above IC). This book reviews in two parts (nine chapters in all) the capabilities of integrated nanosystems to match low power and high performance either by using hybrid and heterogeneous CMOS in 2D/3D or by using emerging devices for alternative sensing, actuating, data storage, and processing. Future ICTs will need to take into account their new functional opportunities, their energy efficiency, and their sustainability in the global ecosystem.

Emerging Devices for Low-Power and High-Performance Nanosystems: Physics, Novel Functions, and Data Processing
Edited by Simon Deleonibus

ISBN 978-981-4800-11-2 (Hardcover), 978-0-429-45873-6 (eBook)
www.panstanford.com

I.1 The More Integrated Sustainable Functions Era Yet to Come?

Since the foundation of the International Technology Roadmap for Semiconductors (ITRS) [1], its action has been driven by integrated circuits' progress based on complementary metal-oxide semiconductor (CMOS) technology scaling. The dominating idea was summarized in the famous 1965 paper by Gordon E. Moore on the increase of the number of components integrated on a chip [2]. The title of Moore's paper, "Cramming More Devices on a Chip," became a motto for all microelectronics engineers and scientists as a basic guideline to boost their research and development. The domains More Moore, More than Moore, and Beyond CMOS (or Beyond Moore) [1, 3] have been defined to address the different device categories in application areas and possible alternatives.

I.1.1 New Virtuous Paradigm to Proceed the Historical Trend of Breakthroughs

Increasing the level of abstraction has always guided progress in advanced technologies in the past. New challenges are relevant to the integration of pure information science outputs into a wider scope, and their pervasion in the real world. The new routes ahead of us are marked by mottos such as "Cramming More Functions in Integrated Systems," overtaking Gordon Moore's famous words of 1965 [2]. If one considers the evolution of technologies since prehistory, many examples demonstrate the capability of human beings to increase the level of abstraction each time a technological breakthrough has been introduced to be embedded or cointegrated with another technology heterogeneously. For example, the invention of flint tools preceding the invention of the hammer or the knife was possible thanks to the mastering of heterogeneous integration of wood and stone technology and, further on, thanks to the discovery and mastering of iron technology. One can make the same assumption regarding the invention of mechanical calculators followed by the inventions of electric cash registers, mainframe computers, pocket calculators, personal computers, mobile handset telephones, and smart telephones. At each breakthrough new functionalities have been added to the former technology.

A new paradigm could thus be defined as more integrated sustainable functions (MISFs). This is guided by the historical necessity to introduce breakthroughs as well as by social pressure defining new usages. Progress in information and communication technology (ICT) cannot be measured anymore only by implementing on a chip an increasing number of faster devices; it has to be measured by more diversified functions defined in a sustainable way. This new definition of progress takes into account ecoresponsibility, on both environmental and sustainability sides, overtaking the introduction of functional diversification [1, 2]. Today, technical solutions are very often proposed as optimal choices in terms of materials and fabrication processes as well as for computation and data processing. The proposed solutions that result from these choices do not take into account the availability/scarcity of materials and the potential impact on social or ecological environments. These aspects become also the researcher's responsibility because of their scientific and technical consequences. If an optimal choice cannot be adopted, how can it be replaced without any detrimental consequences for performance and expected outputs? In the field of materials science or chemistry, for example, the adopted solutions are brought for the sake of thermodynamic optimization. Envisaging a new solution that is socially acceptable, without increasing the cost, involves efforts in terms of research and development. Hopefully, in some cases, fabrication techniques that are not commonly used by a branch of the sector can be adopted by giving more credit to a new viewpoint. This is the case for catalysis [4] in the field of chemistry, which gains in popularity thanks to its performance (mass efficiency) and impact on waste management. In the field of electronics, it is timely to take such routes and start investigations into new solutions. A new roadmap needs to be built that will take into account at the same time *ecologically sustainable technological solutions and energy efficiency in ICT.*

I.1.2 Ecologically Sustainable Technological Solutions

Scarcity of materials, especially spice or noble metals, is a threat to the earth's economically exploitable reserves and, in fine, the whole nanoelectronics activity! For example, Zn, Au, In, Te, and other rare earths [5] commonly used in our electronic devices or apparatuses are candidates for severe rarefaction of known reserves by 2025. According to different sources, the more commonly used materials, such as Sn, Cu, Ta, and Pb, would pose the same type of issues by 2030–2040 and Ni by 2050. It is relevant to draw new routes for sustainable electronics. This would entail either alternative solutions inspired by less material greedy green chemistry [4] or process solutions to promote a circular economy [6]. A circular economy was traditionally seen as the "poor" people's solution to survive. Today, it has become a major tool to master sustainability in the main ecosystem [6]. Nevertheless, this should not be a good/bad opportunity to increase "programmed obsolescence," a great temptation to boost business based strictly on financial considerations.

I.1.3 Energy Efficiency in ICT

Energy efficiency has always been a major request for decision making in the ICT world. The priority for scaling digital electronics has been balanced by the urgent necessity to reduce power consumption along with increasing their speed, also designated as energy efficiency. Information processing and storage has been considered separately or in association (combination), depending on the applications.

In order to address the future of information processing, a critical assessment has been carried out, by the Micro/Nanoelectronics scientific community, through the different aspects of state variable, material science, device physics, data representation and computing architecture (see for example the Emerging Devices (ERD) ITRS chapter in the 2007 version of Ref. [1]). In this respect, the pertinent figure of merit to consider in order to make fair comparisons is the bandwidth delivered per unit of consumed energy.

As far as the CMOS is concerned, this was a difficult paradigm that the ITRS encountered at the beginning of the 2000s and it finally defined three product families for silicon-based integrated circuits (ICs): high performance (HP), low operating power (LOP), and low standby power (LSTP). The increasing popularity of mobile devices pushed for power saving but requested a HP for fast video Internet access. Thus, combining LSTP or LOP devices in systems on a chip (SOC) became a necessity.

Meanwhile, other architectures have been considered or sometimes reconsidered as alternatives to a CMOS by many research teams worldwide and the ERD Working Group of ITRS [1, 7–10]. In all the various benchmarks made since 2001, the CMOS appeared to be "not too bad" in terms of essential figures of merit comparing speed versus switching energy, size, and potential cost. Among all charge transfer devices, tunnel field-effect transistors (TFETs) demonstrate by far the best opportunity to scale down device size and applied voltage. The supply voltage is the strongest tuning parameter to reduce standby and active power dissipation. TFET would be accepted as a solution because it could be directly derived from the CMOS by changing the type of one of the current injecting electrodes (source or drain). This would be possible if, and only if, short channel effects (mainly charge sharing and drain-induced barrier lowering) are drastically reduced. On this latter point, nanowire channels with a gate all around the architecture are the ultimate way to ensure scalability [7–9, 11]. Once these architectures have been accounted for, it has been possible to consider other candidates using different state variables, such as spin, magnetization, strong electron correlation, phase, and polarization, that would potentially be much more energy efficient. The ERD group of the ITRS made this classification as early as 2007. The technological demonstration and one-to-one comparison are not always obvious due to the differences in maturity of the different technologies or their known process integration.

To date, no new Holy Grail has been reported to replace the CMOS straightforwardly once we have exhausted all the possibilities offered by the dominating digital computing–based technology. This is why the association of different technologies would also be

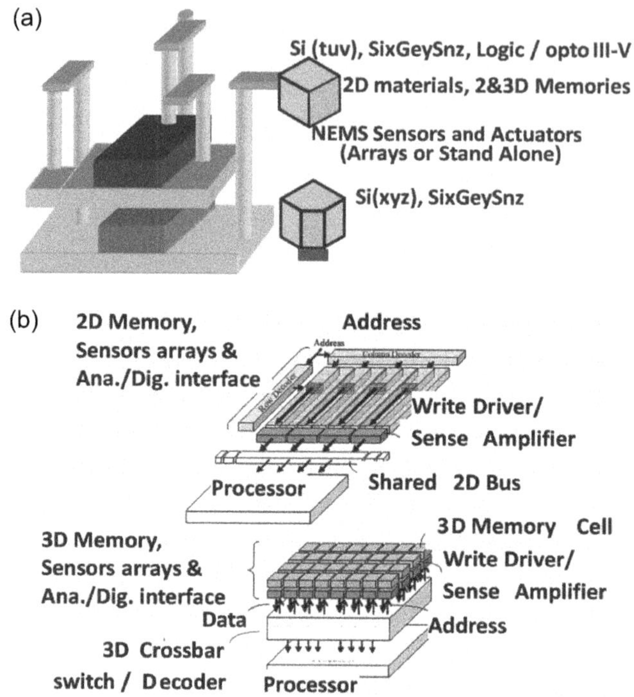

Figure I.1 (Modified and adapted from Ref. [7]) Monolithic/Sequential 3D integration makes possible heterogeneous cointegration of (a) materials and sensors, thanks to cold end processing in midprocess. Improvements on layout area and energy-delay product can be obtained on partitioned Si CMOS FPGAs. (b) High-density or 3D memories (DRAM, RERAM, RERAM/SRAM) embedded with a digital circuit can deliver more bandwidth, reduce latency, and introduce neuromimetic and programmable architectures. This scheme can be used to implement massively parallel based computing architecture necessary for quantum computing and highly resolvant sensors as well. Passive components can be cointegrated as usual, above the IC, at the backend of the line (or in package) after the sequential process.

eligible for evaluation once the priority issue of CMOS compatibility has been settled. However, very little heterogeneous cointegration of the different technologies and their applications has been achieved. Today, thanks to the reduced thermal budget, it is possible to envisage such scenarios by using high-density 3D integration (Fig. I.1) in many ways (see also Refs. [7, 8, 12–14]). Different semiconducting or metallic materials and devices using different actuation and sensing modes (electrostatic, mechanical, spin based, magnetic, RF waves, etc.) can be cointegrated with a CMOS in midprocess sequential 3D) or at the backend of the line or stacked on top of an already processed chip. New opportunities arise to bring in new applications of or add-ons to a CMOS in many ways, such as adiabatic ultra-low-power logic or sensing based on cointegration of the CMOS and mechanical resonators, spin manipulation in CMOS devices for quantum computing applications [15], and electric charge- or magnetic-based quantum cellular automata.

In this book, hybrid integration and heterogeneous integration are emphasized as eligible approaches for an improved or augmented CMOS, emerging device architecture, and enablers for alternative computing or sensing data storage paradigms.

I.2 The Third Volume of the Pan Stanford Series on Intelligent Nanosystems

Volume 3 of the Pan Stanford Series on Intelligent Nanosystems is titled *Emerging Devices for Low-Power and High-Performance Nanosystems*. Historically, emerging devices have always posed new issues regarding data processing, because of their physics and the possible new functions they might enable. That is why the subtitle of Volume 3 is *Physics, Novel Functions, and Data Processing*. Charge transfer devices have been dominating computer architectures since the early development of ICs. With the scaling of supply voltages and electrical signals, mechanisms different from charge diffusion could be eligible, such as band-to-band tunneling and ionic transport, provided no major device degradation is brought. Still, electrostatic integrity is a major concern to master their energy efficiency: the choice of device architecture and heterogeneous association of materials with other devices are interesting questions to evaluate. In the context of low applied supply voltages and electric fields, different state variables might be eligible to compute information besides electric charge, such as spin, magnetic waves, and magnetic coupling. Finally, new data processing paradigms at the device level or circuit/system level can be envisaged with the advent of new architectures offered by emerging devices: adiabatic logic, neuromorphic data storage and processing, quantum information processing, close to zero energy consumption management with nanowires, mechanical and magnetic switches, and new sensing capabilities. In all the cases exposed in this book, the authors demonstrate that the proposed solutions are compatible with cointegration or coexistence with silicon CMOS technology.

This volume features nine chapters in two parts: **Part 1**, Hybrid and Heterogeneous CMOS for Ultra-Low-Power Data Processing (Chapters 1 to 4), and **Part 2**, Revised Actuation, Sensing, and Data Storage Modes in Emerging Devices for Alternative Computing (Chapters 5 to 9).

In **Part 1**, junctionless transistors, III-V/Si heterostructure–based TFETs, nanoelectromechanical (NEM) switches for logic and memory, and NEM switches for ultra-low-power adiabatic architecture design are reviewed.

In **Chapter 1**, Jean-Pierre Colinge of CEA-LETI gives a thorough review of the junctionless transistor architecture. The transistor integration process simplification opens up new opportunities to cointegrate with Si other semiconductors, such as Ge, whether crystalline or noncrystalline, in a much more efficient way. The junctionless architecture is straightforwardly suited for 3D sequential/monolithic cold-end integration of a CMOS and sensors/actuators. The physics and applications of junctionless transistors are described, and a list of models developed to simulate their electrical characteristics is given.

In **Chapter 2**, Katsuhiro Tomioka of Hokkaido University reviews the challenges and demonstrates achievements of vertical TFETs using III-V/Si heterostructures with steep subthreshold slopes lower than 21 mV/decade. This device is a good illustration of what a hybrid CMOS could be. Process integration issues are analyzed, and solutions to reduce misfit dislocations, thanks to the small-diameter III-V/Si nanowire channel, are proposed. TFETs are serious candidates to achieve a HP CMOS at a low voltage.

In **Chapter 3**, Chuang Qian and Tsu-Jae King of the University of California, Berkeley, investigate the opportunities in the cointegration of CMOS and NEM switches in 2D and 3D. They have demonstrated the superior performances of body-biased NEM relays and sub-100-mV switching low-power possibilities. Several switch designs are studied that take into account their process integration capabilities and issues, such as contact reliability. Hybrid vertical monolithic integration over a CMOS can offer new alternatives in the field of logic computation, such as look-up tables (LUTs), thanks to switch-based nonvolatile memory integration.

In **Chapter 4**, Hervé Fanet of CEA-LETI reviews the principles of adiabatic logic and its interest compared to the CMOS for ultra-low-power applications. The chapter introduces nanoelectromechanical system (NEMS)-based adiabatic logic combined with the CMOS and demonstrates its ability to drastically reduce power consumption. The different architectures of adiabatic logic gates are reviewed. Their switching energy versus performance figures of merit are compared with those of the CMOS. Finally, capacitance-based adiabatic logic, using NEMS-based integration and alleviating the switch contact resistance limitations, is studied as a promising new paradigm for ultra-low-power logic.

Part 2 features the breakthrough approaches as follows: single spin quantum bit manipulation, memristors and neuromorphic architectures as well as nanomagnetic logic. Quantum bit manipulation is studied either on pMOSFET or electron-based coupled quantum dots, both processed on silicon. Metal/insulator/metal or metal/phase change materials/metal-based memristors potential serve dual data storage or process and bio-inspired neuromorphic 1 transistor-1 resistor circuit architectures. Nanomagnetic logic architectures use multi-layered nanomagnets exchanging bias vertically and horizontally.

Chapter 5, by Romain Maurand et al. is a common effort by CEA, INAC and CEA, LETI on the realization of a single spin quantum bit (Q-bit) based on a silicon CMOS technology. The devices are processed through an already mature technology node available in a 300 mm wafer fab and a research laboratory. This consortium has a lot of experience in single electronics on a CMOS and extends its expertise to single spin manipulation for application in quantum electronics. The spin actuation and readout are obtained on PMOS transistors, submitted to a microwave signal and operating in the Pauli blockade regime.

Chapter 6, by Kodera et al. of the Tokyo Institute of Technology, describes the potential of tunnel-coupled double quantum dots (DQDs) as candidates for large-scale integration of Q-bits by using a CMOS-compatible process flow. A single electron transistor (SET) is used to sense laterally as low as single charge and spin *in tunnel-coupled DQDs*. A well-defined charge stability diagram is obtained for the few-electron regime in the DQD.

In **Chapter 7**, Rao et al., of the University of Massachusetts, review the concept of a memristor and highlight its great ability to dually store and process data. The memristors are based on two terminals devices sandwiching an insulator or a phase change material between two metal electrodes. This is the main class of devices in which the internal resistance depends on the history of applied voltage or current. Ionic and oxygen vacancies motion or thermal phase materials change between electrodes are the various physical phenomena operating in these devices. The authors highlight typical examples, from digital to analog applications and, particularly, spiking neural networks, scalable down to sub-10 nm and integrated in a 3D CMOS.

In **Chapter 8**, Vianello et al., in a joint effort by CEA-LETI and CEA-LIST, show how to implement spike-based neuromorphic circuits by integrating one transistor–one resistor memory (1T1R RRAM) devices and arrays. The resulting synapses can be used for visual pattern classification through a convolutional neural network as well as to perform on-chip unsupervised learning in fully connected neural networks. To mimic bioinspired circuit architectures and algorithms, these functions demonstrate time-dependent plasticity, essential to future impact on the neural deep learning process efficiency and its power consumption. Long- and short-term plasticity of the neural network can provide a data set–free learning process and high background-noise immunity, respectively.

In **Chapter 9**, Dey et al., in a joint effort by the University of Notre Dame and Pazmany University of Budapest, review the principle of magnetically coupled cellular automata logic, their ultra-low-power highlights, and speed and directionality limitations in the historical field coupling–based architectures. Magnetization is the state variable of the coupled cells. The authors propose solutions to enhance the performance of the nanomagnet logic (NML) architectures by adding unidirectional anisotropy capabilities, thanks to an exchange of bias between magnetic multilayers. In addition, speed can be increased through lateral exchange coupling among nanomagnets mediated by a common underlying magnetic film.

References

1. http://www.itrs2.net/itrs-reports.html
2. Moore, G. E. (1965). Cramming more components onto integrated circuits, *Electronics*, **38**(8), pp. 114–117.
3. Brillouët, M. (2003). Heterogeneity on silicon or in a package for future system integration, MEDEA + Scientific Committee, Available: http://www.imec-tw.tw/content/user/File/Heterogeneity_FINAL_1_.pdf
4. Anastas, P., and Warnert, J. C. (eds.) (1998). *Green Chemistry: Theory and Practice*, Oxford University Press, Oxford.
5. Panorama 2010 du marché du tellure, BRGM/RP-60206-FR, September 2011; http://infoterre.brgm.fr/rapports/RP-60206-FR.pdf
6. http://ec.europa.eu/environment/archives/greenweek2014/index.html
7. Deleonibus, S., Faynot, O., Ernst, T., Vinet, M., Batude, P., Andrieu, F., Weber, O., Cooper, D., Bertin, F., Moriceau, H., DiCioccio, L., Signamarcheix, T., Sanquer M., Jehl, X., Cueto, O., Fanet, H., Martin, F., Okuno, H., Nemouchi, F., Poupon, G., Lamy, Y., Gasparutto, D., Baillin, X., Duraffourg, L., Arcamone, J., Perniola, L., De Salvo, B., Vianello, E., Hutin, L., Poulain, C., Beigne, E., Tiron, R., Pain, L., Tedesco, S., Barnola, S., Posseme, N., Le Royer, C., Villalon, A., and Salot, R. (2014). Future challenges and opportunities for heterogeneous process technology. Towards the thin films, zero intrinsic variability devices, zero power era, in *IEEE International Electron Devices Meeting (IEDM) Technical Digest*, pp. 9.2.1–9.2.4.
8. Deleonibus, S. (2016). The energy and variability efficient era (E.V.E.) is ahead of us, *IEEE J. Electron Devices Soc.*, **4**(5), pp. 236–245.
9. Deleonibus, S. (ed.) (2009). *Electronic Device Architectures for the NanoCMOS Era: From Ultimate CMOS Scaling to Beyond CMOS Devices*, Pan Stanford Publishing Pte., Singapore.

10. Nikonov, D. E., and Young, I. A. (2015). Benchmarking of beyond-CMOS exploratory devices for logic integrated circuits, *IEEE J. Explor. Solid-State Computat. Devices Circuits*, **1**, pp. 3–11.
11. Ferrain, I., Colinge, C. A., and Colinge, J.-P. (2011). Multigate transistors as the future classical metal-oxide-semiconductor field-effect transistors, *Nature*, **479**, pp. 310–316.
12. Batude, P., Brunet, L., Fenouillet-Beranger, C., Andrieu, F., Colinge, J.-P., Lattard, D., Vianello, E., Thuries, S., Billoint, O., Vivet, P., Santos, C., Mathieu, B., Sklenard, B., Lu, C.-M. V., Micout, J., Deprat, F., Avelar Mercado, E., Ponthenier, F., Rambal, N., Samson, M.-P., Cassé, M., Hentz, S., Arcamone, J., Sicard, G., Hutin, L., Pasini, L., Ayres, A., Rozeau, O., Berthelon, R., Nemouchi, F., Rodriguez, P., Pin, J.-B., Larmagnac, D., Duboust, A., Ripoche, V., Barraud, S., Allout, N., Barnola, S., Vizioz, C., Hartmann, J.-M., Kerdiles, S., Acosta Alba, P., Beaurepaire, S., Beugin, V., Fournel, F., Besson, P., Loup, V., Gassilloud, R., Martin, F., Garros, X., Mazen, F., Previtali, B., Euvrard-Colnat, C., Balan, V., Comboroure, C., Zussy, M., Mazzocchi, Faynot, O., and Vinet, M. (2017). 3D sequential integration: application driven technological achievements and guidelines, in *IEEE International Electron Devices Meeting (IEDM) Technical Digest*, pp. 52–55.
13. Deleonibus, S. (ed.) (2014). *Intelligent Integrated Systems: Devices, Technologies and Architectures*, vol. 1, Pan Stanford Series on Intelligent Nanosystems, Pan Stanford Publishing Pte., Singapore.
14. Deleonibus, S. (ed.) (2017). *Integrated Nanodevice and Nanosystem Fabrication: Materials, Techniques and New Opportunities*, vol. 2, Pan Stanford Series on Intelligent Nanosystems, Pan Stanford Publishing Pte., Singapore.
15. Veldhorst, M., Eenink, H. G. J., Yang, C. H., and Dzurak, A. S. (2017). Silicon CMOS architecture for a spin-based quantum computer, *Nat. Commun.*, **8**, Article number: 1766.

Part I

Hybrid and Heterogeneous CMOS for Ultra-Low-Power Data Processing

Chapter 1

The Junctionless Transistor

Jean-Pierre Colinge
CEA-LETI, 17 rue des Martyrs, 38054 Grenoble CEDEX 9, France
Jean-pierre.COLINGE@cea.fr

The junctionless transistor consists of a piece of uniformly doped semiconductor with a gate placed between the source and drain contacts and is, therefore, the simplest transistor structure. Because of its simplicity, the junctionless architecture has been used to fabricate devices in a large number of materials, including polycrystalline semiconductors deposited on various types of substrates. The absence of a need to form source and drain junctions enables the fabrication of stacked vertical transistors or low-temperature top tiers in 3D integrated circuits. Here we describe the physics and applications of junctionless transistors and list the models developed to simulate their electrical properties. The advantages and drawbacks of these devices are described at length, and a long list of references is given to help the Reader find easily the articles published on the different properties of junctionless transistors.

Emerging Devices for Low-Power and High-Performance Nanosystems: Physics, Novel Functions, and Data Processing
Edited by Simon Deleonibus

ISBN 978-981-4800-11-2 (Hardcover), 978-0-429-45873-6 (eBook)
www.panstanford.com

1.1 Introduction

Every textbook on semiconductor device physics starts with the same succession of chapters: (1) Crystal Structure, (2) Energy Band Theory, (3) Transport Phenomena, and (4) The PN Junction. It is only after understanding the physics of the PN junction that one can contemplate studying the physics of transistors, since every classical transistor contains PN junctions, including junction field-effect transistors (JFETs), bipolar junction transistors (BJTs), and metal-oxide-semiconductor field-effect transistors (MOSFETs). A transistor without junctions may therefore seem to be heresy. Yet, the very first transistor device ever conceived was junctionless: the first patent for a transistor principle was filed in Canada by Austrian Hungarian physicist Julius Edgar Lilienfeld in 1925 [1]. The Lilienfeld transistor was a field-effect device, much like modern metal-oxide-semiconductor (MOS) devices. Figure 1.1 shows a copy of Lilienfeld's US Patent 1,900,018, "Device for Controlling Electric Current," filed in 1928 [2]. It consists of a thin semiconductor film, 12, deposited on a thin insulator layer, 11, which is itself deposited on a metal electrode, 10. The latter metal electrode is the gate of the device. The current flows in the resistor between contact electrodes, 14 and 15, much the same way drain current flows between source and drain in a modern MOSFET. The initial semiconductor material proposed in Lilienfeld's patent was copper sulfide, and the gate insulator was alumina. The device is a simple resistor, and the application of a gate voltage allows one to deplete the semiconductor film of carriers, thereby modulating its conductivity. Ideally, one should be able to completely deplete the semiconductor film of carriers, in which case the resistance of the device becomes quasi infinite and the device can be turned off. This is why Lilienfeld introduced a V-groove, 13, in the semiconductor film. This allows one to induce full depletion of the copper sulfide at the bottom of the groove, where the film is very thin, to turn the device off. Having a larger semiconductor thickness everywhere else helps reduce the overall resistance of the device when it is turned on. It took 87 years for the Lilienfeld device to be successfully built. In 2012 Migita et al. used a silicon-on-insulator (SOI) substrate and a V-groove gate to fabricate a junctionless MOSFET with a channel length of only 3 nm. The device is fully functional with an on-current/off-current ratio larger than 10^6 at a drain voltage (V_D) of 1 V, a subthreshold slope (SS) of 95 mV/decade, and a drain-induced barrier lowering (DIBL) of 189 mV/V [3, 4]. The thickness of the silicon under the V-groove was only 1 nm. A similar structure made in N-type $In_{0.53}Ga_{0.47}As$ instead of silicon, and with a channel length of 6 nm, was reported a year later [5].

The theoretical operation of a transistor with a uniform doping concentration through the source, channel, and drain was published in 2008 in the *Journal of Computational Electronics* [6] and in 2009 at the EUROSOI Conference [7] and in *Applied Physics Letters* [8]. The first fabricated and electrically characterized junctionless transistor was published in 2009 at the IEEE International Conference [9], but it is only after its publication in the journal *Nature* [10] that the junctionless transistor concept really took off. Soon after, trigate SOI junctionless transistors were demonstrated by INTEL and LETI with gate lengths of 26 and 13 nm, respectively [11, 12].

1.2 Junctionless MOSFET Operation and Properties

1.2.1 Device Physics

The junctionless transistor is essentially a thin-body, heavily doped (typically in the 10^{19} cm^{-3} range) semiconductor resistor with a constant uniform doping concentration from source to drain and a gate electrode that controls the flow of current between source and drain. The device design is extremely simple since there are no PN junctions and the junctionless transistor architecture can easily be adapted to semiconductor materials other than silicon. The device operation relies on fully depleting the semiconductor nanowire/nanoribbon/thin film using the work function difference with the gate material to turn the device off. When the device is turned on, the depletion in the channel disappears and the current is carried in the bulk of the semiconductor, but the current can be

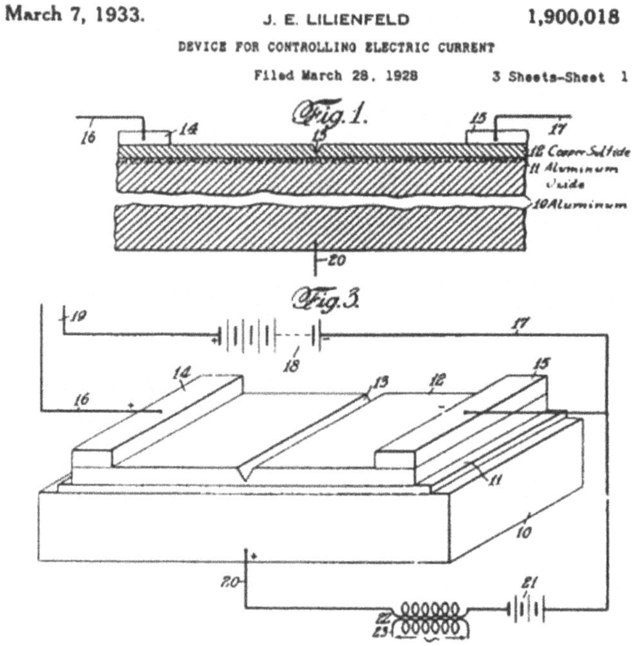

Figure 1.1 Copy of Figs. 1 and 3 of Lilienfeld's US Patent 1,900,018, "Device for Controlling Electric Current." The device contains a metal gate electrode, 10, a gate insulator layer, 11, a semiconductor resistive film, 12, with two contacts, 14 and 15, and a V-groove in the semiconductor film to locally reduce the semiconductor layer thickness, 13. Contacts 14 and 15 are the equivalent of the source and drain contacts in a modern MOSFET.

further increased by increasing the gate voltage and creating an accumulation channel [13].

The operation of a junctionless transistor can be understood using a simple analogy. Let a semiconductor nanowire be represented by the plastic hose in Fig. 1.2. Let water flow through the hose like it does in a garden hose. The flow of water represents the flow of electrons through the nanowire. Since the nanowire is heavily doped the electron density is equal to the doping density and the current is a drift current. Electrons flow from the source to the drain like in any other MOSFET. This corresponds to the on state of a junctionless transistor. If a clamp is used to squeeze the hose, the flow of water decreases and eventually disappears if the

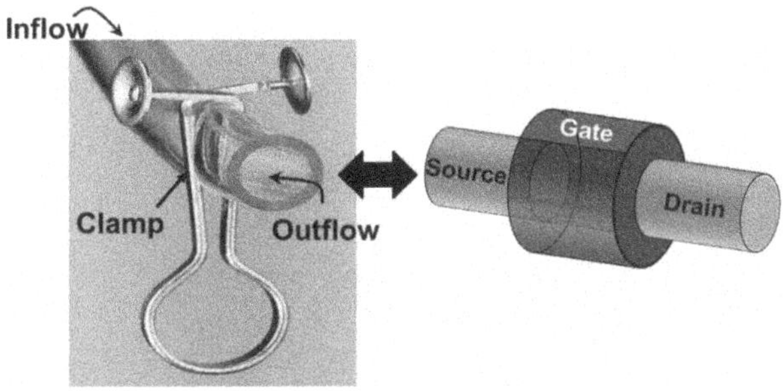

Figure 1.2 Analogy between hose and clamp (left) and junctionless transistor (right).

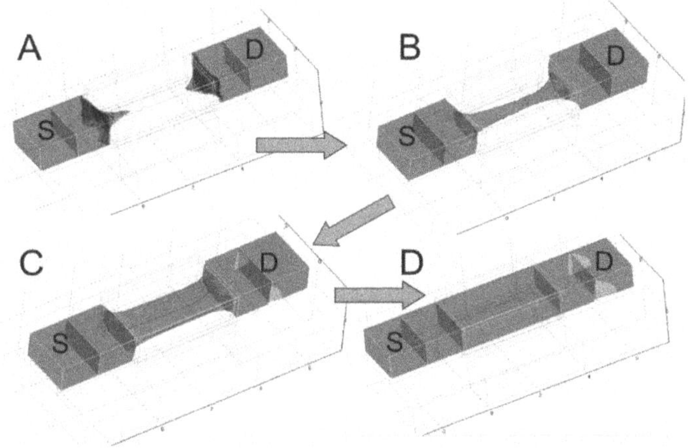

Figure 1.3 Isoconcentration plots of electron density in an N-channel trigate nanowire junctionless transistor for $V_{DS} = 50$ mV. The isoconcentration is equal to the doping concentration, N_D. A: $V_G < V_{THbulk}$; B: $V_{THbulk} < V_G < V_{FB}$; C: Same as B, but at a higher V_G value; D: $V_G = V_{FB}$.

squeezing is strong enough. This corresponds to the off state of the transistor.

3D plots of the neutral (undepleted) silicon in an N-channel trigate nanowire junctionless transistor for $V_{DS} = 50$ mV are shown in Fig. 1.3. The depletion regions are transparent. Below threshold voltage (Fig. 1.3A) the silicon is completely depleted from source to drain. When the threshold is reached a neutral

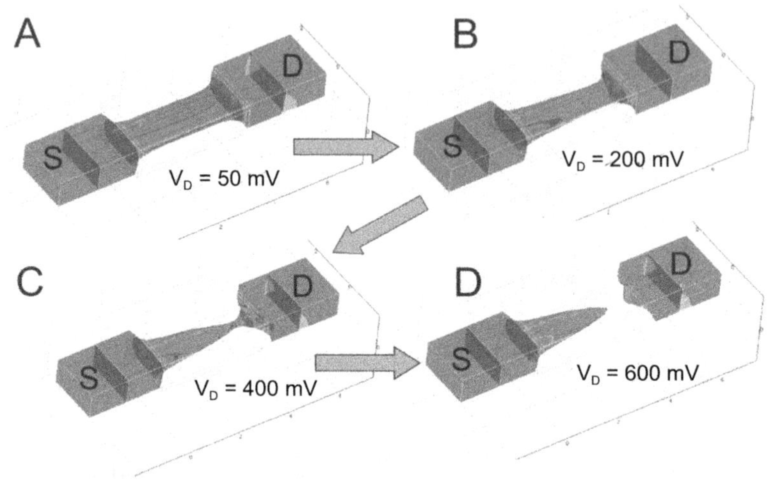

Figure 1.4 Isoconcentration plots of electron density in an N-channel trigate nanowire junctionless transistor for a constant gate voltage corresponding to case C in Fig. 1.3. The isoconcentration is equal to the doping concentration, N_D. A: $V_D = 50$ mV; B: $V_D = 200$ mV; C: $V_D = 400$ mV, which is where channel pinch-off occurs at the drain; D: $V_D = 600$ mV, where the device is in saturation.

(undepleted) channel is formed in the center of the nanowire, connecting source to drain (Fig. 1.3B). As the gate voltage is increased the cross section of the conduction channel increases, allowing for an increase in current flow (Fig. 1.3C). When flatband is reached (Fig. 1.3D) the entire nanowire is neutral. Further increase in gate voltage can be applied to increase the carrier concentration in the channel and create surface accumulation channels.

Figure 1.4 shows similar plots for a constant gate voltage (the same gate voltage as in Fig. 1.3C). One can see the evolution of the neutral (undepleted channel as the drain voltage is increased from 50 mV (Fig. 1.4A) to 200 mV (Fig. 1.4B). The neutral channel becomes pinched off when the drain voltage reaches 400 mV (Fig. 1.4C) and the transistor is in saturation for higher gate voltages (Fig. 1.4D).

1.2.1.1 Effective channel length modulation

When a junctionless transistor is turned off the region of the semiconductor covered by the gate (i.e., the channel region) is depleted, as seen in the previous section. Since the doping in the source and the drain is the same as in the channel, the fringing field lines from the gate edge deplete part of the source and the drain as well. This source and drain depletion effect increases the distance between the neutral (undepleted) parts of the source and the drain such that the effective channel length becomes larger than the physical gate length when the device is turned off [14]. The depletion extends further in the drain than in the source because $V_D > V_S$ (taking the case of an N-channel device) [15, 16]. The source and drain depletion decreases as gate voltage is increased and finally disappears when the gate voltage reaches flatband voltage (Fig. 1.5) [17, 18]. This effect greatly reduces short-channel effects and yields a better SS and DIBL than in inversion-mode transistors for very short channel lengths [8]. Channel length modulation can be boosted by using high-κ spacers at the edges of the gate at

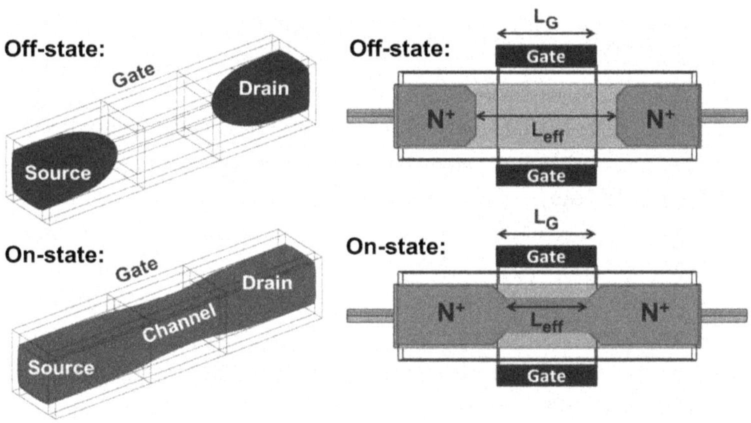

Figure 1.5 (Left) Illustration of effective gate length variation from off state to on state in a junctionless transistor using a plot of carrier isoconcentration. The dark areas are neutral (i.e., not depleted; $n = N_D$). The depleted regions are transparent. (Right) Comparison between effective gate length (L_{eff}) and physical gate length (L_G) in the off and on states. The dark-gray areas are neutral, and the light-gray areas are depleted.

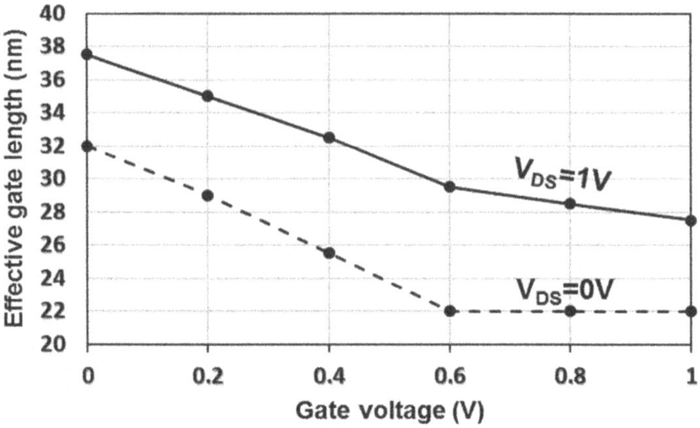

Figure 1.6 Effective gate length in a junctionless transistor as a function of gate voltage. The physical gate length is 22 nm.

the expense of the capacitance between gate and source/drain [19]. The use of dual-κ spacers (i.e., high-κ spacers along the gate sidewalls, which are themselves covered by low-κ spacers) has been shown to improve electrical characteristics. Technology computer-aided design (TCAD) optimization of dual-κ spacer junctionless transistors with a gate length of 20 nm can yield a current drive increase of 72%, a reduction in the SS of 6.5%, and a decrease in the DIBL of 38% [20]. In very short channel inversion-mode devices (L_G = 10 nm and below) direct source-to-drain tunneling becomes a serious problem; when the device is turned off, electrons can tunnel from source to drain though the narrow potential barrier generated by the gate, greatly increasing leakage current. This effect can be greatly reduced by using the gate length modulation effect in a junctionless transistor, since the tunneling current is an exponential function of the width of the potential barrier between source and drain, and thus, of the effective channel length.

The channel length modulation effect is a very effective means to mitigate short-channel effects [21]. Figure 1.6 shows the simulated effective channel length (i.e., the distance between the undepleted regions of the source and the drain) as a function of gate voltage in a junctionless transistor with a physical (printed) gate length of 22 nm. The transistor is a double-gate (DG) device with a thickness of 10 nm and a doping concentration $N_D = 10^{19}$ cm^{-3}. The gate oxide thickness is 1 nm. Under zero gate bias conditions the transistor is turned off and the effective length ranges from 32 to 38 nm. The larger the drain voltage, the longer the effective gate length because the depletion in the drain is longer than that in the source when $V_{DS} > 0$. The effective gate length decreases when the gate voltage is increased and becomes equal to the physical gate length when the transistor is turned on and $V_{DS} = 0$. It remains longer than the physical gate length when the device is turned on if $V_{DS} = 1$ V [22].

It is also worth pointing out that the extension of the channel depletion zone into the drain reduces the gate-induced drain leakage (GIDL) when the transistor is turned off. GIDL current is caused by band-to-band tunneling (BTBT) between the channel in inversion (for a large reverse gate current) and the drain. The extension of the channel depletion in the drain increases the tunneling distance and decreases the electric field, which in turn deceases BTBT current. This is one of the reasons why junctionless transistors usually offer lower leakage current and a better I_{ON}/I_{OFF} ratio than inversion-mode or accumulation-mode transistors [23, 24].

1.2.1.2 Mobility

The behavior of mobility in the channel of inversion-mode MOSFETs is well known and has extensively been described by Takagi et al. in 1994 in the so-called universal mobility model [25]. The mobility is usually plotted as a function of the effective (vertical) electric field (MV/cm) at the silicon/gate insulator interface, on a log scale, but it is also frequently plotted as a function of the inversion charge density (cm^{-2}). For understanding the physical mechanisms involved, we will use a plot of mobility as a function of gate voltage, since in a junctionless transistor we will be dealing with both bulk mobility and surface accumulation mobility. The scattering mechanisms limiting mobility in the channel of an inversion-mode MOSFET as a function of increasing gate voltage (Fig. 1.7) are the following:

- Coulomb scattering due to the presence of doping impurities. This scattering increases with doping concentration

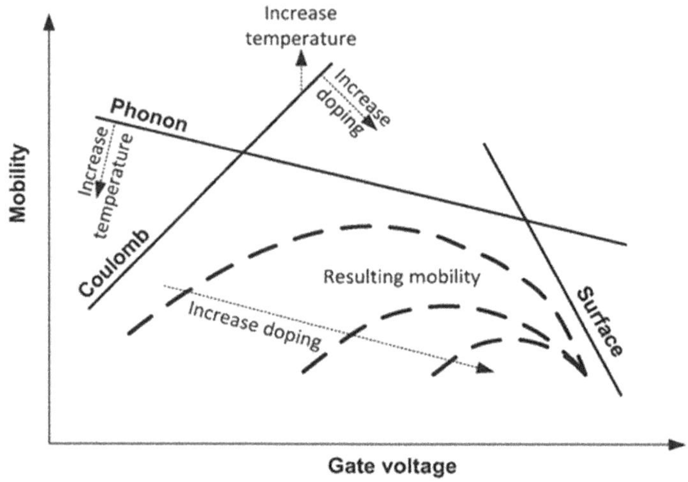

Figure 1.7 Different scattering mechanisms limiting the mobility in an inversion or accumulation layer (Coulomb scattering, phonon scattering, and surface roughness scattering) as a function of gate voltage. The resulting mobility is shown by the dashed lines for different doping concentrations.

and decreases with temperature such that the Coulomb scattering mobility varies as $\frac{T^{3/2}}{N}$, where N is the doping concentration.

- Phonon scattering due to the vibrations in the crystal lattice. Phonon scattering increases with temperature such that the phonon scattering mobility varies as $T^{-3/2}$.
- Surface roughness scattering (SRS) due to the interactions between the carriers in the channel and the semiconductor/gate oxide interface. Since the carriers in the channel are ever more "pressed" against the interface as the gate voltage is increased, SRS increases with gate voltage, which decreases mobility. As a result, carrier mobility in an inversion channel is always lower than bulk mobility where only Coulomb and phonon scattering mechanisms are present.

In a junctionless transistor, part of the current is due to purely resistive conduction in the bulk of the device. The bulk current experiences no SRS as the electric field perpendicular to the current flow in the neutral bulk conduction channel is essentially equal to zero [26, 27] but Coulomb scattering is high because of the high doping concentration. There is an exception to this rule, however; if the variance of the roughness step height becomes comparable to or larger than the diameter of the nanowire, then SRS can become the dominant scattering mechanism [28]. The other part of the current is flowing in surface accumulation channels if the gate voltage is larger than the flatband voltage. Early experiments show that the carrier mobility in an accumulation channel is larger than that in an inversion channel and that the accumulation mobility can become even larger than bulk mobility [29–31]. This is caused by the screening of the ionized impurity charges (that dominate overall scattering) by the high concentration of majority carriers in the accumulation channel [32]. As a result, accumulation mobility in a junctionless transistor can become larger than bulk mobility and larger than the mobility in a lightly doped channel of an inversion-mode device [11]. Furthermore, the accumulation mobility stays constant or increases with gate voltage until SRS starts to appear and channel mobility eventually decreases, as it does in an inversion channel. Accumulation mobility can reach values 20% higher than bulk mobility for a doping concentration of 10^{19} cm^{-3} in planar junctionless transistors [33] and up to 46% in nanowire junctionless devices [34]. Furthermore, for an equal amount of inversion or accumulation charge, the electric field is lower in an accumulation layer than in an inversion layer, which reduces SRS scattering [27, 35]. In the case of nanowires, mobility increases when the width of the nanowire is decreased. This is attributed to the presence of a majority carrier concentration that is higher than the doping concentration in the bulk of the nanowires [36]. These majority carriers partially screen the electric field generated by the doping impurity atoms, which reduces Coulomb scattering and, therefore, increases bulk mobility [37]. Figure 1.8 shows the measured mobility in germanium P-channel gate-all-around (GAA) junctionless nanowire MOSFETs [38]. The measured mobility is compared to that of a lightly doped inversion-mode device. At low gate voltage overdrive ($V_G - V_{TH}$), inversion mobility is larger than the apparent mobility in junctionless devices. The apparent mobility results in the combination of bulk conduction

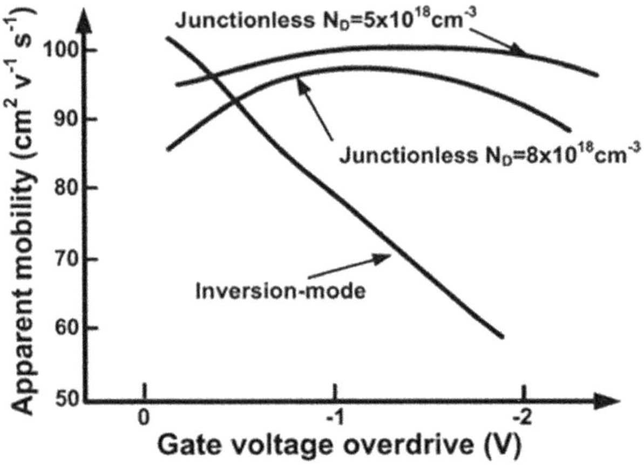

Figure 1.8 Mobility versus gate voltage overdrive in P-channel germanium junctionless and inversion-mode GAA MOSFETs [38].

and accumulation channel conduction. At larger gate voltages, the junctionless mobility slightly increases due to the screening effect, while the inversion mobility steadily decreases due to SRS. Junctionless mobility eventually starts to decrease for large values of gate voltage overdrive due to SRS. It is worth noting that strain can be used to increase mobility in junctionless transistors, as it is the case in standard inversion-mode field-effect transistors (FETs) [39, 40].

A method has been developed to extract separately bulk conduction and accumulation layer conduction [41–43]. The method is based on the use of both a front-gate and a back-gate voltage on planar junctionless SOI FETs. Modulating the back-gate voltage and using the $Y = I_D/\sqrt{g_m}$ function, where I_D is the drain current and g_m is the front transconductance allows one to separate the two types of conduction and extract the corresponding mobilities [44]. The study, carried out on P-channel devices, reveals that both bulk mobility and accumulation mobility increase as the (negative) front-gate voltage is increased. Bulk mobility enhancement under front-accumulation conditions can be understood through a reduction of the doping impurity scattering due to the screening of Coulomb centers. In a junctionless transistor the ionized impurities and the charged interface states are gradually screened by the build-up of the accumulation layer when the gate voltage is increased, which reduces the scattering of majority carriers [33]. The carrier mobility in the accumulation layer is larger than the bulk mobility and increases with gate voltage, confirming the results obtained by Rudenko et al. on junctionless nanowire transistors [34].

Nagai et al. simulated the impurity scattering mechanisms in junctionless transistors using a Monte Carlo technique [45]. The study reveals that forward scattering (i.e., in the current flow direction) is dominant when majority carriers are scattered by an ionized impurity in the channel. Impurity scattering is basically isotropic for low-energy electrons, but scattering becomes increasingly forward-oriented as electron energy is increased. For instance, 81% of electrons with an energy of 0.5 eV are forward scattered when interacting with an impurity. As a result, channel mobility is decreased (isotropic scattering) for small gate voltage overdrives. When the gate voltage is increased, however, the current drive of a junctionless transistor approaches that of a conventional inversion-mode transistor because of the forward scattering nature by ionized impurities in the high-electric-field regime. In addition, as channel length is decreased, the forward scattering nature of impurity scattering appears at lower gate voltage overdrive values than in long-channel devices, due to the formation of a high-field region, such that junctionless transistors have essentially the same current drive as inversion-mode FETs with low doping concentrations.

Finally, a study by Ueda et al. based on atomistic $sp^3d^5s^*$ tight-binding simulations of heavily doped junctionless GAA silicon nanowire transistors shows that if the doping atoms can be periodically placed along the channel length, high bulk values mobility values can be obtained [47]. The evenly spaced doping atoms form periodic potential wells for the majority carriers, which can easily tunnel from one impurity well to the next. Placing doping atoms in an evenly distributed manner constitutes an important challenge for device processing, but it can be achieved [46]. An increase of doping concentration increases the transmission of carriers from source to drain, such that electron mobility can in principle increase from 50 to 300 $cm^2V^{-1}s^{-1}$ when the doping concentration is increased from 2×10^{19} to 8×10^{19} cm^{-3}, respectively [47].

1.2.1.3 Subthreshold slope

Junctionless nanowire transistors with a trigate, pi-gate, omega-gate, or GAA geometry have excellent subthreshold characteristics. Even the earliest junctionless transistors with a trigate geometry were exhibiting a room temperature SS of 60 mV/decade over 5 orders of magnitude subthreshold current and a DIBL of 7 mV/V for a gate length of 50 nm [48]. Early devices with a gate length of 30 nm were featuring a SS of 68 mV/decade over 6 orders of magnitude subthreshold current and a DIBL of 15 mV/V [12]. Ab initio simulations of GAA junctionless devices with a gate length of 3 nm still show a SS of 70 mV/V [49]. The subthreshold current of a nanowire junctionless transistor is essentially a diffusion current. It is independent of drain voltage, as is the case in inversion-mode transistors. If the width of the transistor is increased in such a way that it becomes a planar, single-gate transistor, the electrostatic control of the gate over the channel is degraded, as reported by Park et al. [50] in measurements on 80-nm-wide SOI junctionless transistors with a doping concentration of $1\text{–}2 \times 10^{19}$ cm^{-3}. In such a case a drift current component is added to the diffusion current and the SS reaches low values for only a limited range of subthreshold current values.

1.2.1.4 Threshold voltage and transconductance

A junctionless transistor has two threshold voltages: a threshold for bulk conduction, V_{TH}, which is the gate voltage below which the channel is fully depleted and above which a neutral conduction region is formed, and an accumulation threshold voltage, which is basically equal to the flatband voltage, V_{FB}. In the case of an N-channel device, $V_{TH} < V_{FB}$. The device is in subthreshold operation when $V_G < V_{TH}$. It is in bulk (resistive) conduction mode when $V_{TH} < V_G < V_{FB}$, and an additional accumulation channel is added to bulk conduction when $V_G > V_{FB}$. This operation is similar to that of an accumulation-mode FET [99]. In a typical accumulation-

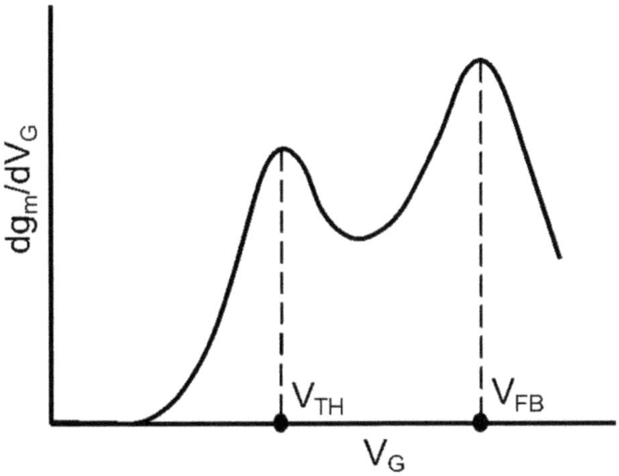

Figure 1.9 Bulk conduction threshold voltage (V_{TH}) and accumulation threshold voltage (V_{FB}) extracted from the dg_m/dV_G plot of a junctionless transistor.

mode transistor, however, the channel concentration is low and the bulk conduction current is, therefore, much smaller than the accumulation current. In a junctionless transistor, the channel doping concentration is typically equal to or higher than 10^{19} cm^{-3}, such that the bulk channel current is much larger than in a classical accumulation-mode FET. If low-voltage operation is sought the device can be operated below flatband, in which case the entirety of the drain current flows through the bulk channel [51]. The drain current can, however, be boosted by increasing the gate voltage and forming an accumulation channel. Both V_{TH} and V_{FB} can easily be extracted by plotting the derivative of the transconductance versus gate voltage, dg_m/dV_G [50, 52, 53]. The dg_m/dV_G plot of a junctionless transistor shows two distinct peaks, corresponding to V_{TH} and V_{FB} (Fig. 1.9). The relative heights of the peaks depend on the doping concentration and on the geometry of the transistor (channel length and height and width of the channel in the case of a nanowire device). An exhaustive review of threshold voltages and mobility extraction techniques applied to junctionless transistors can be found in the *Journal of Nano Research* [54].

1.2.1.5 Gate capacitance

As the gate voltage is increased below threshold in an inversion-mode FET the depth of the depletion width decreases and the concentration of weak inversion carrier increases underneath the gate stack. When the threshold is reached, a strong inversion channel is formed and then the gate capacitance reaches a constant value equal to $C_{ox}WL$, where C_{ox} is the gate capacitance per unit area $C_{ox} = \varepsilon_{ox}/t_{ox}$ and W and L are the gate width and gate length, respectively. For simplicity we are neglecting quantum effects that cause the channel to be located at some small but finite distance below the semiconductor/insulator interface, as well as fringing capacitances. We also assume V_{DS} is small. In a junctionless transistor, the channel region is fully depleted below the threshold. When the gate voltage reaches the bulk conduction channel, most of the channel is still fully depleted, and depletion completely disappears when the gate voltage reaches flatband voltage, at which point an accumulation layer starts forming. As long as accumulation is not reached there is a depletion layer between the conducting (neutral) channel and the semiconductor/insulator interface, which reduces the gate capacitance, when compared to an inversion-mode FET [55]. This relative reduction increases with channel doping concentration [11, 14]. Figure 1.10 shows the gate capacitance in inversion-mode and junctionless trigate transistors with different doping concentrations. A model for the transcapacitances C_{GG}, C_{GD}, C_{DG}, C_{SG}, C_{SD}, and C_{DD} has been developed by Jazaeri et al. for DG and cylindrical GAA junctionless transistors. Each transcapacitance is defined as $C_{ij} = \frac{\partial Q_i}{\partial V_j}$, where i and j represent the source, gate, or drain [56]. The reduction of gate capacitance can give a speed advantage to junctionless complementary metal-oxide semiconductor (CMOS) logic circuits, despite the reduced mobility due to Coulomb scattering. Recent work suggests that junctionless FinFETs can increase speed performance by 12% over inversion-mode CMOS FinFETs at the 10-nm-node level. Dynamic power consumption is deceased as well [57].

Some studies indicate that despite having a lower current drive due to lower carrier mobility, a CMOS junctionless circuit can operate at a higher frequency and with lower dynamic

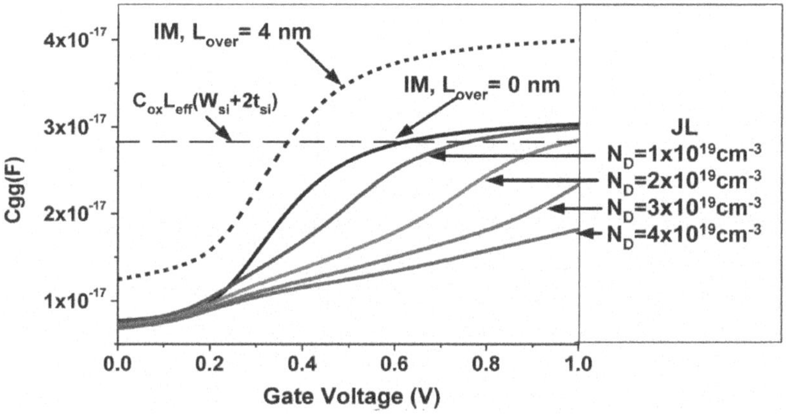

Figure 1.10 Gate capacitance, C_{gg}, in inversion-mode (IM) and junctionless (JL) trigate FETs with an effective gate length, L_{eff}, equal to 25 nm and a cross section of 10 nm × 10 nm. $V_D = 50$ mV and $t_{ox} = 1$ nm. The IM device with 4 nm S&D overlap ($L_{over} = 4$ nm) has a physical gate length of 25 nm + (2 × 4 nm) = 33 nm. The horizontal dashed line represents the intrinsic gate capacitance $C_{ox}WL$ [14]. The gate capacitance can become larger than the intrinsic gate capacitance because of the overlap between source/drain and gate in the IM device and because of fringing capacitances between gate and source/drain in the junctionless transistor.

power consumption than inversion-mode devices, simply because of the lower gate capacitance. For example, simulations indicate an improvement of ~28% and ~10% in rise time and fall time, respectively, for a CMOS inverter as well as an ~12% increase in the oscillation frequency of oscillation of a CMOS ring oscillator (10 nm technology node) when junctionless transistors are used instead of inversion-mode devices [58].

1.2.1.6 Miller capacitance

In a junctionless transistor in the off state, the depletion region in the channel extends beyond the edges of the gate to some distance inside the source and the drain. As mentioned previously, this effect modulates the channel length [14]. It also decreases the gate/source (C_{gs}) and gate/drain (C_{gd}) capacitances. The extension in the drain is larger than that in the source because $V_D > V_S$ (we take here the case of an N-channel device), such that $C_{gd} < C_{gs}$. This effect has been observed in inversion-mode FETs with source and drain underlap [59, 60] and is systematically observed in junctionless FETs. The gate/drain capacitance is also known as the Miller capacitance, which is an important limiting factor in radio-frequency (RF) circuits. Junctionless nanowire transistors show lower Miller capacitances than inversion-mode and accumulation-mode transistors [61]. Because of the depletion in the source and the drain, junctionless transistors present a C_{gs}/C_{gd} ratio two times smaller than inversion-mode devices or, in other words, a Miller capacitance that is half that of the inversion-mode MOSFET. Coupling this effect with the reduction in gate capacitance and an increase in the early voltage due to the extension of the depletion in the source and the drain, one can contemplate a cutoff frequency (f_T) and a maximum oscillation frequency (f_{max}) two times higher in junctionless than in inversion-mode FETs, along with a 65% improvement of intrinsic transistor gain, A_{vo}, despite having a lower mobility (transconductance) handicap [62, 63].

1.2.1.7 Variability

Variability of the electrical parameters of a MOSFET plays an important role in a manufacturing environment, since the threshold voltage, the on-current and the off-current of all devices must fall within a predefined set of specifications. Variability is affected by the variation of many fabrication parameters: gate length variations, gate edge roughness, nanowire thickness and width variations, work function fluctuations in polycrystalline metal gates, gate oxide thickness, charge fluctuations, etc. [64, 65]. An important contribution is due to random dopant fluctuations (RDFs). This contribution is represented by the parameter A_{VT} of a Pelgrom plot, in which the width of distribution of the threshold voltage mismatch is plotted as a function of the inverse of the square root of the gate length × gate width product of a population of transistors. In mathematical terms this can be written as $\sigma \Delta V_T = \frac{A_{VT}}{\sqrt{WL}}$. In an inversion-mode transistor, $\sigma \Delta V_T = \frac{A_{VT}}{\sqrt{WL}} = \frac{q\sqrt{2Nx_{depl}}}{C_{ox}\sqrt{WL}}$. One can see that the variability increases with gate oxide thickness and doping concentration [66–68]. Let us compare junctionless and inversion-mode transistors in terms of different sources of variability:

- Random dopant fluctuation (RDF): Because of its high channel doping concentration, variability is worse in junctionless transistors than in other types of transistors that all have lighter channel doping concentrations [69, 70]. In some instances, undoped channels can be used, which reduces RDF variability to a minimum. In a junctionless transistor variability is most severe when the device is turned off or operating in the subthreshold regime [71], but screening of the doping impurities reduces RDF variability when a junctionless is turned on and driven into accumulation [72]. According to Leung and Chui, the impact of RDF on junctionless device variability is alleviated to some extend when the devices are scaled down in dimensions [73]. One solution for eliminating RDF in junctionless transistors would be to reduce the doping concentration to zero. The major problem with this approach is the Schottky barriers formed between metal and semiconductor at source and drain contacts. According to Chen et al. [74], it is possible to eliminate that problem by using metal-insulator-semiconductor (MIS) tunnel contacts instead of metal-semiconductor (MS) contacts. Simulation studies show that in undoped and unintentionally doped Si (doping concentration ranging from 10^{10} to 10^{17} cm^{-3} for both N- and P-type dopants), junctionless FETs with MIS contacts may exhibit higher on-currents than those with MS contacts and show almost the same transfer characteristics. The later result implies that junctionless FETs with MIS contacts possess good immunity against RDF [74].
- Channel length variations: The channel length modulation effect renders junctionless less prone to short-channel effects such as DIBL and SS degradation, which reduces device variability as a function of gate length variations.
- Device architecture: Planar junctionless transistors suffer from large RDF variability issues [75]. This effect can, however, be mitigated by using multigate architectures

(trigate, omega-gate, GAA), which are known to improve the electrostatic control of the channel and reduce A_{VT} [76].

One obvious way of reducing RDF variability would be to reduce doping concentration and, in an optimal case, use undoped channel, source, and drain. If the doping is reduced between the usual 10^{19} cm^{-3} value (taking silicon as an example), the contacts between metal and source and drain become high-resistance Schottky contacts. One study suggests that these MS contacts can be advantageously replaced by MIS contacts, in which the current tunnels through the thin (typically 0.5 nm thick) insulator. MIS contacts eliminate Fermi-level pinning, reduce contact resistance, and inject free carriers in the source and the drain. Simulations results suggest that by using tunnel insulators with a high dielectric constant ($\kappa = 15$) one can contemplate driving a current of 830 μA/μm for a supply voltage of 0.7 V, an equivalent oxide thickness (EOT) of 1 nm, and an off-current of 20 pA/μm in a silicon junctionless transistor with a low doping concentration (10^{15} cm^{-3}) [77].

1.2.1.8 Reliability and noise

The main elements of reliability of an FET are due to the injection of carriers from the semiconductor into the gate oxide. These charges induce threshold voltage shifts and, in some instances, channel carrier mobility degradation. Negative bias temperature instability (NBTI) and positive bias temperature instability (PBTI) and hot-carrier injection (HCI) are the main degradation mechanisms. HCI is caused by the injection of high-energy electrons into the gate dielectric. These electrons are accelerated in the high-field pinch-off region where the channel meets the drain. This electric field is lower in junctionless transistors [78], resulting in a lower hot-carrier generation rate and less HCI degradation. Furthermore, in an inversion-mode transistor, the peak electric field occurs underneath the channel, underneath the gate dielectric. In a junctionless transistor, it occurs in the drain, at some distance from the gate dielectric. This contributes to reducing the injection of carriers into the gate oxide [79–81].

NBTI and PBTI effects are lower in junctionless transistors than in inversion-mode FETs because the electric field across the gate oxide is lower and there are fewer interactions between the carriers and interface traps during device operation (device carriers are mostly within the bulk of a nanowire in a junctionless transistor, while they are located at the semiconductor/gate dielectric interface in an inversion-mode device) [82, 83].

There are two main noise components in junctionless transistors: random telegraph noise (RTN) and low-frequency (LF) noise [84, 85]. The RTN measured in the time domain is due to the trapping/detrapping of carriers at the semiconductor/gate insulator interface. These trapping/detrapping events slightly decrease/increase the drain current as a function of time. Because carriers tend to flow in the bulk of the semiconductor in a junctionless transistor and at the semiconductor/gate insulator interface in an inversion-mode device, the amplitude of the RTN is considerably lower in a junctionless transistor than in an inversion-mode FET, unless an accumulation layer is formed, which increases trapping/detrapping rates [86]. Increasing the number of nanowires in parallel during measurement, the RTN can be reduced substantially due to an averaging effect [87].

LF noise is measured in the frequency domain and is also due to the trapping/detrapping of carriers at the semiconductor/gate insulator interface. The LF behavior of a junctionless transistor can be explained by a carrier number fluctuation model, even though conduction occurs in the bulk of the semiconductor. The LF noise can be explained by the trapping/detrapping in slow traps at the semiconductor/gate dielectric interface of majority carriers that have diffused from the bulk region of the channel into the depletion regions surrounding it or that directly tunnel from the bulk channel into the traps. LF noise is slightly lower in junctionless transistors than in inversion-mode FETs [88–91]. The drain current noise spectral density (S_{ID}, $\mathrm{A^2Hz^{-1}}$) in junctionless nanowire transistors increases when gate bias is increased and an accumulation layer is formed. The noise spectrum reveals a $1/f^{\gamma}$ noise behavior and the frequency exponent, γ, is observed close to 1. The value of γ close to 1 signifies a uniform trap distribution within the gate oxide. The drain current noise ($S_{\mathrm{ID}}/I_{\mathrm{D}}^2$), on the other hand, decreases when gate voltage is increased: below flatband, when conduction occurs in the bulk of the nanowire, noise is dominated by bulk mobility fluctuations. In the accumulation regime (above flatband voltage), the noise mechanisms become dominated by carrier number fluctuation, and the drain current noise decreases with gate voltage. At large gate voltages the current noise ceases to decrease and saturates at a constant minimum value due to a noise contribution by the access source and drain regions [87].

By applying different front- and the back-gate bias, it is possible to place the bulk conduction channel of a trigate nanowire junctionless transistor closer to or further from the semiconductor/insulator interfaces. Using this technique, it has been demonstrated that the LF noise increases when the channel moves closer to these interfaces [92]. P-channel junctionless transistors have been fabricated in a bottom-up fabrication technique using in situ highly doped nanowires grown by low-pressure chemical vapor deposition (CVD) vapor-liquid-solid (VLS) growth. LF noise characterization of these devices biased in saturation was conducted at different gate biases. Wet thermal oxidation was then carried out at 650°C to grow a 5–10 nm gate oxide. The final diameter of the silicon nanowire was 10–12 nm. The noise spectrum indicates either a Lorentzian or $1/f$ noise depending on the gate bias. Analysis of the results indicates very low trap densities on the order of 10^{16} $cm^{-3}eV^{-1}$. The low trap densities in these devices are associated with the VLS fabrication technique with in situ doping and the absence of the semiconductor junction and the ion implantation step in the process [93].

Analysis carried out on nanowire silicon-oxide-nitride-oxide-silicon (SONOS) flash memory devices submitted to 1000 program/erase cycles shows a LF noise power spectral density that is over three times lower in junctionless than in inversion-mode transistors. This is attributed to the bulk conduction mechanism in junctionless transistors, which is less sensitive to the presence of interface traps than the channel(s) of an inversion-mode device, which are closer to the silicon/gate insulator interface. The reported results are consistent with the carrier number fluctuation model [94].

1.2.1.9 Bipolar effect

Although a junctionless transistor does not contain PN junctions, it is possible to bias the device in such a way that bipolar amplification can be observed. There are three examples of junctionless bipolar amplification in the literature. We will consider the case of an N-channel device.

- If the gate bias is sufficiently negative, inversion channels can be created under the gate. The work function difference between the gate material and the channel is included in the gate bias, such that the actual gate voltage can be positive. In that case the source, inversion channels, and drain can then form the emitter, base collector, and emitter of an NPN transistor, respectively [95]. The device has an electrically floating base.
- Using a higher gate voltage (but one still below the threshold), the injection of holes into the channel region can create a hole-rich region in the channel that can forward-bias the channel region with respect to the source and create a "parasitic" NPN bipolar transistor. Such injection of carriers can be achieved by impact ionization in the drain region [96] or by a heavy ion strike (single-event upset) [97].

Bipolar gain can reach values over 100 and depends on semiconductor thickness, gate oxide thickness, and doping concentration.

1.2.1.10 Models for the junctionless transistor

A simple but useful physically based model that encompasses both the bulk current and the accumulation current of a junctionless transistor can be found in Table 1.1 [98]. It is derived from an earlier model developed for the accumulation-mode transistors [99]. It is valid only above the threshold but accounts for saturation of either the bulk or the accumulation current, or both. It allows one to use different mobility values for the bulk current and the accumulation current.

Table 1.1 Expressions[a] for the drain current in a junctionless transistor

Bias conditions	Drain current
$V_{GS} > V_{po}$ $V_{GS} < V_{FB}$ $V_{DS} < V_{DSat1}$	$I_D = \frac{q\mu_b N_D}{L_{effb}}\left(\frac{1}{n+1}\frac{S_{max}-S_{min}}{(V_{FB}-V_{po})^n}\left((V_{GS}-V_{po})^{n+1}-(V_{GS}-V_{DS}-V_{po})^{n+1}\right)+S_{min}V_{DS}\right)$
$V_{GS} > V_{po}$ $V_{GS} < V_{FB}$ $V_{DS} > V_{DSat1}$	$I_D = \frac{q\mu_b N_D}{L_{effb}}\left(\frac{1}{n+1}\frac{S_{max}-S_{min}}{(V_{FB}-V_{po})^n}(V_{GS}-V_{po})^{n+1}+S_{min}V_{DS}\right)$
$V_{GS} > V_{FB}$ $V_{DS} < V_{DSat2}$	$I_D = \frac{q\mu_b N_D}{L_{effb}}S_{max}C_{ox}+\frac{\mu_{acc}C_{ox}W_{eff}}{L_{effacc}}\left(V_{DS}(V_{GS}-V_{FB})-\frac{1}{2}V_{DS}^2\right)$
$V_{GS} > V_{FB}$ $V_{DS} < V_{DSat1}$ $V_{DS} > V_{DSat2}$	$I_D = \frac{q\mu_b N_D}{L_{effb}}\left((S_{max}-S_{min})(V_{GS}-V_{FB})+S_{min}V_{DS}+\frac{S_{max}-S_{min}}{n+1}\frac{\left((V_{FB}-V_{po})^{n+1}-(V_{GS}-V_{DS}-V_{po})^{n+1}\right)}{(V_{FB}-V_{po})^n}\right)$ $+\frac{1}{2}\frac{\mu_{acc}C_{ox}W_{eff}}{L_{effacc}}(V_{GS}-V_{FB})^2$
$V_{GS} > V_{FB}$ $V_{DS} > V_{DSat1}$	$I_D = \frac{q\mu_b N_D}{L_{effb}}\left(S_{max}(V_{GS}-V_{FB})+\frac{S_{max}+nS_{min}}{n+1}(V_{FB}-V_{po})\right)+\frac{1}{2}\frac{\mu_{acc}C_{ox}W_{eff}}{L_{effacc}}(V_{GS}-V_{FB})^2$

[a]These expressions include both bulk channel and accumulation channel currents. The different symbols used in the equations are defined in Table 1.2.

Table 1.2 Definitions of the symbols used in Table 1.1

Symbol	Definition
V_{FB}	Flatband voltage
V_{po0}	Linear pinch-off voltage at $V_D = 0V$
V_{po}	Pinch-off voltage $V_{po} = V_{po0} - \eta V_{DS}$
η	DIBL coefficient
C_{ox}	Gate oxide capacitance normalized by the channel perimeter
W_{eff}	Channel perimeter
S	Neutral (nondepleted) cross section of the channel ($S = S_{min}$ when the surface is inverted, and $S = S_{max}$ when the surface is accumulated.)
V_{DSat1}	Drain saturation voltage for the neutral bulk channel $V_{DSat1} = \frac{V_{GS} - V_{po0}}{1-\eta}$
V_{DSat2}	Drain saturation voltage for the accumulation channel $V_{DSat2} = V_{GS} - V_{FB}$
L_{effacc}; L_{effb}	Effective length of the accumulation channel; effective length of the neutral bulk channel
μ_{acc}; μ_b	Effective carrier mobility in the accumulation channel; effective carrier mobility in the neutral bulk channel

A number of more sophisticated models based on solving Poisson's equation for a simple-gate, a DG, a trigate, and a GAA geometry have been developed since and are available in the literature. Let us look at some of them, sorted in chronological order:

- Model for threshold voltage in trigate junctionless transistors [100].
- Charge-based model for DG junctionless transistor, valid in deep depletion (subthreshold) to accumulation and from linear saturation [101].
- Theory of the junctionless transistor, applied to a cylindrical GAA nanowire device transistor, including subthreshold operation [102].
- Atomic-scale simulation of the short-channel (3 nm) cylindrical GAA nanowire transistor [103].

- Model for a junctionless thin-film SOI FET with a single gate [104].
- Surface-potential-based model for trigate junctionless transistors. This model accounts for short-channel effects [105].
- Surface-potential-based model for long-channel DG MOSFETs, valid in deep depletion (subthreshold) to accumulation and from linear to saturation [106].
- Physically based threshold voltage definition for trigate junctionless transistors [107].
- Model including short-channel effects in trigate junctionless transistors [108].
- Continuous model for junctionless cylindrical GAA nanowire MOSFETs [109].
- Model for the DG junctionless transistor in the subthreshold region, including short-channel effects [110].
- Model for ultrathin junctionless DG MOSFETs in the subthreshold regime [111].
- First-principle-based analysis of silicon GAA nanowire transistors with a gate length of 3 nm [112].
- Charge-based model for DG junctionless FETs, including inversion when the gate bias is sufficiently negative (N-channel device) [113].
- A 2D closed form analytical compact model for long- and short-channel junctionless accumulation mode DG MOSFETs developed using conformal mapping techniques [114].
- Compact model of subthreshold current, including source/drain depletion effects, for short-channel cylindrical GAA junctionless MOSFETs [17].
- Model for short-channel triple-gate junctionless transistors [115].
- Charge-based model for DG junctionless FETs, including charge trapped at the interface (interface charge trapping is at the core of junctionless nanowire chemical and biochemical sensor operation) [116].
- Charge-based compact analytical model for trigate junctionless nanowire transistors, validated by a comparison with both numerical simulations and measured data [117].

- Analytical model for short-channel trigate junctionless MOSFETs, calibrated with numerical simulations, and capable to accurately predict the DIBL and the SS as a function of gate length [118].
- Analytical model for the dynamic behavior of triple-gate junctionless transistors that includes short-channel effects, doping concentration, device dimensions, and quantum effects. It predicts gate capacitances (C_{GG}, C_{GS}, C_{GD}) and transconductance [119].
- Charge-based compact model for triple-gate junctionless MOSFETs that includes short-channel effects, saturation velocity overshoot, series resistance, and mobility degradation effects. The model consists of a single equation that covers subthreshold and above-threshold operation, as long as $V_G \leq V_{FB}$ [120].
- A threshold voltage model for trigate junctionless transistor, including substrate bias effects [121].
- A continuous charge-based compact model derived using Pao–Sah's dual integral applying to all operation regions (fully depleted, partly depleted, and accumulation) and including charge modeling for AC and transient circuit simulation. This model has been implemented in the HSPICE circuit simulator with Verilog-A language and used to simulate a junctionless transistor inverter and oscillator without any convergence problems [122].

1.2.1.11 Carrier confinement effects

Nanowire transistors exhibit carrier confinement effects when their section is small enough and both temperature and drain voltage are low enough. In nanowire devices, carriers are confined in the directions perpendicular to carrier transport, which gives rise to effects such ad formation of energy sub-bands and population of valleys that can be different than in bulk material [123]. These effects manifest themselves in the form of oscillations of transconductance when gate voltage is increased. In inversion-mode nanowire transistors these effects appear above the inversion threshold, in which case the inversion layer may be close to the nanowire center when the cross section is small or close to the gate oxide when the cross section is larger. In a junctionless transistor, the oscillations appear as soon as a neutral channel is formed, and the carrier centroid is always near the center of the nanowire, which yields clearly visible oscillations [124–126]. Oscillations have even been observed at room temperature in silicon GAA junctionless nanowires transistors with a diameter of 5 nm [127], in V-groove germanium junctionless transistors with a channel thickness of 14 nm [128], and in planar SOI junctionless FETs with a silicon thickness of 2.5 and 5 nm [129]. Subthreshold oscillations can be observed in some cases. These are attributed to Coulomb blockade effects. The Coulomb oscillations in the subthreshold transport are due to the transistor channel breaking up into one or more charge islands that are tunnel-coupled to the source and drain electrodes. The formation of these islands may be either due to disorder within the channel, for example, surface roughness or RDFs, or due to strain- or confinement-induced tunnel barriers at the extremities of the transistor channel [130].

1.2.2 Junctionless Transistor Architectures

In the literature one can find many different architectures that have been used to make junctionless transistors (Fig. 1.11).

1.2.2.1 SOI trigate

The first published junctionless transistors were trigate (aka pi-gate or omega-gate) devices (Fig. 1.11A). The fabrication of such a device is quite simple: nanowires are etched in an SOI film and uniformly doped using ion implantation and thermal annealing. A gate stack is then grown or deposited, and the gate electrode is patterned to form the final structure. The gate material can either be doped polysilicon or a midgap metal [10–12].

1.2.2.2 SOI planar transistors

It is possible to fabricate planar SOI junctionless transistors (Fig. 1.11B), but the silicon film must be very thin in order to be able to turn the device off. Very thin film devices operate

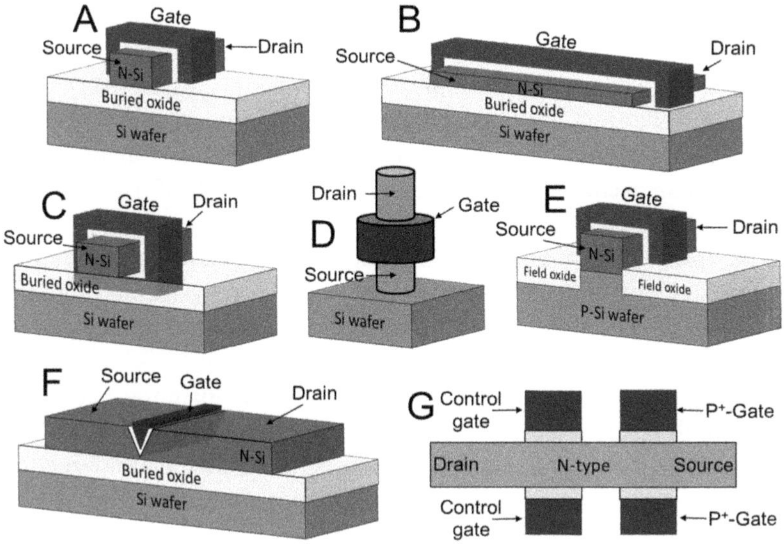

Figure 1.11 Junctionless transistor architectures. (A) Trigate/omega-gate SOI device (cross section perpendicular to channel); (B) thin-film SOI device (cross section perpendicular to channel); (C) horizontal GAA device (cross section perpendicular to channel); (D) vertical GAA device; (E) bulk trigate junctionless transistor with junction isolation underneath the transistor; (F) V-groove gate SOI device; (G) junctionless TFET with electrostatic doping control.

properly, but the SS and the off-current degrade rapidly if the film thickness is increased [131, 132]. Using very thin films is rather inconvenient because the source and drain resistance is high. This can be somehow corrected by thinning the film locally using a V-groove (Fig. 1.11F), which allows for fabricating both a thin channel and relatively thick source and drain [3, 4]. By placing metallic oxide nanoparticles in the gate stack of the device at the bottom of the V-groove, memory effect can be obtained. Using a gate oxide composed of a thin (3 nm) tunnel SiO_2 layer, ferritin nanoparticles (FeO_x, CoO_x, or InO_x nanoparticles), and an Al_2O_3 (10–20 nm) top layer, hysteresis of the $I_D(V_G)$ curves with an amplitude of 2 V can be obtained. This corresponds to an estimated injection/removal of 10 electrons per nanoparticle [133]. Such a device has a potential for high-density flash storage applications.

Planar strings of junctionless transistors have been used to demonstrate the operation a nonvolatile NAND memory. Each string consists of 32 junctionless transistors in series (bit line [BL]), each having a floating gate as the storage element, in addition to a control gate (word line [WL]). The WLs and BLs are arranged in a crossbar architecture. Devices with a half pitch of 15 nm have been fabricated, and simulations show they can be scaled down to a half pitch of 10 nm. The gate stack is composed of a junctionless transistor, a 7-nm-thick SiO_2 tunnel oxide, a polysilicon/titanium nitride (TiN) floating gate, a HfAlO/SiO_2/HfAlO intergate dielectric, and a polysilicon control gate [134].

Because of the absence of the need for forming junctions, junctionless transistors can be formed using a relatively low temperature process, which makes them attractive candidates for 3D monolithic integration (sequential monolithic integration). A predoped SOI layer can be transferred to the top of an already processed wafer using wafer bonding and handle wafer removal. Junctionless transistors can be made in that layer using low-temperature gate stack formation, without the need for junction implantation and annealing. Such a process was demonstrated using a gate-first approach process using an ALD HfO_2 dielectric deposited at 300°C and a CVD TiN metal gate deposited at 430°C, which was the maximum temperature used in the process. Such a low thermal budget makes junctionless transistor fabrication compatible with the low-κ dielectric Cu BEOL structures used in the bottom device layer below the (top) junctionless transistors [135, 233]. Polysilicon junctionless transistors can be used as a low-cost, low-performance alternative for low-temperature top-tier device fabrication [83].

1.2.2.3 GAA

The GAA architecture offers better electrostatics and hence better short-channel effect control than any other gate configuration (Figs. 1.11C and 1.11D). GAA junctionless transistors have been fabricated both in single-crystal and polycrystalline silicon [136, 137]. Several horizontal GAA nanowires can be stacked on top of each other in order to increase the current drive per device footprint [138, 139]. Single-crystal and polycrystalline silicon GAA junctionless transistors can be used as flash memory devices using either a SONOS gate stack [140, 141] or a gate oxide containing silicon nanocrystals [142]. Such devices have a potential for 3D-stacked memory applications [143].

1.2.2.4 Vertical nanowire FETs

Vertical nanowire junctionless transistors have been reported by several groups (Fig. 1.11D). The junctionless architecture is particularly well suited to the fabrication of vertical GAA nanowire transistors since the gate is automatically self-aligned to source and drain. Arrays of vertical GAA nanowire transistors with a gate length of 14 nm have shown to have overall good electrical characteristics, with excellent electrostatic behavior (SS = 95 mV/decade and DIBL = 7 mV/V) [144, 145]. These excellent short-channel characteristics are maintained for nanowire diameters up to 40 nm owing to the channel modulation effect [16, 146].

Vertical nanowire junctionless transistors with gate lengths down to 20 nm and a SONOS gate stack have been shown to have comparable electrical characteristics to those of junction-based nanowire SONOS devices (SS $<$ 70 mV/decade, leakage current $<$1 pA). Because there is no need to form PN junctions, the process complexity is significantly reduced and this device becomes a suitable platform for vertically stacked ultra-high-density memory applications [143]. Operation of a junctionless vertical silicon nanowire GAA SONOS memory (JL-SONOS) with two physical storage nodes per cell was first reported in 2011 [147]. In this device, charges are separately stored in the nitride layer near the top/bottom region of the vertical wire channel, that is, either near the source or near the drain. Measurements of the program/erase speed, endurance, and retention reveal that a robust 2-bit-per cell storage can be achieved. This structure and the vertical nature of the transistor relax channel length and integration density limitations associated with traditional horizontal devices. In addition, the absence of junctions makes this device highly manufacturable at a low cost and within a relatively low thermal budget.

Vertical junctionless nanowires also look promising for static random-access memory (SRAM) applications. The junctionless architecture greatly relaxes process constrains by avoiding the need for precise junction formation and the need to align the gate to the source and drain junctions. A SRAM cell design was proposed by Veloso et al. in which two vertical junctionless transistors of the same polarity are stacked on top of each other (i.e., N-channel on N-channel and P-channel on P-channel), such that a six-transistor SRAM cell is stacked on top of another six-transistor SRAM cell. Both cells share supply voltage (VDD), ground (VSS), bit line (BL), and inverted bit line (BLB), while access of the full 2-bit 12-transistor SRAM is controlled by two separate WLs. This design offers a 39% increase in density compared to a standard SRAM design, mainly due to a simplified interconnect scheme [148]. Vertical MOS transistors are indeed well adapted to the formation of memory arrays because the area penalty associated with interconnects is lessened compared to horizontal devices.

Vertical GAA transistors offer promises for RF applications, mainly because of their reduced gate and Miller capacitances. Several groups have investigated the matter using both silicon and II-V nanowires [149, 150]. Finally, one can mention the fabrication of P-channel vertical Ge/Si core/shell nanowire transistors. Because of the excellent gate coupling of the GAA architecture, a 1D hole gas is formed in the nanowire core, enabling the devices to deliver a high on-current (750 μA/μm for $L_G = 544$ nm and $V_{DS} = 2$ V) [151].

1.2.2.5 Bulk and stacked PN junctionless transistors

The bulk junctionless transistor is junctionless in the sense that there are no source and drain junctions, but a PN junction is needed underneath the device to isolate the source, drain, and channel from the substrate (Fig. 11.E) [152]. This type of device reportedly offers a better on-current/off-current ratio and better short-channel characteristics than its SOI counterpart because the effective channel thickness is reduced by the channel/substrate PN junction [153]. Single-crystal SiGe junctionless bulk trigate FETs were successfully fabricated on a silicon substrate by Kim et al., and it was shown that reverse-biasing the PN junction underneath the channel can modulate the effective channel thickness and, therefore, modulate the threshold voltage [154]. Polysilicon SOI junctionless

FETs have been fabricated using a similar structure; the device is a P-channel device made out of P-type polysilicon. However, there is a layer of N-type polysilicon between the P-type layer and the buried oxide. This configuration seems to increase performance and relaxes constraints on the (poly)silicon film thickness, that is, thicker films can be used while retaining good turn-off characteristics. As thicker films can be used, source and drain resistance are decreased [155–157]. The comparative analysis of germanium- and silicon-stacked PN or NP inversion-mode, accumulation-mode, and junctionless transistors with a gate length of 3 nm confirms the superior short-channel effect immunity of the junctionless architecture [158].

Furthermore, several N-type and P-type polysilicon layers can be stacked on each other to form a P/N/P/N/P/N ... P/N structure. This type of structure is still junctionless in the direction of current flow and offers a large current drive per footprint, as well as good short-channel effect control [159].

1.2.2.6 Tunnel FET

It is possible to replace physical doping of a transistor by field-induced or electrostatic doping. This has been first described by de Marchi et al. in 2012, in which the polarity (i.e., either N-channel or P-channel) of a GAA nanowire FET can be changed at will [160, 161]. This is accomplished by means of applying a bias to "polarization gates," which turns the source and drain regions N-type if one applies a positive bias to the polarization gates or P-type if one applies a negative bias to the polarization gates. The concept has been extended to junctionless transistors in which the N^+ doping of the device can locally be turned to P^+ using a gate with high work function, such as platinum, called the P-gate, as shown in Fig. 1.11G. According to simulations, the proposed device can operate as an N-channel P^+-N-N^+ tunnel FET (TFET), where the P^+ region is electrostatically induced by the P-gate and the P^+-N barrier (tunnel barrier amplitude) is varied using the control gate [162]. When the control gate is positively biased BTBT occurs at the P^+-N barrier and an electron current flows through the device. If parameters such as gate oxide thickness and dielectric constant, and gate metal work functions, are properly adjusted, a SS below 60 mV/decade at room temperature is predicted. Likewise, one can design a P-channel TFET using a P^+-doped junctionless architecture and a low work function, thereby replacing the P-gate by an N-gate [163]. It is worth noting that, according to simulations, a sharp SS can also be obtained in junctionless transistors using impact ionization effects instead of BTBT and using electrostatic doping induced by appropriate gate work functions and applied biases [164, 165].

A polysilicon GAA junctionless FET with an ultrathin channel (0.65 nm thick) was shown to have a measured SS of 43 mV/decade at room temperature. Simulations using a 3D quantum transport simulation tool show that a SS this low is due to BTBT in the channel-drain "junction." The device has an on-current/off-current ratio in excess of 10^8 [166].

1.3 "Nonsilicon" Junctionless Transistors

Because they do not require the formation of PN junctions and because the flow of current is in the bulk of the semiconductor material rather than on its surface, junctionless transistors have an architecture and process simplicity that lend themselves particularly well to the fabrication of FETs in many materials other than single-crystal silicon. Although polysilicon is technically made out of silicon, it will be included in this section because its properties are quite different from those of single-crystal silicon.

1.3.1 Germanium

P-channel junctionless GAA germanium nanowire transistors with a gate length of 250 nm, an I_{ON} of 390 μA/μm, and an I_{ON}/I_{OFF} ratio larger than 10^6 were demonstrated in 2015 and compared with inversion-mode nanowires [38]. At low gate voltage overdrive (V_G-V_{TH}) the junctionless transistors have a lower mobility than their inversion-mode counterparts due to higher impurity scattering. As the gate voltage is increased, however, the junctionless transistors show a much better immunity to SRS as the conduction channel is near the center of the nanowire in the junctionless device and near the germanium/gate dielectric interface in the inversion-mode devices, yielding higher mobility and current drive (Fig. 1.8). The mobility in the junctionless transistors increases with temperature, which is another indicator of impurity scattering–limited mobility.

In contrast to P-channel FETs, N-channel germanium FETs have a notoriously poor performance despite the high mobility of electrons in bulk germanium. This is due to the presence of a high density of interface states located in the upper part of the bandgap (i.e., near the conduction band edge) at the germanium/insulator interface [167]. These interface states are due to the formation of substoichiometric GeO_x compounds resulting from a reaction from a stoichiometric GeO_2 and oxygen vacancies from the Ge during thermal processing. A high density of acceptor-like interface states is also found in the upper part of the Ge bandgap when high-κ dielectrics are used in the gate stack [168]. It is, therefore, logical to consider the use of junctionless devices where surface scattering is reduced and where the channel is in the center of a nanowire.

N-channel germanium GAA junctionless transistors were first demonstrated by Wong et al. in 2014 [169]. The devices featured tensile strain and phosphorous doping concentrations ranging from 10^{19} to 8×10^{19} cm^{-3}. Mobility in transistors with a gate length of 280 nm reached 125 cm^{-2}V^{-1}s^{-1}. As in the case of P-channel junctionless transistors, surface scattering is negligible and the mobility does not decrease with gate voltage overdrive. Furthermore, electron mobility increases with temperature, as expected for a device where impurity Coulomb scattering is dominant. Improvement in processing and reduction of gate length to 60 nm yielded N-channel GAA Ge junctionless nanowire transistors with an I_{ON} of 1235 μA/μm at $V_G - V_{TH} = V_{DS} = 1$ V, a SS of 95 mV/decade, an I_{ON}/I_{OFF} ratio of 2×10^6, and reduced noise power density [170]. Here again, mobility shows little dependence on gate voltage overdrive (V_G-V_{TH}) and an increase in temperature. The channel mobility reached 250 and 100 cm^{-2}V^{-1}s^{-1} in nanowires with a width of 10 and 7 nm, respectively. It is worth noting that the density of states (DoS) in germanium nanowires is lower than in silicon. As a result, the quantum capacitance, and thus the carrier concentration, is lower in germanium nanowire devices than in the equivalent silicon devices, no matter the operation mode (inversion mode, accumulation mode, or junctionless). As a result, the performances of Ge and Si devices are comparable in scaled devices operating in the quantum capacitance limit, depending on operating conditions [21].

Planar (non-nanowire) N- and P-channel junctionless germanium transistors have been fabricated on germanium-on-insulator (GeOI) wafers [171]. The device structure is similar to that published by Migita et al. in 2012 [3]. Sub-10-nm channel lengths were successfully demonstrated [172]. Fully depleted, junctionless P-channel transistors with fully silicided (NiGe) source, and drains have been made in 10-nm-thick GeOI films. These transistors offer smaller source and drain resistance than conventional (NiGe on Ge) devices and have an I_{ON}/I_{OFF} ratio of 10^5, indicating that there is no increase in leakage due to the full silicidation of the source and the drain [173].

1.3.2 Polycrystalline Si and Ge

Transistors made in polycrystalline silicon or polycrystalline germanium have much lower performance than transistors made in single-crystal semiconductors and are typically used in applications such as flat-panel displays and are usually referred to as thin-film transistors (TFTs). Polycrystalline semiconductor films are composed of small crystallites separated by grain boundaries. The grain boundaries trap free carriers, such as electrons in N-type doped polysilicon. These traps are caused by dangling bonds at the edges of the crystallites. The trapped carriers cause the formation of a potential barrier at the boundary between grains, which in turn reduces the overall carrier mobility. This effect is at the origin of the unusual behavior of the resistance of polysilicon as a function of doping concentration: the resistance is high and is quite independent of dopant concentration until a concentration of a few 10^{18} cm^{-3} is reached, at which point the resistance suddenly decreases and becomes comparable to that of single-crystal silicon. To be more specific, the potential barrier increases with doping concentration until the traps at the grain boundaries become saturated with majority carriers, and any further increase in doping concentration decreases the height of the potential barriers [174, 175]. In polysilicon, this occurs at a doping concentration of a few 10^{18} cm^{-3}. In a classical TFT, carriers from the inversion channel are trapped at grain boundaries, causing the formation of potential barriers that need to be overcome by the rest of the carriers in the channel. This increases the threshold voltage and reduces the effective mobility [176]. In junctionless transistors, the channel is doped with a high impurity concentration, typically around 10^{19} cm^{-3}. The potential barriers are thus smaller than in TFTs, and any increase of gate voltage above accumulation threshold further increases the concentration of majority carriers, giving rise to a relatively high carrier mobility. N-channel GAA polysilicon junctionless transistors have a transconductance (and thus an apparent mobility) that is up to five times larger than that of inversion-mode GAA polysilicon TFTs, due to the abundance of carriers flowing through the body of the channel [177]. The pi-gate architecture provides good electrostatic control of the channel as well, and polysilicon junctionless nanowire transistors with pi-gates can reach a SS as low as 61 mV/decade (67 mV/decade average) [178, 179]. As is the case with single-crystal devices, the electrostatic control of polycrystalline junctionless transistors improves with the reduction of the device width (i.e., when the cross section evolves from planar to nanoribbon to nanowire). This, in turn, improves the on-current/off-current ratio, the SS, and the DIBL [180, 181]. Both current drive and on-current/off-current ratio can be increased in a junctionless transistor by using a hybrid structure that is still junctionless from source to drain but is constituted of a polysilicon channel that is N^+ at its bottom and P^+ at its top (to form a P-channel FET); the gate material is N^+ polysilicon. This technique allows one to utilize thicker nanowire channels and to increase the channel concentration without jeopardizing turn-off characteristics. This hybrid polysilicon P/N junctionless TFT reportedly exhibits an on-current/off-current ratio larger than 10^7, a SS of 64 mV/decade, and a DIBL of 3 mV/V. The channel thickness is 25 nm, and the current drive is nine times larger than that of a standard junctionless TFT with similar dimensions [156].

Vertical transistors are increasingly being used to fabricate 3D NAND flash memories. Each transistor has a polycrystalline silicon channel, and these transistors are used to make ultra-high-density, low-bit-cost memories. In vertical nanowire transistors it is extremely difficult, if not impossible, to align source and drain junctions to the gate. In a junctionless transistor, there is no need for such an alignment. Among several approaches the bit-cost-scalable (BiCS) flash technology has been demonstrated by several groups. The BiCS cell array consists of a multiple control gate placed in series around polysilicon "nanopipes" or "nanopillars." Typically, the control gate electrodes are made by etching holes in stacked conductive plates. These are subsequently coated with an oxide-nitride-oxide (ONO) gate dielectric stack or small floating gates and filled with the polysilicon channel material. Usually the polysilicon does not completely fill the holes but rather coats their insides, thereby forming a "macaroni-shaped" semiconductor layer, the inside of which is subsequently filled with a dielectric material. Each intersection of a control gate plate and a polysilicon "tube" forms a flash memory cell. In the case where the gate dielectric is ONO, electrons from the channel can tunnel though the bottom oxide layer and be trapped in the nitride, thereby changing the threshold voltage of the transistor and thus storing information. The resulting transistor is called a SONOS device. A string of such transistors forms a NAND flash memory structure [182–186]. As an alternative to ONO, silicon nanocrystals embedded in the gate oxide can trap charge and be used for nonvolatile memory devices. N- and P-channel junctionless polycrystalline GAA nanowires made using this technique have been shown to retain 95% of stored charge for a period of 10 years at a temperature of 85°C, making them contenders for system-on-panel and 3D-stacked memory applications [187]. To probe further, we recommend the excellent book by Prince on vertical 3D memory technologies [188].

Polycrystalline germanium has the potential for higher current drive than polysilicon transistors. P-channel poly-Ge junctionless transistors with a gate length smaller than 100 nm have been reported to carry a current 50% to 70% lower than that of their single-crystal counterparts. The poly-Ge nanowires are formed at a relatively low temperature (550°C) with an atomic layer deposition (ALD) gate oxide formed at 250°C and a back-end-of-line (BEOL) process with a maximum thermal budget of 300°C. This makes such devices good candidates for 3D monolithic integration [189]. Flash annealing of polycrystalline germanium allows one to increase grain size and, therefore, improve mobility. N- and P-channel junctionless transistors made using this technique with a maximum processing temperature of 400°C have shown mobilities of 200 and 140 $cm^2V^{-1}s^{-1}$ for electrons and holes, respectively. This translates into drive currents of 311 and 119 μA/μm for a gate length of 69 nm, which is comparable with the current drive of single-crystal silicon MOSFETs with a gate length of 60 nm [190]. Laser-recrystallized, large-grain polycrystalline germanium has been used to fabricate N-channel junctionless transistors. These devices were made using plasma in-situ-doped nanocrystalline Ge recrystallized using visible pulsed laser annealing. The germanium film exhibits a high electrically active concentration of 2×10^{19} cm^{-3}. The fabricated N-type poly-Ge junctionless transistors show an I_{ON}/I_{OFF} ratio of 6×10^4, a subthreshold swing of 237 mV/decade at $V_D = 1$ V, and a DIBL of 101 mV/V for $L_G = 50$ nm. Such low-thermal-budget devices could be used in a 3D sequential-layer integration process and flexible electronics [191].

1.3.3 III-V Semiconductors

III-V transistors are more prone to short-channel and direct source-to-drain tunneling than silicon MOSFETs because they have a higher dielectric constant and a lower effective mass (at least for electrons) than silicon. As a result, both SS and DIBL degrade rapidly in III-V nanowire transistors as the gate length is reduced to sub-10-nm dimensions. Using the effective gate length modulation property of junctionless transistors, this problem can, however, be mitigated. Improvements in the SS and DIBL on the order of 20% and 40% can be obtained in InP, GaAs, and GaSb GAA MOSFETs, respectively, with a gate length of 10 nm and a cross section of 5×5 nm^2 by using junctionless architecture instead of the conventional inversion-mode architecture [192].

The junctionless architecture is well suited to III-V materials. The high dopant concentration present in the channel of a junctionless MOSFET is less problematic for a semiconductor such as $In_{0.53}Ga_{0.47}As$ than it is for silicon. Indeed, the bulk electron mobility in silicon is lower than 100 $cm^2V^{-1}s^{-1}$ for a donor doping concentration, N_D, equal to 10^{19} cm^{-3}, while in the mobility $In_{0.53}Ga_{0.47}As$ is ~4000 $cm^2V^{-1}s^{-1}$ for a similar doping level. Moreover, the junctionless architecture circumvents the difficulties associated with source and drain implantation and thermal activation techniques that are generally used to form the source and drain regions of III-V inversion-mode MOSFETs. Junctionless $In_{0.53}Ga_{0.47}As$ MOSFETs have been demonstrated with an I_{ON}/I_{OFF} ratio of 10^6, a SS of 88 mV/decade, and a flatband effective mobility of 1200 $cm^2V^{-1}s^{-1}$ [193, 194]. As in the case of silicon devices, both volume conduction and surface accumulation are observed. Because the InGaAs/gate oxide interface is of lesser quality than the Si/gate oxide interface, however, the surface accumulation mobility is slightly lower than the bulk mobility [195]. Trigate gallium nitride (GaN) junctionless transistors have been reported as well. The devices were made in N-type GaN fins and nanowires sitting on top of a semi-insulating GaN layer grown on a sapphire substrate [196]. InP junctionless FinFETs with an extremely high fin aspect ratio (H_{fin}/W_{fin} = 600 nm/20 nm = 30) have been shown to reach an I_{ON}/I_{OFF} ratio of 3×10^5 [197]. AlGaN/GaN junctionless nanowire transistors were fabricated with an omega-gate structure. The devices show superior electrostatics with an I_{ON}/I_{OFF} ratio of 10^{10} and a SS of 62 mV/decade [198]. GAA GaN nanowire transistors were fabricated on GaN-on-insulator wafers. These achieve an I_{ON}/I_{OFF} ratio of 10^8 and an off-current of 0.1 pA/μm [199]. GAA indium arsenide (InAs) junctionless transistors fabricated using the VLS technique have been reported. These devices show an electron mobility of 2900 and 2200 $cm^2V^{-1}s^{-1}$ at temperatures of 50 K and 300 K, respectively. At low temperatures, measurement shows activated transport due to donors present in the bulk as well as at the nanowire/gate oxide interface. At room temperature, the carrier transport is limited by the interplay of remote Coulomb scattering and acoustic phonon scattering [200].

1.3.4 Other Materials

Because there is no requirement to form junctions or use doping gradients, the junctionless architecture can be used to make MOSFETs in a variety of materials, often using a low-thermal-budget process. Here are some examples.

1.3.4.1 Transition metal dichalcogenides

Junctionless transistors have been fabricated in 2D transition metal dichalcogenides (TMDs) such as molybdenum disulfide (MoS_2) [201–203]. TMDs are usually used in monolayers or a few monolayers to fabricate MOSFETs that have excellent short-channel characteristics, because the film of the TMD can be as thin as a fraction of a nanometer and because TMD materials have a low dielectric constant. When comparing inversion-mode (with junctions) and junctionless TMD transistors, the usual properties encountered in other types of FETs are still observed: a better SS and DIBL in junctionless devices due to modulation of the effective channel length with gate voltage [204], a better I_{ON}/I_{OFF} ratio, and a lower drain electric field [205]. The drawbacks are also the same as in other types of FETs, that is, higher variability and a higher threshold voltage control due to the high doping concentration. The advantage of the junctionless architecture over the inversion-mode structure is quite evident when it comes to processing TMD films into transistors. One study in particular highlights the following points [201]: (i) junctionless $MoSi_2$ FETs outperform their Si counterparts, (ii) the junctionless FET structure further improves the device electrostatics and on-current in MoS_2-based FETs, (iii) the doping density can be optimized to reach the highest on-current, and (iv) the threshold voltage of MoS_2-based junctionless FETs can always be turned to be positive by using proper gate metals. This allow one to use higher doping concentrations, which is desired to reduce the resistance of contacts. The scattering effect due to high doping may be an issue for high-performance circuits but not for low-power applications since the latter do not rely heavily on carrier mobility.

1.3.4.2 Carbon nanotube transistors

Junctionless carbon nanotube transistors (CNTs) have been fabricated and tested as sensors for cholesterol [206]. The transistors show a linear response of drain current versus cholesterol concentration in liquid concentrations between 0 and 20 micromoles, in which range the current varies from 0 to 1400 μA. Sensing is made possible by functionalizing the ZnO_2 gate dielectric with polyaniline, which produces electrical charges when in contact with cholesterol molecules. The gate electrode is a solution containing the species (cholesterol) to be detected.

1.3.4.3 Metal oxide transistors

Junctionless transistors have been made in indium-tin oxide (ITO) and zinc oxide (ZnO). The ITO transistors have been formed by a low-temperature process on both polymer and paper substrates. This approach opens the door to manufacturing cheap, flexible transistor circuits for wearables and other low-cost circuits. Transparent junctionless transistors made on polymer films exhibit a SS of 84 mV/decade, an I_{ON}/I_{OFF} ratio of 5.5×10^7, and a current drive of several microamps [207]. ITO junctionless transistors made on paper substrates have been shown to deliver a drain current of tens of microamps and an I_{ON}/I_{OFF} ratio of 2×10^6 [208]. ITO is a semiconductor that requires no doping because the free electron concentration in the as-deposited films is higher than 10^{20} cm^{-3}. ZnO is another semiconducting metal oxide with a bulk bandgap of 3.37 eV. Like ITO films, ZnO films are naturally N-type. SOI-like ZnO-on-oxide junctionless transistors have been proposed. Simulations suggest that, because of the large bandgap of ZnO, I_{ON}/I_{OFF} ratios as high as 10^{10} can be obtained with a current drive up to 800 μA/μm for a gate length of 20 nm [209]. Transparent indium-zinc oxide (IZO) junctionless transistors have been fabricated on glass substrates using a simple sputtering process through a shadow mask. The gate dielectric is an SiO_2-based solid electrolyte film. The transistors have a SS smaller than 200 mV/decade and an I_{ON}/I_{OFF} ratio of 10^6 for a supply voltage of 1 V. Such junctionless transparent low-voltage FETs could have potential applications in low-cost portable sensors [210]. "Biodegradable" junctionless transistors have been made in Al:ZnO films deposited on a sodium alginate membrane. Under 1 V operation the devices have an I_{ON}/I_{OFF} ratio of 10^5. The devices are biodegradable because they completely dissolve in water in a period of time as short as 1 h [211]. Junctionless transistors have also been made in IZO deposited on SiO_2-coated paper, using Al_2O_3 as the gate oxide. These transistors show a low-voltage operation of 1.5 V and a high performance with a reasonable field-effect mobility (>18 cm^2/V·s), a small subthreshold swing (<130 mV/decade), and a high current on/off ratio ($>10^6$) [212]. Junctionless TFTs made in transparent amorphous InGaZnO with thicknesses ranging from 10 to 50 nm have been shown to have I_{ON}/I_{OFF} ratios equal to or better than 10^7 [213].

1.3.4.4 Metallic transistors

Some ab initio studies indicate the possibility of forming junctionless nanowire transistors in metals. This can be made possible by to a semimetal-semiconductor transition in the case of tin (Sn) [214]. In that device, the diameters of the source and the drain are large enough for tin to be semimetallic and the diameter of the channel is small enough to have a bandgap. No doping is necessary since the channel carriers are supplied by the semimetal source. Decent simulated $I_D(V_G)$ and $I_D(V_D)$ characteristics were obtained for gate lengths as short as 2.3 nm for a channel diameter of 1 nm. The field effect has been observed experimentally in oxygen-rich TiN thin films. TiN is normally a metal with a free electron concentration of 10^{22} cm^{-3} or higher, but using a combination of quantum confinement to open a bandgap and high oxygen doping, the nitride can undergo a transition to a semiconductor state. Using this technique gate modulation of channel current has been observed at room temperature in junctionless transistors made in thin (4.8 nm) TiN films doped with a high oxygen concentration [215]. Another study suggests making a transistor using a gold (Au) nanowire with very small cross section (0.58 nm $\times$ 0.87 nm) combined with a GAA architecture. An energy gap forms in the channel, which enables the formation of a transistor. The current drive is quite large (up to 10 mA/μm) for a gate length of 3 nm. The I_{ON}/I_{OFF} ratio is, however, pretty poor ($I_{ON}/I_{OFF} = 6$ for $L_G = 3$ nm and $I_{ON}/I_{OFF} = 30$ for $L_G = 5$ nm), because the DoS in the nanowire bandgap is not equal to zero (the nanowire still shows a metallic behavior) [216].

1.4 Junctionless Nanowire Sensors

In some applications, frequently called "More than Moore" applications, it is suitable to combine pure microelectronic circuitry with a layer of devices such as sensors or actuators to create compact monolithic devices with high levels of functionality. For example, one can contemplate making intelligent, implantable medical biosensors that can be remotely accessed to monitor outpatients at home instead of having to keep them in a hospital environment. Nanowire-based biosensors are actively being developed to detect early warning signs of a variety of medical conditions, such as heart disease and cancer [217]. Semiconductor nanowires and nanowire transistors are particularly well adapted to the fabrication of chemical or biochemical sensors for several reasons:

- Nanowires have an extremely high surface-to-volume ratio, which is ideal for sensing minute quantities of a chemical compound [218].
- The liquid/solid interface of a nanowire transistor has a strong influence on the electronic properties of the nanowire and the presence of electrically charged molecules can easily be detected through changes in nanowire resistivity.
- It is possible to create dense arrays of nanowire sensors and connect them to readout electronics [219, 220].
- Nanowire sensors can be reusable [221].
- Silicon nanowire sensor devices are compatible with CMOS processing and can, therefore, offer biomedical and biochemical system-on-chip (SoC) solutions [222].

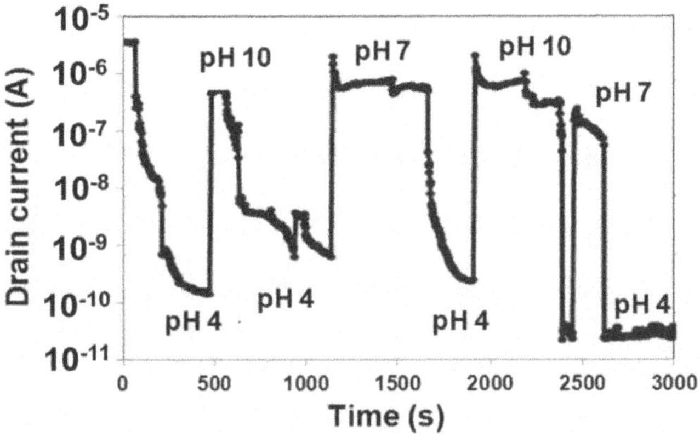

Figure 1.12 Drain current variation with pH level at $V_D = 1$ V and $V_{BG} = 0$ V for a back-gated silicon nanowire transistor of 20 nm width and 1 μm length.

Furthermore, the use of top-down silicon fabrication techniques can ensure wafer-scale device integration and repeatable device fabrication based on reliable process techniques established over the years by the semiconductor industry.

As a general rule, the sensitivity of a nanowire sensor increases when reduced doping, smaller diameter, and longer nanowires are used. For maximum sensitivity, the sensors should be operated in the depletion mode (i.e., the doping of the sensor and the molecule should have the same polarity) [223]. It is, however, worth pointing out that nanowire transistors with high doping concentrations can achieve high sensitivity, provided they are operated in the subthreshold region (i.e., in depletion); back-gated junctionless nanowire transistors have demonstrated very high sensitivity to sensing pH levels. When such a device operates in the subthreshold region, it exhibits a 3 orders of magnitude difference in current, responding to pH values changing from 4 to 7 or 10 (Fig. 1.12) [224]. Nanowire sensors can also be made sensitive to ionic concentrations in pH-neutral solutions obtained by diluting phthalate in buffered Fisher pH 7 solutions [225] (Fig. 1.13). To reduce fabrication cost, work is also ongoing to develop polysilicon junctionless nanowire transistors to detect biochemicals [226].

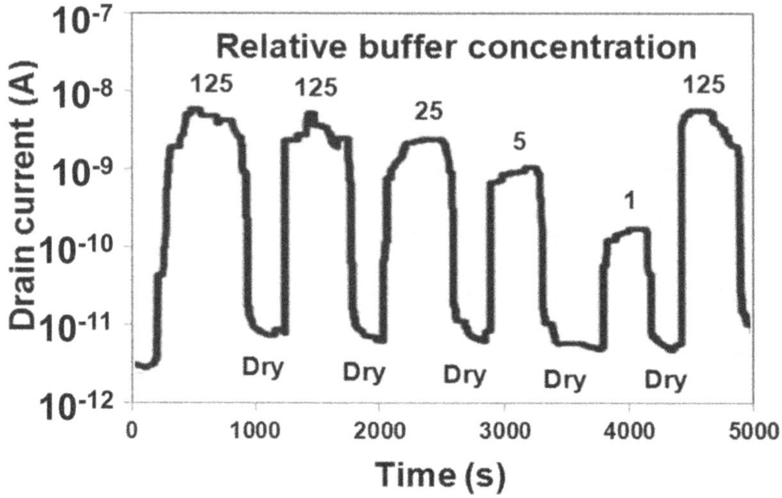

Figure 1.13 Time dependence of the drain current demonstrating the sensitivity of back-gated nanowire transistors to phthalate diluted at different concentrations in buffered Fisher pH 7 solutions.

Low-power junctionless nanowire sensors have been designed and fabricated, showing high sensitivity in the subthreshold regime [227]. Junctionless nanowire transistors can also be used for sensing subterahertz radiations [228], gases [229–231], and light [232].

1.5 Conclusions

The junctionless transistor architecture has been adopted by many research institutes worldwide. It allows one to fabricate MOSFETs in new materials much more easily than if the conventional inversion-mode were used. The junctionless transistor fabrication requires a relatively low thermal budget and is, therefore, suited for 3D monolithic integration. The performances of junctionless polycrystalline transistors are better than those of conventional TFTs, which can be useful for low-cost and wearable electronics. Some devices, such as N-channel germanium MOSFETs, show much better electrical properties when realized using the junctionless architecture than the classical inversion-mode structure. The

Table 1.3 Comparison of inversion-mode, accumulation-mode, and junctionless nanowire MOSFETs

	Inversion mode	Accumulation mode	Junctionless
Ease of fabrication	Ref	0	+
Applicability to new materials	Ref	0	+
Applicability to polycrystalline semiconductors	Ref	0	+
Coulomb scattering	Ref	+	-
Surface roughness scattering	Ref	+	+
Mobility	Ref	0	-
Channel length modulation	Ref	0	+
Applicability 3D monolithic	Ref	0	+
integration Variability	Ref	0	- (in subthreshold)
Subthreshold slope	Ref	0	0
Matching	Ref	0	-
Source and drain resistance	Ref	0	-
Current drive	Ref	0	-
I_{ON}/I_{OFF} ratio	Ref	0	+

Ref = reference, "0" = similar to reference, "+" = better than reference, and "-" = worse than reference.

relative merits of junctionless transistors are compared to those of inversion-mode and accumulation-mode transistors in Table 1.3.

References

1. Lilienfeld, J. E. (1925). Method and apparatus for controlling electric current, Canada patent 1745175.
2. Lilienfeld, J. E. (filed 1928, granted 1933). Device for controlling electric current, US patent 1900018.
3. Migita, S., Morita, Y., Masahara, M., and Ota, H. (2012). Electrical performances of junctionless-FETs at the scaling limit (L_{CH} = 3 nm), in *IEEE International Electron Devices Meeting (IEDM) Technical Digest*, pp. 191–194.
4. Migita, S., Morita, Y., Masahara, M., and Ota, H. (2013). Fabrication and demonstration of 3nm-channel-length junctionless field-effect transistors on silicon-on-insulator substrates using anisotropic wet etching and lateral diffusion of dopants, *Jpn. J. Appl. Phys.*, **52**, p. 04CA01.

5. Goh, K. H., Guo, Y., Gong, X., Liang, G. C., and Yeo, Y. C. (2013). Near ballistic sub-7nm junctionless FET featuring 1nm extremely-thin channel and raised S/D structure, in *IEEE International Electron Devices Meeting (IEDM) Technical Digest*, pp. 433–436.
6. Sorée, B., Magnus, W., and Pourtois, G. (2008). Analytical and self-consistent quantum mechanical model for a surrounding gate MOS nanowire operated in JFET mode, *J. Comput. Electron.*, **7**, pp. 380–383.
7. Lee, C. W., Afzalian, A., Yan, R., Dehdashti, N., and Colinge, J. P. (2009). Junctionless MuGFETs, *Proc. EUROSOI Conf.*, pp. 21–22.
8. Lee, C. W., Afzalian, A., Dehdashti Akhavan, N., Yan, Y., Ferain, I., and Colinge, J. P. (2009). Junctionless multigate field-effect transistor, *Appl. Phys. Lett.*, **94**, p. 053511.
9. Colinge, J. P., Lee, C. W., Afzalian, A., Dehdashti, N., Yan, R., Ferain, I., Razavi, P., O'Neill, B., Blake, A., White, M., Kelleher, A. M., McCarthy, B., and Murphy, R. (2009). SOI gated resistor: CMOS without junctions, *Proc. IEEE Int. SOI Conf.*, paper 11.1.
10. Colinge, J. P., Lee, C. W., Afzalian, A., Dehdashti, N., Yan, R., Ferain, I., Razavi, P., O'Neill, B., Blake, A., White, M., Kelleher, A. M., McCarthy, B., and Murphy, R. (2010). Nanowire transistors without junctions, *Nat. Nanotechnol.*, **5**(3), pp. 225–229.
11. Rios, R., Cappellani, A., Armstrong, M., Budrevich, A., Gomez, H., Pai, R., Rahhal-orabi, N., and Kuhn, K. (2011). Comparison of junctionless and conventional trigate transistors with L_g down to 26 nm, *IEEE Electron Device Lett.*, **32**(9), pp. 1170–1172.
12. Barraud, S., Berthomé, M., Coquand, R., Cassé, M., Ernst, T., Samson, M. P., Perreau, P., Bourdelle, K. K., Faynot, O., and Poiroux, T. (2012). Scaling of trigate junctionless nanowire MOSFET with gate length down to 13 nm, *IEEE Electron Device Lett.*, **33**(9), pp. 1225–1227.
13. Georgiev, V. P., Mirza, M. M., Dochioiu, A. I., Lema, F. A., Amoroso, S. M., Towie, E., Riddet, C., MacLaren, D. A., Asenov, A., and Paul, D. J. (2016). Experimental and simulation study of a high current 1D silicon nanowire transistors using heavily doped channels, *IEEE Nanotechnol. Mater. Devices Conf. (NMDC)*, pp. 1–3; and (2017). Experimental and simulation study of 1D silicon nanowire transistors using heavily doped channels, *IEEE Trans. Nanotechnol.*, **16**(5), pp. 727–735.
14. Yu, T. H., Hsu, E., Liu, C. W., Colinge, J. P., Sheu, Y. M., Wu, J., and Diaz, C. H. (2013). Electrostatics and ballistic transport studies in junctionless nanowire transistors, in *Proceedings of the International Conference on Simulation of Semiconductor Processes and Devices (SISPAD)*, pp. 85–88.

15. Colinge, J. P., Kranti, A., Yan, R., Ferain, I., Dehdashti Akhavan, N., Razavi, P., Lee, C. W., Yu, R., and Colinge, C. A. (2011). A simulation comparison between junctionless and inversion-mode MuGFETs, *Electrochem. Soc. Trans.*, **35**(5), pp. 63–72.
16. Trevisoli, R., Doria, R. T., de Souza, M., and Pavanello, M. A. (2015). Effective channel length in junctionless nanowire transistors, *30th Symposium on Microelectronics Technology and Devices (SBMicro)*, pp. 1–4.
17. Razavi, P., Fagas, G., Ferain, I., Yu, R., Das, S., and Colinge, J. P. (2012). Influence of channel material properties on performance of nanowire transistors, *J. Appl. Phys.*, **111**, p. 124509.
18. Xiao, Y., Zhang, B., Lou, H., Zhang, L., and Lin, X. (2016). A compact model of subthreshold current with source/drain depletion effect for the short-channel junctionless cylindrical surrounding-gate MOSFETs, *IEEE Trans. Electron Devices*, **60**(5), pp. 2176–2181.
19. Gundapaneni, S., Ganguly, S., and Kottantharayil, A. (2011). Enhanced electrostatic integrity of short-channel junctionless transistor with high-k spacers, *IEEE Electron Device Lett.*, **32**(10), pp. 1325–1327.
20. Saini, G., and Choudhary, S. (2017). Improving the subthreshold performance of junctionless transistor using spacer engineering, *Microelectron. J.*, **59**, pp. 55–58.
21. Thirunavukkarasu, V., Lee, J., Sadi, T., Georgiev, V. P., Lema, F. A., Pandian Soundarapandian, K., Jhan, Y. R., Yang, S. Y., Lin, Y. R., Kurniawan, E. D., Wu, Y. C., and Asenov, A. (2017). Investigation of inversion, accumulation and junctionless mode bulk Germanium FinFETs, *Superlattices Microstruct.*, **111**, pp. 649–655.
22. Xiao, Y., Lin, X., Loi, H., Zhang, B., Zhang, L., and Chan, M. (2016). A short channel double-gate junctionless transistor model including the dynamic channel boundary effect, *IEEE Trans. Electron Devices*, **63**(12), pp. 4661–4667.
23. Hur, J., Lee, B. H., Kang, M. H., Ahn, D. C., Bang, T., Jeon, S. B., and Choi, Y. K. (2016). Comprehensive analysis of gate-induced drain leakage in vertically stacked nanowire FETs: inversion-mode versus junctionless mode, *IEEE Electron Device Lett.*, **37**(5), pp. 541–544.
24. Sahay, S., and Kumar, M. J. (2017). Physical insights into the nature of gate-induced drain leakage in ultrashort channel nanowire FETs, *IEEE Trans. Electron Devices*, **64**(6), pp. 2604–2610.
25. Takagi, S. I., Toriumi, A., Iwase, M., and Tango, H. (1994). On the universality of inversion layer mobility in Si MOSFETs: part I – effects

of substrate impurity concentration, *IEEE Trans. Electron Devices*, **41**(12), pp. 2357–2362.

26. Colinge, J. P., Lee, C. W., Ferain, I., Dehdashti Akhavan, N., Yan, R., Razavi, P., Yu, R., Nazarov, A. N., and Doria, R. T. (2010). Reduced electric field in junctionless transistors, *Appl. Phys. Lett.*, **96**, p. 073510.
27. Park, S. J., Jeon, D. Y., Montès, L., Mouis, M., Barraud, S., Kim, G. T., and Ghibaudo, G. (2014). Less mobility degradation induced by transverse electric-field in junctionless transistors, *Appl. Phys. Lett.*, **105**, p. 213504.
28. Sorée, B., Magnus, W., and Vandenberghe, W. (2011). Low-field mobility in ultrathin silicon nanowire junctionless transistors, *Appl. Phys. Lett.*, **99**, p. 233509.
29. Sun, S. C., and Plummer, J. D. (1980). Electron mobility in inversion and accumulation layers on thermally oxidized silicon surfaces, *IEEE Trans. Electron Devices*, **27**(8), pp. 1497–1508.
30. Chindalore, G., Mudanai, S., Shih, W. K., Tash, A. F., Jr., and Maziar, C. M. (1999). Temperature dependence characterization of effective electron and hole mobilities in the accumulation layers of N- and P-type MOSFET's, *IEEE Trans. Electron Devices*, **46**(6), pp. 1290–1294.
31. Bhuvaneshwari, Y. V., and Kranti, A. (2017). Analysis of mobility in double gate junctionless transistors, in *30th International Conference on VLSI Design and 16th International Conference on Embedded Systems*, pp. 283–288.
32. Ye, H. Y., Lan, H. S., and Liu, C. W. (2016). Electron mobility in junctionless Ge nanowire NFETs, *IEEE Trans. Electron Devices*, **63**(11), pp. 4191–4195.
33. Goto, K. I., Yu, T. H., Wu, J., Diaz, C. H., and Colinge, J. P. (2012). Mobility and screening effect in heavily doped accumulation-mode metal-oxide-semiconductor field-effect transistors, *Appl. Phys. Lett.*, **101**, p. 073503.
34. Rudenko, T., Nazarov, A., Ferain, I., Das, S., Yu, R., Barraud, S., and Razavi, P. (2012). Mobility enhancement in heavily doped junctionless nanowire silicon-on-insulator metal-oxide-semiconductor field-effect transistors, *Appl. Phys. Lett.*, **101**, p. 213502.
35. Park, S. J., Jeon, D. Y., Montès, L., Mouis, M., Barraud, S., Kim, G. T., and Ghibaudo, G. (2014). Low-temperature operation of junctionless nanowire transistors: Less surface roughness scattering effects and dominant scattering mechanisms, *Appl. Phys. Lett.*, **105**, p. 263505.

36. Wei, K., Zeng, L., Wang, J., Du, G., and Liu, X. (2016). Physically based evaluation of electron mobility in ultrathin-body double-gate junctionless transistors, *IEEE Electron Device Lett.*, **35**(8), pp. 817–819.
37. Rudenko, T., Nazarov, A., Yu, R., Barraud, S., Cherkaoui, K., Razavi, P., and Fagas, G. (2013). Electron mobility in heavily doped junctionless nanowire SOI MOSFETs, *Microelectron. Eng.*, **109**, pp. 326–329.
38. Wong, I. H., Chen, Y. T., Huang, S. H., Tu, W. H., Chen, Y. S., and Liu, C. W. (2015). Junctionless gate-all-around pFETs using in-situ boron-doped Ge channel on Si, *IEEE Trans. Nanotechnol.*, **14** (5), pp. 878–882.
39. Raskin, J. P., Colinge, J. P., Ferain, I., Kranti, A., Lee, C. W., Dedashti Akhavan, N., Yan, R., Razavi, P., and Yu, R. (2010). Mobility improvement in nanowire junctionless transistors by uniaxial strain, *Appl. Phys. Lett.*, **97**, p. 042114.
40. Sung, P. J., Cho, T. C., Hou, F. J., Hsueh, F. K., Chung, S. T., Lee, Y. J., Current, M. I., and Chao, T. S. (2017). High-performance uniaxial tensile strained n-channel JL SOI FETs and triangular JL bulk FinFETs for nanoscaled applications, *IEEE Trans. Electron Devices*, **64**(5), pp. 2054–2059.
41. Jeon, D. Y., Park, S. J., Mouis, M., Joo, M. K., Barraud, S., Kim, G. T., and Ghibaudo, G. (2014). Separation of surface accumulation and bulk neutral channel in junctionless transistors, *Appl. Phys. Lett.*, **104**, p. 263510.
42. Singh Parihar, M., Liu, F., Navarro, C., Barraud, S., Badewin, M., Ionica, I., Kranti, A., and Cristoloveanu, S. (2015). Back-gate effects and mobility characterization in junctionless transistor, in *Proceedings of the European Solid-State Research Conference (ESSDERC)*, pp. 282–285.
43. Singh Parihar, M., Liu, F., Navarro, C., Barraud, S., Badewin, M., Ionica, I., Kranti, A., and Cristoloveanu, S. (2015). Back-gate effects and mobility characterization in junctionless transistor, *Solid-State Electron.*, **125**, pp. 154–160.
44. Ghibaudo, G. (1988). New method for extraction of MOSFET parameters, *Electron. Lett.*, **24**(9), pp. 543–545.
45. Nagai, K., Tsuchiya, H., and Ogawa, M. (2013). Channel length scaling effects on device performance of junctionless field-effect transistor, *Jpn. J. Appl. Phys.*, **52**, p. 044302.
46. Shinada, T., Hori, M., Guagliardo, F., Ferrari, G., Komatubara, A., Kumagai, K., Tanii, T., Endo, T., Ono, Y., and Prati, E. (2011). Quantum transport in deterministically implanted single-donors in Si FETs, in *IEEE International Electron Devices Meeting (IEDM) Technical Digest*, pp. 697–700.

47. Ueda, A., Luisier, M., and Sano, Y. (2015). Enhanced impurity-limited mobility in ultra-scaled Si nanowires junctionless field-effect transistors, *Appl. Phys. Lett.*, **107**, p. 253501.
48. Lee, C. W., Ferain, I., Kranti, A., Dehdashti Akhavan, N., Razavi, P., Yan, R., Yu, R., O'Neill, B., Blake, A., White, M., Kelleher, A. M., McCarthy, B., Gheorghe, S., Murphy, R., and Colinge, J. P. (2010). Short-channel junctionless nanowire transistors, in *International Conference on Solid-State Devices and Materials Conference (SSDM)*, Tokyo, pp. 1044–1045.
49. Ansari, L., Feldman, B., Fagas, G., Colinge, J. P., and Greer, J. C. (2010). Simulation of junctionless Si nanowire transistors with 3 nm gate length, *Appl. Phys. Lett.*, **97**, p. 062105.
50. Park, S. J., Jeon, D. Y., Montès, L., Mouis, M., Barraud, S., Kim, G. T., and Ghibaudo, G. (2016). Behavior of subthreshold conduction in junctionless transistors, *Solid-State Electron.*, **124**, pp. 58–63.
51. Sing Parihar, M., Ghosh, D., and Kranti, A. (2013). Ultra low power junctionless MOSFETs for subthreshold logic applications, *IEEE Trans. Electron Devices*, **60**(5), pp. 1540–1546.
52. Jeon, D. Y., Park, S. J., Mouis, M., Berthomé, M., Barraud, S., Kim, G. T., and Ghibaudo, G. (2013). Revisited extraction methodology for electrical characterization of junctionless transistors, *Solid-State Electron.*, **90**, pp. 86–93.
53. Bhuvaneshwary, Y. V., and Kranti, A. (2017). Extraction and analysis of mobility in double-gate junctionless transistors, in *30th International Conference on VLSI Design and 16th International Conference on Embedded Systems (VLSID)*, Hyderabad, India, pp. 283–288.
54. Rudenko, T., Barraud, S., Georgiev, Y. M., Lysenko, V., and Nazarov, A. (2016). Electrical characterization and parameter extraction of junctionless nanowire transistors, *J. Nano Res.*, **39**, pp. 17–33.
55. Qian, X., Yang, Y., Zhu, Z., Zhang, S. L., and Wu, D. (2011). Evaluation of DC and AC performance of junctionless MOSFETs in the presence of variability, in *IEEE International Conference on IC Design & Technology*, pp. 1–4.
56. Jazaeri, F., Barbut, L., and Sallese, J. M. (2014). Trans-capacitance modeling in junctionless gate-all-around nanowire FETs, *Solid-State Electron.*, **96**, pp. 34–37.
57. Guin, S., Sil, M., and Mallik, A. (2017). Comparison of logic performance of CMOS circuits implemented with junctionless and inversion-mode FETs, *IEEE Trans. Electron Devices*, **64**(3), pp. 953–959.

58. Guin, S., Sil, M., and Mallik, A. (2017). Comparison of logic performance of CMOS circuits implemented with junctionless and inversion-mode FinFETs, *IEEE Trans. Electron Devices*, **64**(3), pp. 953–959.

59. Kranti, A., and Armstrong, G. A. (2008). Source/drain extension region engineering in FinFETs for low-voltage analog applications, *IEEE Electron Device Lett.*, **28**(2), pp. 139–141.

60. Kranti, A., and Armstrong, G. A. (2007). Design and optimization of FinFETs for ultra-low-voltage applications, *IEEE Trans. Electron Devices*, **54**(12), pp. 3308–3316.

61. Wang, J., Du, G., Wei, K., Zhao, K., Zeng, L., Zhang, Z., and Liu, X. (2014). Mixed-mode analysis of different mode silicon nanowire transistor-based inverters, *IEEE Trans. Nanotechnol.*, **13**(2), pp. 362–367.

62. Ghosh, D., Singh Parihar, M., Armstrong, G. A., and Kranti, A. (2012). High-performance MOSFETs for ultralow-power analog/RF applications, *IEEE Electron Device Lett.*, **33**(10), pp. 1477–1479.

63. Baral, B., Mohan Biswal, S., Padhee, J., De Debashis, and Sarkar, A. (2016). Effect of gate length downscaling on RF/analog and linearity performance of junctionless double gate MOSFET for analog/mixed signal system-on-chip applications & it's comparative study with conventional MOSFET, *Advances Ind. Eng. Manage.*, **5**(1), pp. 130–137.

64. Kuhn, K. J., Giles, M. D., Becher, D., Kolar, P., Kornfeld, A., Kotlyar, R., Ma, S. T., Maheshari, A., and Mudanai, S. (2011). Process technology variation, *IEEE Trans. Electron Devices*, **58**(8), pp. 2197–2208.

65. Lin, X., Lou, H., Xiao, Y., Wan, W., Zhang, L., and Chan, M. (2017). Silicon-based junctionless MOSFETs: device physics, performance boosters and variations, in *Outlook and Challenges of Nano Devices, Sensors, and MEMS* (Li, T., and Liu, Z., eds), Springer International Publishing AG, pp. 183–236.

66. Pelgrom, M. J. M., Duinmaijer, A. C. J., and Welbers, A. P. G. (1989). Matching properties of MOS transistors, *IEEE J. Solid-State Circuits*, **24**(5), pp. 1433–1440.

67. Pelgrom, M. J. M., Tuinhout, H. P., and Vertregt, M. (1998). Transistor matching in analog CMOS applications, in *IEEE International Electron Devices Meeting (IEDM) Technical Digest*, pp. 915–918.

68. Ghibaudo, G. (2012). Evaluation of variability performance of junctionless and conventional trigate transistors, *Solid-State Electron.*, **75**, pp. 13–15.

69. Chen, C. Y., Lin, J. T., and Chiang, M. H. (2017). Threshold-voltage variability analysis and modeling for junctionless double-gate transistors, *Microelectron. Reliab.*, **74**, pp. 22–26.

70. Dehdashti Akhavan, N., Umana-Membreno, G., Gu, R., Antoszewski, J., and Faraone, L. (2017). Random dopant fluctuations and statistical variability in n-channel junctionless FETs, *Nanotechnology*, **29**(2), p. 025203.

71. Aldegunde, M., Martinez, A., Barker, J. R. (2012). Study of discrete doping-induced variability in junctionless nanowire MOSFETs using dissipative quantum transport simulations, *IEEE Electron Device Lett.*, **33**(2), pp. 194–196.

72. Han, M. H., Chang, C. Y., Jhan, Y. R., Wu, J. J., Chen, H. B., Cheng, Y. C., and Wu, Y. C. (2013). Characteristic of p-type gate-all-around nanowire transistor and sensitivity analysis, *IEEE Electron Device Lett.*, **34**(2), pp. 157–159.

73. Leung, G., and Chui, C. O. (2012). Variability impact of random doping fluctuations on nanoscale junctionless FinFETs, *IEEE Electron Device Lett.*, **33**(6), pp. 767–769.

74. Chen, L. Y., Hsieh, Y. F., and Kao, K. H. (2017). Undoped and doped junctionless FETs: source/drain contacts and immunity to random dopant fluctuation, *IEEE Electron Device Lett.*, **38**(6), pp. 708–711.

75. Migita, S., Matsukawa, T., Morita, Y., Masahara, M., and Ota, H. (2013). Variability of short channel junctionless field-effect transistors caused by fluctuation of dopant concentration, in *Proceedings of the 14th International Conference on Ultimate Integration (ULIS)*, pp. 137–140.

76. Kuhn, K. J. (2012). Considerations for ultimate CMOS scaling, *IEEE Trans. Electron Devices*, **59**(7), pp. 1813–1828.

77. Chen, L. Y., Hsieh, Y. F., and Kao, K. H. (2017). Undoped and doped junctionless FETs: source/drain contacts and immunity to random doping fluctuations, *IEEE Electron Device Lett.*, **38**(6), pp. 708–711.

78. Cheng, Y. C., Wu, Y. C., Chen, H. B., Han, M. H., Lu, N. H., Su, J. S., and Chang, C. Y. (2013). High voltage characteristics of junctionless polysilicon thin film transistors, *Appl. Phys. Lett.*, **103**, p. 123510.

79. Park, J. T., Kim, J. Y., and Colinge, J. P. (2012). Negative-bias temperature instability and hot carrier effects in nanowire junctionless p-channel multigate transistors, *Appl. Phys. Lett.*, **100**, p. 083504.

80. Park, C. H., Ko, M. D., Kim, K. H., Lee, S. H., Yoon, J. S., Lee, J. S., and Jeong, Y. H. (2012). Investigation of low-frequency noise behavior after

hot-carrier stress in an n-channel junctionless nanowire MOSFET, *IEEE Electron Device Lett.*, **33**(11), pp. 1538–1540.

81. Cho, M., Hellings, G., Veloso, A., Simoen, E., Roussel, Ph., Kaczer, B., Arimura, H., Fang, W., Franco, J., Matagne, P., Collaert, N., Linten, D., and Thean, A. (2015). On and off state hot carrier reliability in junctionless high-k MG gate-all-around transistors, in *IEEE International Electron Devices Meeting (IEDM) Technical Digest*, pp. 366–369.
82. Toledano-Luque, M., Matagne, P., Sibaja-Hernández, A., Ciarella, T., Ragnarsson, L. A., Sorée, B., Cho, M., Mocuta, A., and Thean, A. (2014). Superior reliability of junctionless pFinFETs by reduced oxide electric field, *IEEE Electron Device Lett.*, **35**(12), pp. 1179–1181.
83. Lin, J. Y., Puttaveerappa Vijay Kumar, M., and Chao, T. S. (2018). Junctionless nanosheet (3 nm) poly-Si TFT: electrical characteristics and superior positive gate bias stress reliability, *IEEE Electron Device Lett.*, **39**(1), pp. 8–11.
84. Clément, N., Han, X. L., and Larrieu, G. (2013). Electronic transport in scaled gate-all-around silicon nanowire transistor arrays, *Appl. Phys. Lett.*, **103**, p. 263504.
85. Larrieu, G., Guerfi, Y., Han, X. L., and Clément, N. (2015). Vertical field effect transistor with sub-15nm gate-all-around in Si nanowire array, in *Proceedings of the European Solid-State Device Conference (ESSDERC)*, Graz, Austria, pp. 202–205.
86. Nazarov, A. N., Ferain, I., Dehdashti Akhavan, N., Razavi, P., Yu, R., and Colinge, J. P. (2011). Random telegraph-signal noise in junctionless transistors, *Appl. Phys. Lett.*, **98**, p. 092111.
87. Mukherjee, C., and Maneux, C. (2017). 1/f noise in 3D vertical gate-all-around junction-less silicon nanowire transistors, in *Proceedings of the European Solid-State Device Research Conference (ESSDERC)*, pp. 34–37.
88. Jang, D., Lee, J. W., Lee, C. W., Colinge, J. P., Montès, L., Lee, J. I., Lim, G. T., and Ghibaudo, G. (2011). Low-frequency noise in junctionless multigate transistors, *Appl. Phys. Lett.*, **96**, p. 133502.
89. Jeon, D. Y., Park, S. J., Mouis, M., Barraud, S., Kim, G. T., and Ghibaudo, G. (2013). Low-frequency noise behavior of junctionless transistors compared to inversion-mode transistors, *Solid-State Electron.*, **81**, pp. 101–104.
90. Doria, R. T., Trevisoli, R. D., de Souza, M., and Pavanello, M. A. (2013). Trap density characterization through low-frequency noise in junctionless transistors, *Microelectron. Eng.*, **109**, pp. 79–82.

91. Doria, R. T., Trevisoli, R. D., de Souza, M., and Pavanello, M. A. (2014). Low-frequency noise and effective trap density of short channel p- and n-types junctionless nanowire transistors, *Solid-State Electron.*, **96**, pp. 22–26.

92. Trevisoli Doria, R., Trevisoli, R., de Souza, M., Barraud, S., Vinet, M., Faynot, O., and Pavanello, M. A. (2017). Analysis of the substrate bias effect on the interface trapped charges in junctionless nanowire transistors through low-frequency noise characterization, *Microelectron. J.*, **178**, pp. 17–20.

93. Opondo, N., Ramadurgam, S., Yang, C., and Mohammadi, S. (2016). Trap studies in silicon nanowire junctionless transistors using low-frequency noise, *J. Vac. Sci. Technol. B*, **34**, p. 011804,

94. Bang, T., Lee, B. H., Kim, C. K., Ahn, D. C., Jeon, S. B., Kang, M. H., Oh, J. S., and Choi, Y. K. (2017). Low-frequency noise characteristics in SONOS flash memory with vertically stacked nanowire FETs, *IEEE Electron Device Lett.*, **38**(1), pp. 40–43.

95. Parihar Singh, M., Ghosh, D., Armstrong, G. A., and Kranti, A. (2012). Bipolar snapback in junctionless transistors for capacitorless dynamic random access memory, *Appl. Phys. Lett.*, **101**, p. 263503.

96. Parihar Singh, M., Ghosh, D., Armstrong, G. A., Yu, R., Razavi, P., and Kranti, A. (2012). Bipolar effects in unipolar junctionless transistors, *Appl. Phys. Lett.*, **101**, p. 093507.

97. Munteanu, D., and Autran, J. L. (2012). 3-D numerical simulation of bipolar amplification in junctionless double-gate MOSFETs under heavy-ion radiation, *IEEE Trans. Nucl. Sci.*, **59**(4), pp. 773–780.

98. Berthomé, M., Barraud, S., Ionescu, A., and Ernst, T. (2011). Physically-based, multi-architecture, analytical model for junctionless transistors, in *Proceedings of the 12th International Conference on Ultimate Integration on Silicon (ULIS 2011)*, Cork, Ireland, pp. 11–14.

99. Colinge, J. P. (1990). Conduction mechanisms in thin-film accumulation-mode SOI p-channel MOSFETs, *IEEE Trans. Electron Devices*, **37**(3), pp. 718–723.

100. Trevisoli, R. D., Doria, R. T., de Souza, M., and Pavanello, M. A. (2011). Threshold voltage in junctionless nanowire transistors, *Semicond. Sci. Technol.*, **26**, p. 105009.

101. Sallese, J. M., Chevillon, N., Lallement, C., Iñiguez, B., and Prégaldiny, F. (2011). Charge-based modeling of junctionless double-gate field-effect transistor, *IEEE Trans. Electron Devices*, **58**(8), pp. 2628–2636.

102. Gnani, E., Gnudi, A., Reggiani, S., and Baccarani, G. (2011). Theory of the junctionless transistor nanowire FET, *IEEE Trans. Electron Devices*, **58**(9), pp. 2903–2910.

103. Ansari, L., Feldman, B., Fagas, G., Colinge, J. P., and Greer, J. C. (2012). Subthreshold behavior of junctionless silicon nanowire transistors from atomic simulations, *Solid-State Electron.*, **71**, pp. 58–62.

104. Gnani, E., Gnudi, A., Reggiani, S., and Baccarani, G. (2012). Physical model of the junctionless UTB SOI-FET, *IEEE Trans. Electron Devices*, **59**(4), pp. 941–948.

105. Trevisoli, R. D., Doria, R. T., de Souza, M., Das, S., Ferain, I., and Pavanello, M. A. (2012). Surface-potential-based drain current analytical model for triple-gate junctionless nanowire transistors, *IEEE Trans. Electron Devices*, **59**(12), pp. 3510–3518.

106. Chen, Z., Xiao, Y., Tang, M., Xiong, Y., Huang, J., Li, J., Gu, X., and Zhou, Y. (2012). Surface-potential-based drain current model for long-channel junctionless double-gate MOSFETs, *IEEE Trans. Electron Devices*, **59**(12), pp. 3292–3298.

107. Trevisoli, R. D., Doria, R. T., de Souza, M., and Pavanello, M. (2013). A physically-based threshold voltage definition, extraction and analytical model for junctionless nanowire transistors, *Solid-State Electron.*, **90**, pp. 12–17.

108. Trevisoli, R. D., Doria, R. T., de Souza, M., and Pavanello, M. (2013). Drain current and short channel effects modeling in junctionless nanowire transistors, *J. Integr. Circuits Syst.*, **8**(2), pp. 116–124.

109. Jin, X. S., Liu, X., Kwon, H. I., and Lee, J. H. (2013). A continuous model of accumulation mode (junctionless) cylindrical surrounding-gate nanowire MOSFETs, *Chin. Phys. Lett.*, **30**(3), p. 028502.

110. Holtij, T., Schwarz, M., Kloes, A., and Iñiguez, B. (2013). Threshold voltage, and 2D potential modeling within short-channel junctionless DG MOSFETs in subthreshold region, *Solid-State Electron.*, **90**, pp. 107–115.

111. Jazaeri, F., Barbut, L., Koubab, A., and Sallese, J. L. (2013). Analytical model for ultra-thin body junctionless symmetric double gate MOSFETs in subthreshold regime, *Solid-State Electron.*, **82**, pp. 103–110.

112. Ansari, L., Feldman, B., Fagas, G., Martinez Lacambra, C., Haverty, M. G., Kuhn, K. J., Shankar, S., and Greer, J. C. (2013). First principle-based analysis of single-walled carbon nanotube and silicon nanowire junctionless transistors, *IEEE Trans. Nanotechnol.*, **12**(6), pp. 1075–1081.

113. Jazaeri, F., Barbut, L., and Sallese, J. L. (2014). Generalized charge-based model for double-gate junctionless FETs, including inversion, *IEEE Trans. Electron Devices*, **61**(10), pp. 3553–3557.

114. Holtij, T., Graef, M., Hain, F. M., Kloes, A., and Iñíguez, B. (2014). Compact model fort short-channel junctionless accumulation mode double gate MOSFETs, *IEEE Trans. Electron Devices*, **61**(2), pp. 288–299.

115. Paz, B. C., Cassé, M., Barraud, S., Reimbold, G., Faynot, O., Ávila-Herrera, F., Cerdeira, A., and Pavanello, M. A. (2016). Drain current model for short-channel triple gate junctionless nanowire transistors, *Microelectron. Reliab.*, **63**, pp. 1–10.

116. Yesayan, A., Jazaeri, F., and Sallese, J. M. (2016). Charge-based modelling of double-gate and nanowire junctionless FETs including interface-trapped charges, *IEEE Trans. Electron Devices*, **63**(3), pp. 1368–1374.

117. Ávila-Herrera, F., Paz, B. C., Cerdeira, A., Estrada, M., and Pavanello, M. A. (2016). Charge-based analyctical compact model for triple-gate junctionless nanowire transistors, *Solid-State Electron.*, **122**, pp. 23–31.

118. Guo, Z., Zhang, J., Ye, Z., and Wang, Y. (2016). 3-D analytical model for short-channel triple-gate junctionless MOSFETs, *IEEE Trans. Electron Devices*, **63**(10), pp. 3857–3863.

119. Trevisoli, R., Trevisoli, Doria, R., de Souza, M., Barraud, S., Vinet, M., and Pavanello, M. A. (2016) Analytical model for the dynamic behavior of triple-gate junctionless nanowire transistors, *IEEE Trans. Electron Devices*, **63**(2), pp. 856–863.

120. Oproglidis, T. A., Tsormpatzoglou, A., Tassis, D. H., Karatsori, T. A., Barraud, S., Ghibaudo, G., and Dimitriadis, C. A. (2017). Analytical drain current compact model in the depletion operation region of short-channel triple-gate junctionless transistors, *IEEE Trans. Electron Devices*, **64**(1), pp. 66–72.

121. Gola, D., Singh, B., Kumar Tiwari, P. (2017). A threshold voltage model of tri-gate junctionless field-effect transistors including substrate bias effects, *IEEE Trans. Electron Devices*, **64**(9), pp. 3534–3540.

122. Jiang, C., Liang, R., Xu, J., and Ashraful Alam, M. (2017). A compact quasi-static terminal charge and drain current model for double-gate junctionless transistors and its circuit validation, *IEEE Trans. Electron Devices*, **64**(12), pp. 4823–4830.

123. Colinge, J. P., and Greer, J. C. (2016). *Nanowire Transistors*, Cambridge University Press, ISBN: 9781107052406, pp. 29–35.

124. Park, J. T., Kim, J. Y., Lee, C. W., and Colinge, J. P. (2010). Low-temperature conductance oscillations in junctionless nanowire transistors, *Appl. Phys. Lett.*, **96**, p. 102106.
125. Li, X., Han, W., Ma, L., Wang, H., Zhang, Y., and Yang, F. (2013). Low-temperature quantum transport characteristics in single n-channel junctionless nanowire transistors, *IEEE Electron Device Lett.*, **34**(5), pp. 581–583.
126. Ma, L., Han, W., Wang, H., Li, X., and Yang, F. (2013). Temperature dependence of electronic behaviors in n-type multiple-channel junctionless transistors, *J. Appl. Phys.*, **114**, p. 124507.
127. Yi, K. S., Trivedi, K., Floresca, H. C., Yuk, H., Hu, W., and Kim, M. J. (2011). Room-temperature quantum confinement effects in transport properties of ultrathin Si nanowire field-effect transistors, *Nano Lett.*, **11**(12), pp. 5465–5470.
128. Wu, H., Zhang, J. Y., Gu, J. J., Dong, L., Conrad, N. J., and Ye, P. D. (2013). Room-temperature quantum oscillations in Ge junctionless MOSFETs at the scaling limit, in *71st Device Research Conference*, pp. 1–2.
129. Kim, S. Y., Choi, S. Y., Hwang, W. S., and Cho, B. J. (2016). Valley-engineered ultra-thin silicon for high-performance junctionless transistors, *Nat. Sci. Rep.*, **6**, p. 29354.
130. Mirza, M. M., Schupp, F. J., Mol, J. A., MacLaren, D. A., Briggs, G. A., and Paul, D. J. (2017). One dimensional transport in silicon nanowire junction-less field effect transistors, *Nat. Sci. Rep.*, **7**, p. 3004.
131. Lin, H. C., Lin, C. I., Lin, Z. M., Shie, B. S., and Huang, T. Y. (2013). Characteristics of planar junctionless poly-Si thin-film transistors with various channel thickness, *IEEE Trans. Electron Devices*, **60**(3), pp. 1142–1148.
132. Chou, C. H., Lee, I. C., Lei, D. C., and Cheng, H. C. (2014). Planar junctionless poly-Si thin-film transistors with single-gate and double gate, *Jpn. J. Appl. Phys.*, **53**(6S), p. 06JE07.
133. Ban, T., Migita, S., Uenuma, M., Okamoto, N., Ishikawa, Y., Uraoka, Y., Yamashita, I., and Yamamoto, S. I. (2017). Estimation of charge effects of ultrafine channel utilizing junctionless transistor with nanodot-type floating gate, *Jpn. J. Appl. Phys.*, **56**, p. 03BB05.
134. Blomme, P., Versluis, J., Ercken, M., Sourieau, L., Hody, H., Vecchio, G., Paraschiv, V., Tal, C. L., Van den Bosch, G., and Van Houdt, J. (2016). Junctionless array with ultrathin poly\TiN floating gate and HfAlO based integrate dielectric for sub-15nm planar NAND flash, in *8th IEEE International Memory Workshop*, Paris, France, pp. 1–4.

135. Vandooren, A., Witters, L., Vecchio, E., Kunnen, E., Hellings, G., Peng, L., Inoue, F., Li, W., Waldron, N., Mocuta, D., and Collaert, N. (2017). Double-gate Si junction-less n-type transistor for high performance Cu-BEOL compatible applications using 3D sequential integration, in *IEEE SOI-3D-Subthreshold (S3S) Microelectronics Technology Unified Conference*, Burlingame, CA, USA, pp. 1–2.

136. Chen, H. B., Chang, C. Y., Lu, N. H., Wu, J. J., Han, M. H., Cheng, Y. C., and Wu, Y. C. (2013) Characteristics of gate-all-around junctionless poly-Si TFTs with an ultrathin channel, *IEEE Electron Device Lett.*, **34**(7), pp. 897–899.

137. Liu, T. Y., Pan, F. M., and Sheu, J. T. (2015). Characteristics of gate-all-around junctionless polysilicon nanowire transistors with twin 20-nm gates, *IEEE J. Electron Device Soc.*, **3**(5), pp. 405–409.

138. Lee, B. H., Hur, J., Kang, M. H., Bang, T., Ahn, D. C., Lee, D., Kim, K. H., and Choi, Y. K. (2016). A vertically integrated junctionless nanowire transistor, *Nano Lett.*, **16**(3), pp. 1840–1847.

139. Ahn, D. C., Lee, B. H., Kang, M. H., Hue, J., Bang, T., and Choi, Y. K. (2016). Impact of crystalline damage on a vertically integrated junctionless nanowire transistor, *Appl. Phys. Lett.*, **109**, p. 183108.

140. Su, C. J., Su, T. K., Tsai, T. I., Lin, H. C., and Huang, T. Y. (2012). A junctionless SONOS nonvolatile memory device constructed with in-situ doped polycrystalline silicon nanowires, *Nanoscale Res. Lett.*, **7**, p. 162.

141. Choi, S. J., Moon, D. I., Duarte, J. P., Kim, S., and Choi, Y. K. (2011). A novel junctionless all-around-gate SONOS device with a quantum nanowire on a bulk substrate for 3D stack NAND flash memory, in *Symposium on VLSI Technology Digest of Technical Papers*, pp. 74–75.

142. Yeh, M. S., Wu, Y. C., Chung, M. H., Jhan, Y. R., Chang-Liao, K. S., Liu, K. C., Wu, M. H., and Hung, M. F. (2014). Investigation of p-channel and n-channel junctionles gate-all-all-around polycrystalline silicon nanowires with silicon nanocrystals nonvolatile memory, *Appl. Phys. Lett.*, **105**, p. 042109.

143. Sun, Y., Yu, H. Y., Singh, N., Le, T. T., Gnani, E., Baccarani, G., Leong, K. C., Lo, G. Q., and Kwong, D. L. (2011). Junction-less stackable SONOS memory realized on vertical-Si-nanowire for 3-D application, in *Proceedings of International Symposium on VLSI Technology, Systems and Applications (VLSI-TSA)*, pp. 154–155.

144. Guerfi, Y., and Larrieu, G. (2016). Vertical silicon nanowire field-effect transistors with nanoscale gate-all-around, *Nanoscale Res. Lett.*, **11**, p. 210.

145. Larrieu, G., Guerfi, Y., Han, X. L., and Clément, N. (2017). Sub-15 nm gate-all-around field effect transistors on vertical silicon nanowires, *Solid-State Electron.*, **130**, pp. 9–10.

146. Larrieu, G., and Han, X. L. (2013). Vertical nanowire array-based field-effect transistors for ultimate scaling, *Nanoscale*, **5**, pp. 2437–2441.

147. Sun, Y., Yu, H. Y., Singh, N., Leong, K. C., Lo, G. Q., and Kwong, D. L. (2011). Junctionless vertical-Si-nanowire-channel-based SONOS memory with 2-bit storage per cell, *IEEE Electron Device Lett.*, **32**(6), pp. 725–727.

148. Veloso, A., Parvais, B., Matagne, P., Simoen, E., Huynh-Bao, T., Paraschiv, V., Vecchio, E., Devriendt, K., Rosseel, E., Ercken, M., Chan, B. T., Delvaux, C., Altamirano-Sánchez, E., Versluijs, J. J., Tao, Z., Suhard, S., Brus, S., Sibaja-Hernandez, A., Waldron, N., Lagrain, P., Richard, O., Bender, H., Chasin, A., Kaczer, B., Ivanov, T., Ramesh, S., De Meyer, K., Ryckaert, J., Collaert, N., and Thean, A. (2016). Junctionless gate-all-around lateral and vertical nanowire FETs with simplified processing for advanced logic and logic/RF applications and scaled SRAM cells, in *Digest of Technical Papers of the Symposium on VLSI Technology*, pp. 138–139.

149. Shin, S., Kang, I. M., and Kim, R. (2012). Extraction method for substrate-related components of vertical junctionless silicon nanowire field-effect transistors and its verification on radio frequency characteristics, *Jpn. J. Appl. Phys.*, **51**, p. 06FE20.

150. Yoon, Y. J., Seo, J. H., Kwon, H. I., Lee, J. H., and Kang, I. M. (2016). Enhancement-mode GAN-based junctionless vertical surrounding-gate transistors with dual-material gate structure for high-frequency applications, *J. Nanosci. Nanotechnol.*, **16**(10), pp. 10204–10209.

151. Chen, L., Cai, F., Otuonye, U., and Lu, W. D. (2016). Vertical Ge/Si core/shell nanowire junctionless transistor, *Nano Lett.*, **16**(1), pp. 420–426.

152. Colinge, J. P., Kranti, A., Yan, R., Lee, C. W., Ferain, I., Yu, R., Dehdashti Akhavan, N., and Razavi, P. (2011). Junctionless nanowire transistor (JNT): properties and design guidelines, *Solid State Electron.*, **65–66**, pp. 33–37.

153. Han, M. H., Chang, C. Y., Chen, H. B., Wu, J. J., Cheng, Y. C., and Wu, Y. C. (2013). Performance comparison between bulk and SOI junctionless transistors, *IEEE Electron Device Lett.*, **34**(2), pp. 169–171.

154. Kim, D. H., Kim, T. K., Yoon, Y. G., Hwang, B. W., Choi, Y. K., Chi, B. J., and Lee, S. H. (2014). First demonstration of ultra-thin SiGe-channel junctionless accumulation-mode (JAM) bulk FinFETs on Si substrate with PN junction-isolation scheme, *J. Electron Devices Soc.*, **2**(5), pp. 123–127.

155. Cheng, Y. C., Chen, H. B., Shao, C.S., Su, J. J., Wu, Y. C., Chang, C. Y., and Chang, T. C. (2014). Performance of a novel P-type junctionless transistor using a hybrid poly-Si fin channel, in *IEEE International Electron Devices Meeting (IEDM) Technical Digest*, pp. 622–625.

156. Cheng, Y. C., Chen, H. B., Su, J. J., Shao, C. S., Thirunavukkarasu, V., Chang, C. Y., and Wu, Y. C. (2015). Characteristics of a novel poly-Si P-channel junctionless transistor with hybrid P/N-substrate, *IEEE Electron Device Lett.*, **36**(2), pp. 159–161.

157. Thirunavukkarasu, V., Cheng, C. H., Lin, Y. R., Kurniawan, E. D., and Wu, Y. C. (2016). Performance of hybrid p-channel trench poly-Si junctionless field-effect gate-all-around transistors, in *Proceedings of the Semiconductor Interfaces Specialists Conference (SISC)*, p. 11.2.

158. Thirunavukkarasu, V., Jaehyun Lee, J., Toufik Sadi, T., Georgiev, V. P., Lema, F. A., Pandian Soundarapandian, K., Jhan, Y. R., Yang, S. Y., Lin, Y. R., Kurniawan, E. D., Wu, Y. C., and Asenov, A. (2017). Investigation of inversion, accumulation and junctionless mode bulk Germanium FinFETs, *Superlattices Microstruct.*, **111**, pp. 649–655.

159. Cheng, Y. C., Chen, H. B., Chang, C. Y., Cheng, C. H., Shih, Y. J., and Wu, Y. C. (2016). A highly scalable poly-Si junctionless FETs featuring a novel multi-stacking hybrid P/N layer and vertical gate with very high I_{on}/I_{off} for 3D stacked ICs, in *Symposium on VLSI Technology Digest of Technical Papers*, pp. 188–189.

160. De Marchi, M., Sacchetto, D., Frache, S., Zhang, J., Gaillardon, P. E., Leblebici, Y., and De Michelli, G. (2012). Polarity control in double-gate, gate-all-around vertically stacked silicon nanowire FETs, in *IEEE International Electron Devices Meeting (IEDM) Technical Digest*, pp. 183–186.

161. De Marchi, M., Zhang, J., Frache, S., Sacchetto, D., Gaillardon, P. E., Leblebici, Y., and De Michelli, G. (2014). Configurable logic gates using polarity-controlled silicon nanowire gate-all-around FETs, *IEEE Electron Device Lett.*, **35**(8), pp. 880–882.

162. Ghosh, B., and Waseem Akram, M. (2013). Junctionless tunnel field effect transistor, *IEEE Electron Device Lett.*, **34**(5), pp. 584–586.

163. Akram, M. W., Ghosh, B., Bal, P., and Mondal, P. (2014). P-type double gate junctionless field effect transistor, *J. Semicond.*, **35**(1), p. 014002.

164. Gupta, M., and Kranti, A. (2015). Germanium junctionless MOFET with steep subthreshold swing, *Electrochem. Soc. Trans.*, **66**(5), pp. 79–86.

165. Gupta, M., and Kranti, A. (2016). Transforming gate misalignment into a unique opportunity to facilitate steep switching in junctionless nanotransistors, *Nanotechnology*, **27**, p. 455204.

166. Thirunavukkarasu, V., Jhan, Y. R., Liu, Y. B., Kurniawan, E. D., Lin, Y. R., Yang, S. Y., Cheng, C. H., and Wu, Y. C. (2017). Gate-all-around junctionless silicon transistors with atomically thin nanosheet channel (0.65 nm) and record sub-threshold slope (43 mV/dec), *Appl. Phys. Lett.*, **110**, p. 032101.

167. Gamble, H., Armstrong, B. M., Baine, P. T., Low, Y. H., Rainey, P. V., Mitchell, S. J. N., and McNeill, D. W. (2011). Germanium processing, in *Semiconductor-On-Insulator Materials for Nanoelectronics Applications* (Nazarov, A., Colinge, J. P., Balestra, F., Raskin, J. P., Gamiz, F., and Lysenko, V. S., eds.), Springer, pp. 1–29.

168. Houssa, M., Satta, A., Simoen, E., De Jaeger, B., Meuris, M., Caymax, M., and Heyns, M. (2007). Electrical performance of Ge devices, in *Germanium-Based Technologies* (Claeys, C., and Simoen, E., eds.), Elsevier, pp. 233–265.

169. Wong, I. H., Chen, Y. T., Huang, S. H., Tu, W. H., Chen, Y. S., Shieh, T. C., Lin, T. Y., Lan, H. S., and Liu, C. W. (2014). In-situ doped and tensily strained Ge junctionless gate-all-around nFETs on SOI featuring I_{on} = 828 μA/μm, $I_{on}/I_{off} \sim 1\times10^5$, DIBL = 16–54 mV/V, and 1.4X external strain enhancement, in *IEEE International Electron Devices Meeting (IEDM) Technical Digest*, pp. 239–241.

170. Wong, I. H., Huang, S. H., Lu, F. L., Ye, H. Y., Lu, C. T., Yan, J. Y., Shen, Y. C., Peng, Y. J., Lan, H. S., and Liu, C. W. (2016). Scaled Ge junctionless gate-all-around NFETs with simultaneous I_{on} = 1235 μA/μm at V_{OV} = V_{DS} = 1V, SS = 95 mV/dec, high $I_{on}/I_{off} = 2 \times 10^6$, and reduced noise power density using S/D dopant recovery by selective laser annealing, in *IEEE International Electron Devices Meeting (IEDM) Technical Digest*, pp. 842–845.

171. Ren, S., Bhuiyan, M. A., Wu, H., Jiang, R., Ni, K., Zhang, E. X., Reed, R. A., Fleetwood, D. M., Ye, P., and Ma, T. P. (2016). Total ionizing dose (TID) effects in ultra-thin body Ge-on-insulator (GOI) junctionless CMOSFETs with recessed source/drain and channel, *IEEE Trans. Nucl. Sci.*, **64**(1), pp. 176–180.

172. Wu, H., Li, X. F., Dong, L., Gu, J. J., Conrad, N. J., Zhang, J. Y., and Ye, P. D. (2013). Ultra-scaled junctionless MOSFETs on GeOI Substrates, in *Semiconductor Interface Specialists Conference (SISC)*, Washington D.C., USA.

173. Lee, W. L., Hsu, C. C., Chung, C. T., Lu, Y. H., Luo, G. L., and Chien, C. H. (2017). Fully depleted GeOI-channel junctionless pMOSFET with a low-resistance-raised NiGe alloy S/D, *ECS J. Solid-State Sci. Technol.*, **6**(8), pp. 507–511.
174. Seto, J. Y. W. (1975). The electrical properties of polycrystalline silicon films, *J. Appl. Phys.*, **46**(2), pp. 5247–5254.
175. Colinge, J. P., Demoulin, E., Delannay, F., Lobet, M., and Temerson, J. M. (1981). Grain size and resistivity of LPCVD polysilicon films, *J. Electrochem. Soc.*, **128**(9), pp. 2009–2014.
176. Colinge, J. P., Morel, H., and Chante, J. P. (1983). Field effect in large grain polycrystalline silicon, *IEEE Trans. Electron Devices*, **30**(3), pp. 197–201.
177. Su, C. J., Tsai, T. I., Liou, Y. L., Lin, Z. M., Lin, H. C., and Chao, T. S. (2011). Gate-all-around junctionless transistors with heavily doped polysilicon nanowire channels, *IEEE Electron Device Lett.*, **32**(4), pp. 521–523.
178. Hsieh, D. R., Lin, J. Y., Kuo, P. Y., and Chao, T. S. (2016). High-performance pi-gate poly-Si junctionless and inversion mode FET, *IEEE Trans. Electron Devices*, **63**(11), pp. 4179–4184.
179. Hsieh, D. R., Lin, J. Y., Kuo, P. Y., and Chao, T. S. (2017). Comprehensive analysis on electrical characteristics of Pi-gate poly-Si junctionless FETs, *IEEE Trans. Electron Devices*, **64**(7), pp. 2992–2998.
180. Wu, I. C., Chang, T. C., Liu, P. T., Chen, C. S., Tu, C. H., Zan, H. W., Tai, Y. H., and Chang, C. Y. (2005). Effects of channel width on electrical characteristics of polysilicon TFTs with multiple nanowire channels, *IEEE Trans. Electron Devices*, **52**(10), pp. 2343–2346.
181. Hsieh, D. R., Lin, J. Y., Kuo, P. Y., and Chao, T. S. (2017). Comprehensive analysis on electrical characteristics of Pi-gate poly-Si junctionless FETs, *IEEE Trans. Electron Devices*, **64**(7), pp. 2992–2998.
182. Maeda, T., Itagaki, K., Hishida, T., Katsumata, R., Kito, M., Fukuzumi, Y., Kido, M., Tanaka, H., Komori, Y., Ishiduki, M., Matsunami, J., Fujiwara, T., Aochi, H., Iwata, Y., and Watanabe, Y. (2009). Multi-stacked 1G cell/layer pipe-shaped BiCS flash memory, in *Symposium on VLSI Technology Digest of Technical Papers*, pp. 22–23.
183. Jang, J., Kim, H.-S., Cho, W., Cho, H., Kim, J., Il Shim, S., Younggoan, Jeong, J.-H., Son, B.-K., Kim, D. W., Kihyun, Shim, J.-J., Lim, J. S., Kim, K.-H., Yi, S. Y., Lim, J.-Y., Chung, D., Moon, H.-C., Hwang, S., Lee, J.-W., Son, Y.-H., Chung, U-I., Lee, W.-S. (2009). Vertical cell array using TCAT (Terabit Cell Array Transistor) technology for ultra high density NAND flash

memory, in *Symposium on VLSI Technology Digest of Technical Papers*, pp. 192–193.

184. Kim, W., Choi, S., Sung, J., Lee, T., Park, C., Ko, H., Jung, J., Yoo, I., and Park, Y. (2009). Multi-layered vertical gate NAND flash overcoming stacking limit for terabit density storage, in *Symposium on VLSI Technology Digest of Technical Papers*, pp. 188–189.
185. Jiang, D. D., Xia, Z. L., Jin, L., Zou, W. Q., Zhang, Y., Tang, Z. Y., Li, X. K., and Huo, Z. L. (2016). Analysis of the retention characteristic in three dimensional junction-less charge trapping memory, *ECS Trans.*, **72**(4), pp. 233–238.
186. Li, X. K., Hou, Z. L., Jin, L., Jiang, D. D., Hong, P. Z., Xu, Q., Tang, Z. Y., Li, C. L., and Ye, T. C. (2015). Impact of continuing scaling on the device performance of 3D cylindrical junction-less charge trapping memory, *J. Semicond.*, **36**(9), p. 094008.
187. Yeh, M. S., Wu, Y. C., Chung M. H., Jhan, Y. R., Chang-Liao, K. S., Liu, K. C., Wu, M. S., and Hung, M. F. (2014). Investigation of p-channel and n-channel junctionless gate-all-around polycrystalline silicon nanowires with silicon nanocrystals nonvolatile memory, *Appl. Phys. Lett.*, **105**, p. 042109.
188. Prince, B. (2014). *Vertical 3D Memory Technologies*, John Wiley & Sons, New York.
189. Kamimuta, Y., Ikeda, K., Furuse, K., Irisawa, T., Teuka, T. (2013). Short channel poly-Ge junction-less p-type FinFETs for BEOL transistors, in *International Symposium on VLSI Technology, Systems and Applications (VLSI-TSA)*, pp. 109–110.
190. Usuda, K., Kamata, Y., Kamimuta, Y., Mori, T., Koike, M., and Tezuka, T. (2014). High-performance tri-gate poly-Ge junction-less P- and N-MOSFETs fabricated by flash lamp annealing process, in *IEEE International Electron Devices Meeting (IEDM) Technical Digest*, pp. 422–425.
191. Huang, W. H., Shieh, J. M., Kao, M. H., Shen, C. H., Huang, T. E., Wang, H. H., Yang, C. C., Hsieh, T. Y., Hsieh, J. L., Yu, P., and Yeh, W. K. (2017). Enabling n-type polycrystalline Ge junctionless FinFET of low thermal budget by in situ doping of channel and visible pulsed laser annealing, *Appl. Phys. Express*, **10**, p. 026502.
192. Razavi, P., and Fagas, G. (2013). Electrical performance of III-V gate-all-around nanowire transistors, *Appl. Phys. Lett.*, **103**, p. 063506.
193. Djara, V., Cherkaoui, K., O'Connor, É., Povey, I. M., Thomas, K. K., and Hurley, P. K. (2013). Junctionless InGaAs MOSFETs with InAlAs barrier

isolation and channel thinning by digital wet etch, in *Proceedings of the 71st Annual Device Research Conference (DRC)*, Notre Dame, USA, pp. 131–132.

194. Djara, V., Czornomaz, L., Deshpande, V., Daix, N., Ucelli, E., Caimi, D., Sousa, M., and Fompeyrine, J. (2016). Tri-gate InGaAs-OI junctionless FETs with PE-ALD Al_2O_3 gate dielectric and H_2/Ar anneal, *Solid-State Electron.*, **115**, pp. 103–108.

195. Pirro, L., Park, H. J., Ionica, I., Bawedin, M., Cristoloveanu, S., Czornomaz, L., Djara, V., and Deshpande, V. (2016). Volume and interface conduction in InGaAs junctionless transistors, in *Proceedings of EUROSOI-ULIS*, pp. 104–107.

196. Im, K. S., Seo, J. H., Yoon, Y. J., Jang, Y. I., Kim, J. S., Cho, S., Lee, J. H., Cristoloveanu, S., Lee, H. J., and Kang, I. M. (2014). GaN junctionless trigate field-effect transistors with deep sub-micron gate length: characterization and modeling in RF regime, *Jpn. J. Appl. Phys.*, **53**, p. 118001.

197. Song, Y., Mosheni, K., Kim, S. H., Shin, J. C., Ishihara, T., Adesida, I., and Li, X. (2016). Ultra-high aspect ratio InP junctionless FinFETs by a novel wet etching method, *IEEE Electron Device Lett.*, **37**(8), pp. 970–973.

198. Im, K. S., Sindhuri, V., Jo, Y. W., Son, D. H., Lee, J. H., Cristoloveanu, S., and Lee, J. H. (2015). Fabrication of AlGaN/GaN ?-shaped nanowire fin-shaped FETs by a top-down approach, *Appl. Phys. Express*, **8**, p. 066501.

199. Im, K. S., Won, C. H., Vodapally, S., Caulmilone, R., Cristoloveanu, S., Kim, Y. T., and Lee, J. H. (2016). Fabrication of normally-off GaN nanowire gate-all-around FET with top-down approach, *Appl. Phys. Lett.*, **109**, p. 143106.

200. Konar, A., Mathew, J., Nayak, K., Bajaj, M., Pandey, R. K., Dhara, S., Murali, K. V. R. M., and Deshmukh, M. M. (2015). Carrier transport in high mobility InAs nanowire junctionless transistors, *Nano Lett.*, **15**, pp. 1684–1690.

201. Cao, W., Kang, J., and Banerjee, K. (2015). Junction-less monolayer MoS2 FETs, arXiv:1509.00561 [cond-mat.mes-hall], pp. 1–6.

202. Mirabelli, G., Schmidt, M., Sheehan, B., Cherkaoui, K., Monaghan, S., Povey, I., McCarthy, M., Bell, A. P., Nagle, R., Crupi, F., Hurley, P. K., and Duffy, R. (2016). Back-gated Nb-doped MoS_2 junctionless field-effect transistors, *AIP Adv.*, **6**, p. 025323.

203. Mori, T., Ninomiya, N., Kubo, T., Uchida, N., Watanabe, E., Tsuya, D., Moriyama, S., Tanaka, M., and Ando, A. (2016). Characterization of

effective mobility and its degradation mechanism in MoS_2 MOSFETs, *IEEE Trans. Nanotechnol.*, **15**(4), pp. 651–656.

204. Ghosh, D., Parihar, M., Armstrong, G., and Kranti, A. (2012). High performance junctionless MOSFETs for ultralow power analog/RF applications, *IEEE Electron Device Lett.*, **33**(10), pp. 1477–1479.

205. Agarwal, T., Sorée, B., Radu, I., Raghavan, P., Fiori, G., Iannaccone, G., Thean, A., Heyns, M., and Dehaene, W. (2016). Comparison of short-channel effects in monolayer MoS_2 based junctionless and inversion-mode field-effect transistors, *Appl. Phys. Lett.*, **108**, p. 023506.

206. Barik, A., Sarma, M. K., and Dutta, J. C. (2014). Traditional graphene and junctionless carbon nanotube field effect transistor for cholesterol sensing, in *Proceedings of the International Conference on Emerging Electronics (ICEE)*, Bangalore, India, pp. 1–4.

207. Jiang, J., Wan, Q., and Zhang, Q. (2013). Transparent junctionless electric-double-layer transistors gated by a reinforced chitosan-based biopolymer electrode, *IEEE Trans. Electron Devices*, **60**(6), pp. 1951–1957.

208. Jiang, J., Sun, J., Dou, W., and Wan, Q. (2012). Junctionless flexible oxide-based thin-film transistors on paper substrates, *IEEE Electron Device Lett.*, **33**(1), pp. 65–67.

209. Golve, M., Gundapaneni, S., and Kottantharayil, A. (2012). Novel architectures for zinc-oxide junctionless transistor, in *Proceedings of the International Conference on Emerging Electronics (ICEE)*, Mumbai, India, pp. 1–4.

210. Zhang, G., Wan, Q., Wu, G., and Zhu, L. (2013). Transparent junctionless thin-film transistors with tunable operation mode, *IEEE Electron Device Lett.*, **34**(2), pp. 265–267.

211. Guo, J., Liu, J., Yang, B., Zhan, G., Tang, L., Tian, H., Kang, X., Peng, H., Chen, X., and Yang, C. (2015). Biodegradable junctionless transistors with extremely simple gate structure, *IEEE Electron Device Lett.*, **36**(9), pp. 908–910.

212. Wu, G. D., Zhang, J., and Wan, X. (2014). Junctionless coplanar-gate oxide-based thin-film transistors gated by Al_2O_3 proton conducting films on paper substrates, *Chin. Phys. Lett.*, **31**(10), p. 108505.

213. Kim, S. M., Yu, C. G., Cho, W. J., and Park, J. T. (2017). Device characterization and design guideline of amorphous InGaZnO junctionless thin-film transistor, *IEEE Trans. Electron Devices*, **64**(6), pp. 2526–2532.

214. Ansari, L., Fagas, G., Colinge, J. P., and Greer, J. C. (2012). A proposed confinement modulated gap nanowire transistor based on a metal (tin), *Nano Lett.*, **12**(5), pp. 2222–2227.

215. Cheng, P. H., Wang, C. Y., Chang, T. J., Shen, T. H., Cai, Y. S., and Chen, M. J. (2017). Room-temperature field effect transistors with metallic ultrathin TiN-based channel prepared by atomic layer delta doping and deposition, *Nat. Sci. Rep.*, 7, p. 875.

216. Krotnev, I. P. (2013). Novel metallic field-effect transistors, MS thesis, Univ. of Toronto, https://tspace.library.utoronto.ca/bitstream/1807/42990/3/Krotnev_Ivan_P_201311_MASc_thesis.pdf

217. Chen, H.-C., Qiu, J.-T., Yang, F.-L., Liu, Y.-C., Chen, M.-C., Tsai, R.-Y., Yang, H.-W., Lin, C.-Y., Lin, C.-C., Wu, T.-S., Tu, Y.-M., Xiao, M.-C., Ho, C.-H., Huang, C.-C., Lai, C.-S., and Hua, M.-Y. (2014). Magnetic-composite-modified polycrystalline-silicon nanowire field-effect transistor for vascular endothelial growth factor detection and cancer diagnosis, *Anal. Chem.*, **86**(19), pp. 9443–9450.

218. Shen, M. Y., Li, B. R., and Li, Y. K. (2014). Silicon nanowire field-effect-transistor based biosensors: from sensitive to ultra-sensitive, *Biosens. Bioelectron.*, **60**, pp. 101–111.

219. Ernst, T., Duraffourg, L., Dupré, C., Bernard, E., Andreucci, P., Becu, S., Ollier, E., Hubert, A., Halte, C., Buckley, J., Thomas, O., Delapierre, G., Deleonibus, S., De Salvo, B., Robert, P., and Faynot, O. (2008). Novel Si-based nanowire devices: will they serve ultimate CMOS scaling or ultimate hybrid integration?, in *IEEE International Electron Devices Meeting (IEDM) Technical Digest*, pp. 745–748.

220. Arcamone, J., Philippe, G., Arndt, G., Dupré, C., Savoye, M., Hentz, S., Ernst, T., Colinet, E., Duraffourg, L., and Ollier, E. (2014). Nanosystems monolithically integrated with CMOS: emerging applications and technologies, in *IEEE International Electron Devices Meeting (IEDM) Technical Digest*, pp. 22.1.1-22.1.4.

221. Shen, M. Y., Li, B. R., and Li, Y. K. (2011). Silicon nanowire field-effect transistor-based biosensors for biomedical diagnosis and cellular recording investigation, *Nano Today*, **6**, pp. 131–154.

222. Gao, A., Dai, P., Lu, N., Li, T., and Wang, Y. (2012). CMOS-compatible silicon nanowire based field-effect pH sensor, in *International Conference on Manipulation, Manufacturing and Measurement on the Nanoscale (3M-NANO)*, pp. 113–116.

223. Nair, P. R., and Alam, M. A. (2007). Design considerations of silicon nanowire biosensors, *IEEE Trans. Electron Devices*, **54**(12), pp. 3400–3408.

224. Georgiev, Y. M., Yu, R., Petkov, N., Lotty, O., Nightingale, A. M., deMello, J. C., Duffy, R., and Holmes, J. D. (2014). Silicon and germanium junctionless nanowire transistors for sensing and digital electronics appliations, in *Functional Nanomaterials and Devices for Electronics, Sensors and Energy Harvesting* (Nazarov, A., Balestra, F., Kilchytska, V., and Flandre, D., eds.), Springer, Switzerland, pp. 367–388.

225. Georgiev, Y. M., Petkov, N., McCarthy, B., Yu, R., Djara, V., O'Connell, D., Nightingale, A. M., Thamsumet, N., deMello, J. C., Blake, A., Das, S., and Holmes, J. D. (2014). CMOS-compatible top-down fabrication of sub-50 nm silicon nanowire sensing devices, *Microelectron. Eng.*, **118**, pp. 47–53.

226. Zufilqar, A., Patou, F., Pfreundt, A., Papakonstantinopoulos, C., Svendsen, W. E., and Dimaki, M. (2017). In-situ doped junctionless polysilicon nanowires field effect transistors for low-cost biosensors, *Sens. Biosens. Res.*, **13**, pp. 88–95.

227. Buitrago, E., Fagas, G., Fernández-Bolaños Badia, M., Georgiev, Y. M., Berthomé, M., and Ionescu, A. M. (2013). Junctionless silicon nanowire transistors for the tunable operation of a highly sensitive, low-power sensor, *Sens. Actuators B*, **183**, pp. 1–10.

228. Zaborowski, M., Tomaszewski, D., Malesinska, J., and Grabiec, P. (2014). Fabrication and characterization of junctionless MOSFETs for sensor applications, in *21st International Conference on Mixed Design of Integrated Circuits and Systems (MIXDES)*, Lublin, Poland, pp. 367–371.

229. Asmah, M. T., and Sidek, O. (2014). Junctionless sub-micron silicon wire device fabricated by atomic force microscope lithography and electrically characterized for gas sensing application at room temperature, in *36th International Manufacturing Technology Conference*, Johor, Malaysia, pp. 1–6.

230. Asmah, M. T., Sidek, O., and Hutagalung, S. D. (2015). Junctionless silicon-based device for CO_2 and N_2O gas detection at room temperature, *Procedia Manuf.*, **2**, pp. 385–391.

231. Mohammad Fahad, H., Shiraki, H., Amani, M., Zhang, C., Srinivas Hebbar, V., Gao, W., Ota, H., Hettick, M., Kiriya, D., Chen, Y. Z., Chueh, Y. L., and Javey, A. (2017). Room temperature multiplexed gas sensing using chemical-sensitive 3.5-nm-thin silicon transistors, *Sci. Adv.*, **3**(3), p. e1602557.

232. Roudsari, A. F., Saini, S. S., O N, and Anantram, M. P. (2014). Junctionless phototransistor with nanowire channels, a modeling study, *Opt. Express*, **22**(10), pp. 12573–12582.

233. Vandooren, A., Franco, J., Parvais, B., Wu, Z., Witters, L., Walke, A., Li, W., Peng, L., Desphande, V., Bufler, F. M., Rassoul, N., Hellings, G., Jamieson, G., Inoue, F., Verbinnen, G., Devriendt, K., Teugels, L., Heylen, N., Vecchio, E., Zheng, T., Rosseel, E., Vanherle, W., Hikavyy, A., Chan, B. T., Ritzenthaler, R., Besnard, G., Schwarzenbach, W., Gaudin, G., Radu, I., Nguyen, B. Y., Waldron, N., De Heyn, V., Mocuta, D., and Collaert, N. (2018). 3D sequential stacked planar devices on 300 mm wafers deaturing replacement metal gate junction-less top devices processed at 525°C with improved reliability, *Digest of Technical Papers, Symposium on VLSI Technology*, paper T7-1.

Chapter 2

Several Challenges in Steep-Slope Tunnel Field-Effect Transistors

Katsuhiro Tomioka

Graduate School of Information Science and Technology, and Research Center for Integrated Quantum Electronics (RCIQE), Hokkaido University, Kita 14 Nishi 9, Sapporo 060-0814, Japan

tomioka@rciqe.hokudai.ac.jp

Tunnel FETs (TFETs) with a steep subthreshold slope (SS) have been attracting much attention as building blocks for future low-power integrated circuits and CMOS technology devices. There are, however, several challenges in creating modern TFETs that achieve their utmost performance such as a steeper SS with a higher on-state current. Here we introduce some guidelines for achieving a steep SS and recent advances in vertical TFETs using III-V/Si heterojunctions. These heterojunctions, which are formed by direct integration of III-V nanowires (NWs) on Si, are promising tunnel junctions for achieving a steep subthreshold slope (SS).

2.1 Introduction

The main goal for future electronics is that the power consumption of integrated circuits and internet-of-things (IoT) devices should be

Emerging Devices for Low-Power and High-Performance Nanosystems: Physics, Novel Functions, and Data Processing
Edited by Simon Deleonibus

ISBN 978-981-4800-11-2 (Hardcover), 978-0-429-45873-6 (eBook)
www.panstanford.com

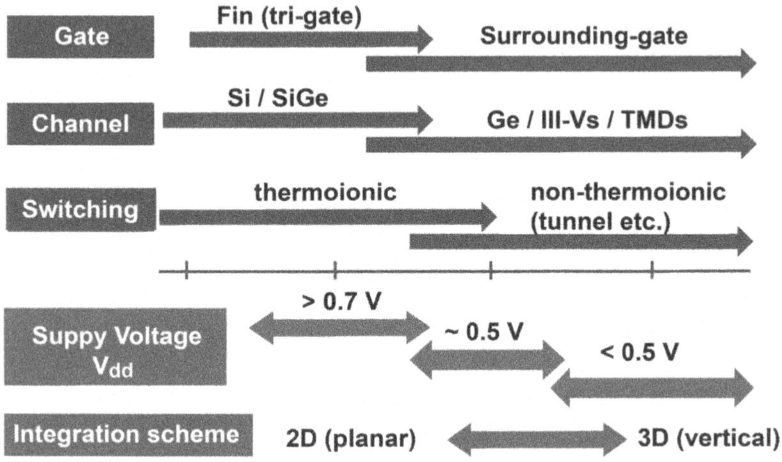

Figure 2.1 Scenario of state-of-the-art CMOS extended technologies.

reduced to ultralow levels and the performance of these circuits and devices should be increased. A simple but difficult approach to decreasing the power consumption of metal-oxide-semiconductor field-effect transistors (MOSFETs) is to lower the supply voltage (V_{dd}) since the active power of integrated circuits is proportional to the square of the V_{dd} and their stand-by power is proportional to the V_{dd} and the off-state leakage current (I_{off}). The V_{dd} is determined by the subthreshold slope (SS) and the difference between the on-state current (I_{on}) and the I_{off}. Therefore, The V_{dd} is expected to be decreased effectively by enhancing the I_{on} under a low bias, reducing the I_{off} by suppressing the short-channel effect, and decreasing the SS. In this regard, Si-based complementary metal-oxide semiconductor (CMOS) technologies with a multigate structure [1, 2], new channel materials [3], and nonclassical switching mechanisms [4] reducing power dissipation are expected to be utilized as extended CMOS technologies. These distinct features should be systematically addressed in extended CMOS technologies before 3D integration takes over (Fig. 2.1).

The I_{off} will be decreased by changing the multigate structure to a gate-all-around (GAA) or a surrounding-gate structure [2, 5, 6], in which the gate-metal/insulator stack is wrapped around the channel in order to provide the best electrostatic control of the channel possible. The vertical surrounding-gate architecture would exhibit geometrical and performance advantages in 7-nm-node devices [7].

Replacing the channel material from strained SiGe to a III-V Ge or two-dimensional transition metal dichalcogenide (2D TMD) is expected to result in a high on-state current under a low bias because these materials have a higher carrier mobility and a lower electron/hole effective mass. Since these materials have a much higher leakage current than silicon due to tunneling leakage, a multigate architecture would suppress the leakage current due to such tunneling and short-channel effect. Furthermore, the inherent mobility mismatch between the n-/p-channels will be a problem in future device integration to match current density. In the near future, the only way to avoid this problem will be to use a vertical channel architecture. In this regard, techniques integrating a vertical III-V channel with p-channel Ge will be important to fabricate Si-CMOS devices. We have already reported that the mobility mismatch problem has been avoided by integrating vertical III-V nanowires (NWs) directly on Ge substrates [8].

The switching mechanism in conventional FETs involves carrier thermal diffusion and drift, limiting the SS at room temperature (RT) to 2.3 $k_B T/q$, or ~60 mV/decade. This physical limitation will stop further scaling of power consumption in integrated circuits. Steep-SS transistors with SS < 60 mV/decade are, therefore, required for further reductions in V_{dd} and power dissipation. Steep-SS transistors such as tunnel field-effect transistors (TFETs) [4] impact ionization FETs [9], and mechanical switches [10] have thus attracted a great deal of attention. Of these steep-SS transistors, TFETs whose switching mechanism involves band-to-band tunneling (BTBT) [11] or Zener tunneling [12] are promising building blocks for future large-scale integrated circuits (LSIs) because their configurations are compatible with those of conventional Si-based FETs and tunneling is induced under an internal electric field.

Using TFETs with a SS strictly lower than 30 mV/decade would reduce the total power consumption of integrated circuits by over 90% because reducing the SS would ease the decrease of V_{dd}. Therefore, various TFETs using Si [13, 14], III-V materials

[15–17], and graphene [18, 19] have been investigated. Semiconductor NWs have also been investigated. These TFETs, however, all had difficulties in reaching sufficient steepness (SS strictly lower than 30 mV/decade) owing to significant challenges, such as those associated with the precise doping techniques needed to form abrupt PN junctions. In this regard, combining Si and III-V materials in one-dimensional (1D) nanostructures could provide a decisive solution to the problem because III-V/Si interfaces formed by integrating III-V NWs on Si have inherently abrupt junctions with band discontinuity [20, 21]. Furthermore, atomically smooth junctions utilizing 2D-TMD/p-Ge heterojunctions have been demonstrated as tunnel junctions [22].

In this chapter, we report on recent advances in vertical TFETs using III-V/Si heterojunctions. We used selective-area metal-organic vapor phase epitaxy (MOVPE) for the direct integration of III-V NWs on a Si(111) substrate. Reducing the opening diameter of the masked substrate decreased the number of misfit dislocations. The TFETs made by using III-V NW/Si heterojunctions demonstrated steep-SS properties at RT. One of the key issues in the vertical TFETs is the formation of intrinsic channels inside the tiny NW, and another is the reduction of NW diameter. We also discuss several current-boosting technologies for the TFETs.

2.2 Challenges in Achieving Steep-Slope TFETs

2.2.1 Benchmarks of Realized Steep-Slope TFETs

TFETs with a steep SS are promising millivolt switches for next-generation nanoelectronics. In our estimation, the SS should be less than 30 mV/decade in order to decrease the supply voltage (currently 0.50 V) of FETs. TFETs with SS $<$ 30 mV/decade would be able to reduce the total power consumption of LSIs by over 90% because the active power consumption of LSIs is proportional to the square of V_{dd}. However, few steep-slope transistors with SS $<$ 30 mV/decade have been demonstrated (Table 2.1). This is because their device architectures and design approaches are quite different from those of conventional MOSFETs and CMOS technologies. Before

Table 2.1 Comparison with experimental TFETs with a steep SS

Tunnel junction	Gate	SS_{min} (mV/decade)	Steep SS range	I_{on} (μA/μm)	I_{off} (A/μm)	Refs.
InAs/GaAsSb	SGT	48	2 dec.	10.6	1×10^{-9}	[23]
InGaAs/InAs	Planar	55	1 dec.	0.3	1×10^{-12}	[24]
Ge/SiGe	SOI	30	1 dec.	0.05	1×10^{-12}	[24]
MoS2/Ge	Planar	4.3	4 dec.	1.0	1×10^{-14}	[22]
InAs/Si	SGT	12	4 dec.	1	1×10^{-12}	[21]
InAs/Si	SGT	30	4 dec.	0.002	8×10^{-14}	[25]
InGaAs/Si	SGT	30	2 dec.	0.1	1×10^{-14}	[26]
InGaAs/Si	SGT	23	2 dec.	0.003	1×10^{-14}	[27]
Si	SGT	30	3 dec.	0.02	2×10^{-13}	[28]
Si	SGT	30	3 dec.	0.44	2×10^{-12}	[29]
Si	SOI	32	2 dec.	1.2	2×10^{-13}	[30]
Ge/Si	SOI	38	3 dec.	0.42	2×10^{-13}	[31]
SiGe	SOI	42	3 dec.	0.02	1×10^{-13}	[32]
Si	SOI	52.8	1 dec.	12.1	1×10^{-9}	[14]
CNT	planar	40	4 dec.	0.01	1×10^{-8}	[33]
Si	GAA	30	2 dec.	2	1×10^{-13}	[34]
Ge	DG	50	2 dec.	0.1	3×10^{-13}	[35]
Si	planar	29	2 dec.	10.2	1×10^{-13}	[36]
SiGe	planar	33	1 dec.	0.05	1×10^{-13}	[37]
InGaAs/InP	GAA	50	2 dec.	2.2	1×10^{-13}	[38]
Ge/SiGe	Planar	28	<1 dec.	0.1	1×10^{-14}	[39]
InGaAs	planar	58	<1 dec.	0.3	1×10^{-11}	[15]

demonstrating vertical TFETs using III-V/Si heterojunctions, we will describe the important challenges of these devices.

2.2.2 Series Resistance

A large internal electric field should be induced at the tunnel junction under a small gate and drain bias in order to achieve a steep SS because the SS of TFETs is ideally determined by the gate and drain bias [24]. In this regard, the arrangement of series resistances is very important for inducing the internal electrical field. The TFET channel architecture can be modeled by three resistances: (i) contact resistance, (ii) junction resistance, and (iii) channel resistance.

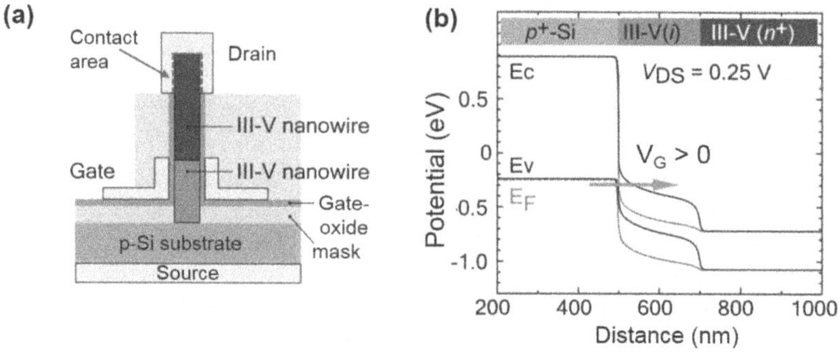

Figure 2.2 (a) Illustration of a vertical TFET using III-V NW/Si heterojunction. (b) Band alignment for the panel (a) at $V_{DS} = 0.25$ V.

The contact resistance should be small and the other resistances should be high enough to localize the internal field at the tunnel junctions.

2.2.2.1 Contact resistance

The contact resistance can be decreased by using a good ohmic contact or controlling the contact area. In the case of III-V compound semiconductors, Mo-related materials are useful in lowering the contact resistance [40]. The effective resistance in III-V NW/Si TFETs can be reduced by enlarging the drain contact area (Fig. 2.2a) [21].

2.2.2.2 Junction resistance

The tunnel junction is the one at which tunnel transport occurs. In the case of homojunction TFETs, controlling this resistance would be difficult because the junction is a *p-i-n* junction. On the other hand, in the case of heterojunction TFETs, the tunnel junction usually consists of staggered Type-II (Fig. 2.2b) or broken-gap band discontinuities. These heterojunctions should possess a high resistance. A heterojunction has a lattice mismatch across the tunnel junction, so crystal defects such as misfit dislocations affect the junction resistance because they form defect levels across the junction that cause trap-assisted tunneling (TAT) [41]. The TAT process enhances the tunneling probability, resulting in

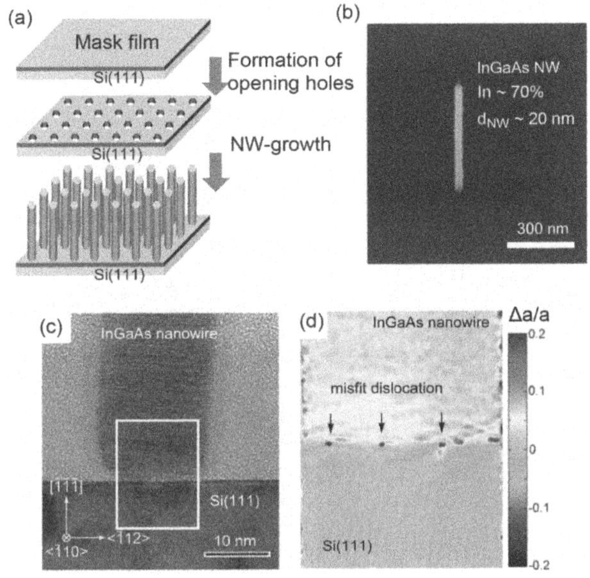

Figure 2.3 (a) Illustration of selective-area growth of NWs on Si. (b) A representative SEM image of an InGaAs NW on Si. The NW diameter is 20 nm. (c) TEM image of the cross section of the NW. (d) Mapping of the lattice constant of the area outlined in white in panel (c).

applications. We have recently reported the formation of a (111)B-oriented surface on the Si(111) surfaces and demonstrated the integration of vertical III-V NWs on Si(111) and Ge(111) substrates by selective-area MOVPE [8, 45, 46]. Representative results for the growth of InGaAs NWs 20 nm in diameter can be seen in Fig. 2.3b. The NWs had no defects originating from the antiphase domain; rather, strain analysis showed misfit dislocations at the InGaAs NW/Si interface. The dislocation periodicity became greater than the value calculated from the lattice mismatch. This is because the lattice relaxation due to the lattice mismatch is suppressed in the nanometer-scaled crystal growth area.

2.3.2 Misfit Dislocations at III-V NW/Si Junctions

The number of misfit dislocations at the III-V NW/Si heterojunction can be controlled by adjusting the NW diameter (or opening

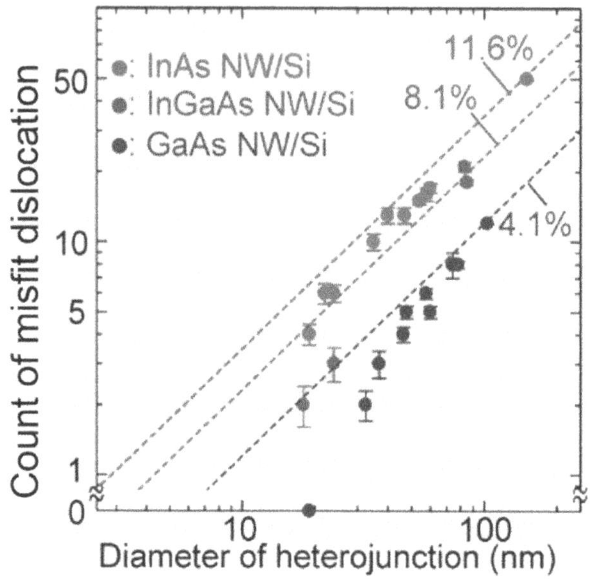

Figure 2.4 Misfit dislocation count at several III-V NW/Si heterojunctions. The heterojunction diameter was measured in TEM images. Broken lines represent estimation from the lattice mismatch noted.

diameter) [25]. Figure 2.4 shows the counts of misfit dislocations as a function of III-V NWs/Si heterojunction diameter, indicating that the number of misfit dislocations falls below the number calculated from the lattice mismatch. InAs NW/Si and InGaAs NW/Si had lattice mismatches of 11.6% and 8.1%, but the number of misfit dislocations decreased with the heterojunction diameter. As for GaAs NW on Si, misfit dislocations were virtually suppressed with a NW diameter of 19 nm, which is due to coherent growth. This coherently grown III-V/Si heterojunction is the ideal tunnel junction to induce only BTBT with a high heterojunction resistance. Note that the misfit dislocations form defect levels across the III-V/Si junction, resulting in the TAT process [41]. When TAT transport is the dominating charge transport process in a TFET, the heterojunction resistance decreases with an increasing SS because of reduction in the internal electrical field.

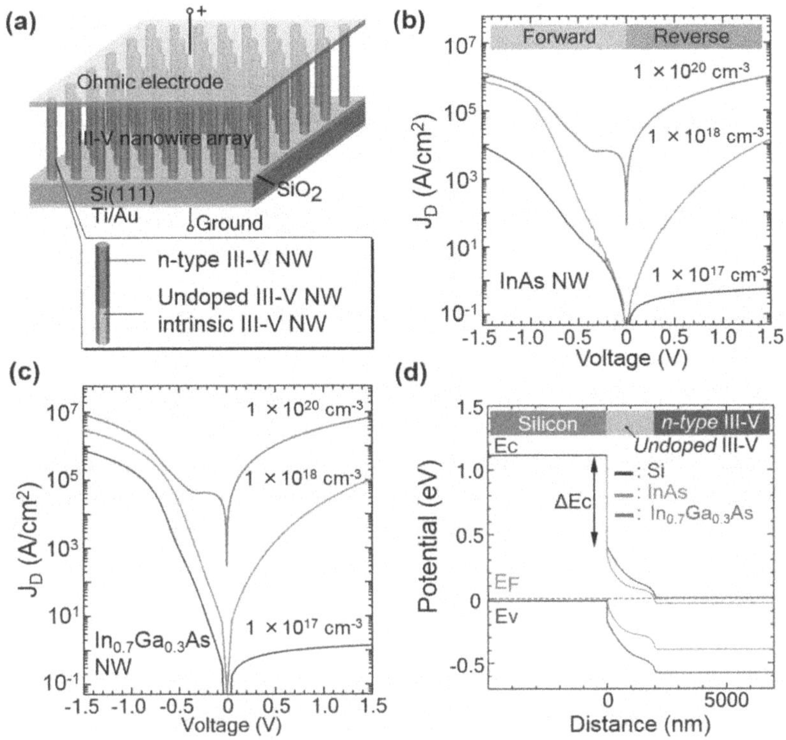

Figure 2.5 Illustration of a vertical diode consisting of 250 NWs on p-Si substrates. (b) Diode properties for InAs NW/p-Si with a different carrier density of the p-Si. Each NW was composed of n-type III-V and undoped III-V segments. (c) InGaAs NW/p-Si. (d) Band alignment estimated by 1D Poisson-Schrödinger equation.

2.3.3 Diode Properties for the III-V NW/Si Junctions

As shown in Fig. 2.5a, a two-terminal device for diode characterization was first fabricated on a p-Si substrate with various carrier concentrations. In this case, the diameter of InGaAs NWs was 70 nm. The device process has been previously reported [20, 26]. The current density (J_D)-voltage (V) curve (the Si substrate was grounded) is shown in Fig. 2.5b. The measured current was divided by the number of NWs and normalized by the surface contact area. In this case, the positive voltage is in the reversed bias direction against the p-i-n junction. The J_D-V curve shows typical rectification properties under a forward bias ($-1.0 < V < 0$), with an ideality factor of 1.9. The J_D under a reverse bias ($0 < V < 1.0$) increases with the carrier concentration of p-Si.

Figure 2.5d shows a band diagram calculated by a 1D Poisson-Schrödinger equation [26, 27], and in that figure the band diagram is staggered Type-II band discontinuity. In this calculation, we set the carrier concentration of p-Si to 1×10^{18} cm^{-3}. The current in the reverse direction is found to be Zener tunneling transport across the InGaAs NW/Si heterojunction, similar to the current in the InAs NW/Si heterojunction (Fig. 2.5b), and the current is increased as the carrier concentration increases. In the case of a heavily doped p-Si substrate ($n = 1 \times 10^{20}$ cm^{-3}), Esaki tunneling transport is also observed under the forward bias direction (at around −0.3 V). This is because the position of the Fermi level lies underneath the valence band in p-Si for $n = 1 \times 10^{20}$ cm^{-3}. These behaviors suggest that controlling the position of the Fermi level by using various electrical fields modulates the tunneling transport induced at the III-V NW/Si heterojunction.

2.4 Vertical Tunnel FETs Using a III-V/SI Junction

2.4.1 MOS Interface in Vertical InGaAs NW FETs

Before demonstrating a vertical TFET using III-V NWs/Si heterojunctions, we show the quality of the MOS structure for the vertical III-V NW architecture because the vertical TFET has a surrounding-gate architecture. Thus, the formation of a high-quality MOS interface is a precondition for achieving a steep SS in TFETs. The idea and the demonstration of SGTs were first reported by Takato et al. [2]. The main advantage of the vertical surrounding-gate structure was to induce the best electrostatic gate-bias control in the channel due to the wrapped gate metal: that enabled the occupied area to be reduced to allow a denser integration.

The SGT structure of a single InGaAs NW on Si is outlined in Fig. 2.6a [47, 48]. The In composition was 70%. We used a gate-first process and etch-back procedures to make the SGT device [49]. Figure 2.6b shows the electrical properties of the InGaAs NW-

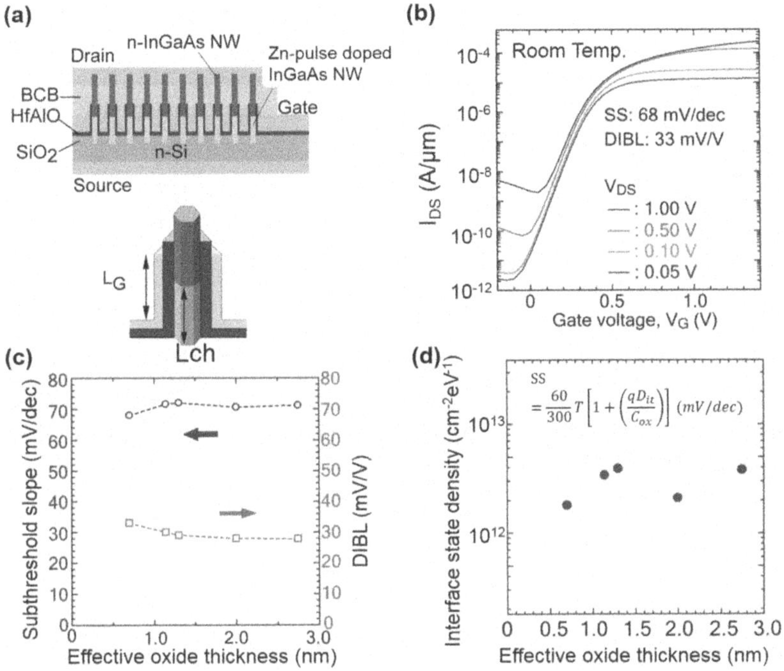

Figure 2.6 (a) Illustration of the SGT structure. The actual device has 10 NWs in parallel with a drain contact. (b) Transfer curves for the InGaAs NW SGTs on Si with 4-nm-thick gate oxide (EOT = 0.75 nm). (c) SS and DIBL versus EOT. (d) Interface state density, D_{it}, estimated from the SS in panel (c).

SGT with an effective oxide thickness (EOT) of 0.75 nm. The NW was 80 nm in diameter. The I_D was reasonably modulated by the V_G with an SS of 68 mV/decade [27]. The drain-induced barrier lowering (DIBL) was 33 mV/V. The small SS close to the ideal SS (~60 mV/decade) and the small DIBL reflect the geometrical advantage of the electrostatic gate bias for the surrounding-gate structure. In Fig. 2.6b, the drain current was increased with the V_{DS} in accumulation mode ($V_G < 0$ V), that is, it originated from gate leakage. The SS was almost completely constant with the variation of EOT in Fig. 2.6c. A simple approximation from the SS estimated the interface state density (D_{it}) to be from 1.8 to 3.9 $\times$ 10^{-12} $cm^{-2}eV^{-1}$. Since the SS for the InGaAs NW SGTs was almost constant

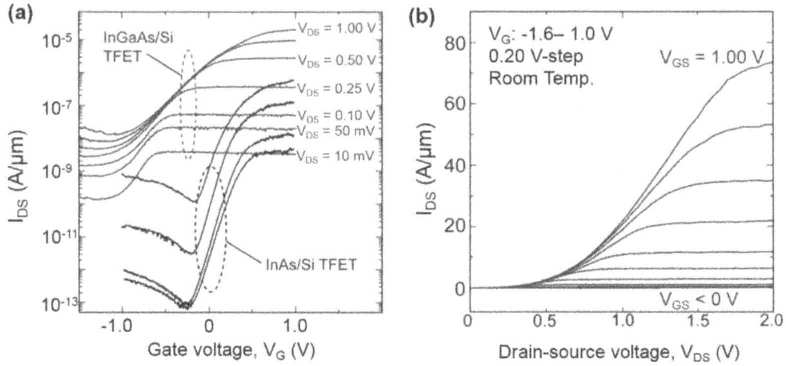

Figure 2.7 (a) Switching properties of a degraded vertical TFET using a single In(Ga)As NW/Si heterojunction. The InGaAs NW was 70 nm in diameter. Black curves are for the InAs NW/Si heterojunction TFET. (b) $I_{DS} - V_{DS}$ curves for the InGaAs NW/Si TFET.

regardless of the EOT, the D_{it} did not depend on the thickness. Thus the extraordinarily low D_{it} for the InGaAs $\{-110\}$ surface/HfAlO interface is thought to be due to the formation of (111)A/B microfacets on the sidewalls of the NW channel. This is because these microfacets form less Ga oxide, which usually increases the D_{it}, and are chemically stable. This specific MOS interface inherently improves electrostatic gate control in the vertical InGaAs/Si TFETs and the InGaAs vertical NW FETs.

2.4.2 Degraded TFETs Using In(Ga)As NW/Si Junctions

Figure 2.7 shows the switching properties of the In(Ga)As NW/Si TFET at drain-source voltages (V_{DS}) from 10 mV to 1 V. The gate voltage (V_G) ranged from -1.50 to 1.50 V. The current values were normalized using a gate perimeter. Current modulations with a SS of 260 mV/decade were observed in the reverse bias direction (V_G is positive for the n-i-p junction). The SS for an InAs NW/Si vertical TFET was 120 mV/decade at RT.

Almost all reports regarding TFETs mention performance degradation due to device architecture and the tunnel junctions. This is because the device structure cannot induce a large internal field at the heterojunction in our case, III-V NW channels, under low V_{DS} and V_G. Since the SS is a function of V_G and V_{DS}, the series resistance of the device must be adjusted to minimize the SS. This device thus requires a large internal field induced under lower V_G and V_{DS}. We discuss the scaling of the NW diameter and doping to achieve a steep SS in vertical TFETs in the next two sections.

2.4.3 Scaling of NW Diameter for a Steep SS

In the case of an InAs NW/Si heterojunction, decreasing the diameter of the NW suppresses the formation of misfit dislocations (Fig. 2.4), so a steep SS can be achieved with InAs NW/Si junction TFETs (Fig. 2.8) [21].

To obtain a steep SS in the III-V NW/Si tunnel junction, III-V NW- diameters must be reduced to suppress the formation of misfit dislocations and the contact area must be increased to reduce the effective contact resistance [21]. Achieving coherent growth without misfit dislocations and TAT was predicted to result in a steep SS

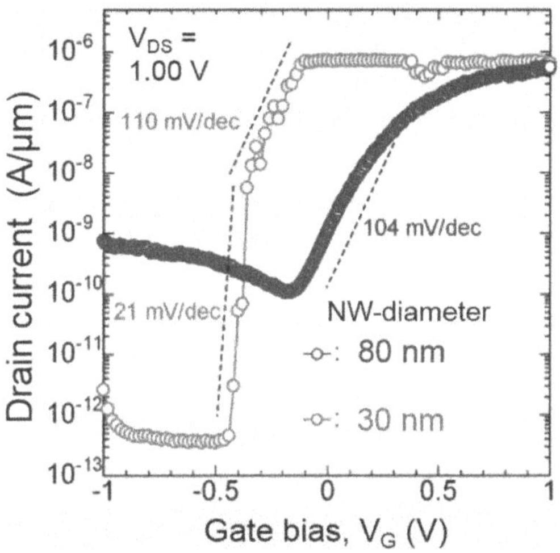

Figure 2.8 NW-diameter dependence of the switching properties of vertical TFETs using InAs NW/Si heterojunctions. The image was modified from Ref. [21].

[25, 41]. However, such a small opening pattern is difficult to form. A device using an InAs NW (30 nm in diameter)/Si heterojunction demonstrated a steep SS (minimum SS = 21 mV/decade) at RT (Fig. 2.8) [21]. There are two SS regions in this device's properties. The SS is very steep in the early stage of gate modulation, but at a higher V_G it takes values close to 110 mV/decade. The physical origin of these SS regions remains under investigation, but two mechanisms (BTBT and TAT) are assumed to contribute in parallel.

2.4.4 Challenges in Doping

Introducing doping in TFETs is a very difficult question, since the TFET is basically a backward device and its channel formation requests PN junctions. Furthermore, the tunneling phenomenon itself occurs in a several-nanometer-scale space. This means the p-i-n junction should ideally be packed within that space. Thus the quality of the doping—in this case, its abruptness—must be better than that in conventional MOSFET at the sub-14-nm nodes. Thus, many challengers stumble over the doping technique for modern TFETs.

The important concept for the doping is how many atoms in the nanochannels are activated. Within the several-nanometer-scale space, a single dopant atom, a contamination or a vacancy, could be activated and contribute to the effective impurity density, eventually increasing it. In the modern MOSFET, this concept is specified in the variability of the device by terms such as "dopant effect" and "random telegraph noise," while in TFETs the doping behavior strongly relates to the device performance. First, the formation of the intrinsic layer by a doping technique is very important for attaining a steep SS in III-V-based TFETs. We will describe the reason and the impact in the next section. Second, heavy doping in the source/drain region for boosting tunnel current or avoiding source starvation effect [50] eventually degrades the abruptness of the tunnel junction because of the formation of an amorphous phase or defect resulting in large dispersion of Urbach tails [51]. Indeed, the heavy doping excess the solubility limit results in the formation of defects. The Urbach tails blur the density of states and allow a contribution of the TAT at the tunnel junction. Thus, excess dopants basically degrade the SS of TFETs.

2.4.4.1 Formation of an intrinsic layer in a III-V NW

As mentioned earlier, III-V materials grown by MOVPE or MBE usually have the problem of unintentional doping. For example, InAs grown in MOVPE is unintentionally doped by carbon dopants derived from metal-organic precursors [42], and the degree of doping depends on the condition of the reactor chamber. So compensation doping is required for forming an intrinsic layer. Other MOVPE-grown materials such as InGaAs and GaAs are also unintentionally doped. Figure 2.9 depicts the transfer curve for the InAs and InGaAs NW–based SGTs on Si. When an undoped layer is used, the concentrations of the InAs and InGaAs NWs are estimated to be 2×10^{17} cm^{-3} and 1×10^{17} cm^{-3}, respectively.

Compensation with counter dopants is difficult to obtaining an intrinsic layer against the NW with small volume. For example, one single Zn atom as a counter dopant has an effective carrier concentration of 4.7×10^{17} cm^{-3} when the InAs NW is 30 nm in diameter and 1 nm long. A III-V NW with an even smaller diameter would have a higher net carrier concentration. Thus the conventional compensation effect changes the conductivity of the NW; that is, a p-type NW is formed. We therefore used a pulse doping technique [25, 26] for forming a nearly intrinsic III-V NW segment in which the carrier concentration, as estimated from the threshold voltage of InAs NW SGTs on Si, was $\sim 2.2 \times 10^{15}$ cm^{-3} (Fig. 2.9b) [26]. In the case of InGaAs NWs, we obtained a nearly intrinsic layer with a carrier concentration of $\sim 5 \times 10^{15}$ cm^{-3} by using the pulse doping technique. This unintentional doping in the InGaAs NW-channel lowers the internal fields due to V_G and V_{DS} because it lowers the channel resistance. This means that a large bias is required for inducing tunnel transport, resulting in a SS of 60 mV/decade.

Figure 2.10 shows the effect of the Zn pulse doping on the properties of TFETs. Formation of a nearly intrinsic NW segment shifted the threshold voltage of the TFET from negative to positive

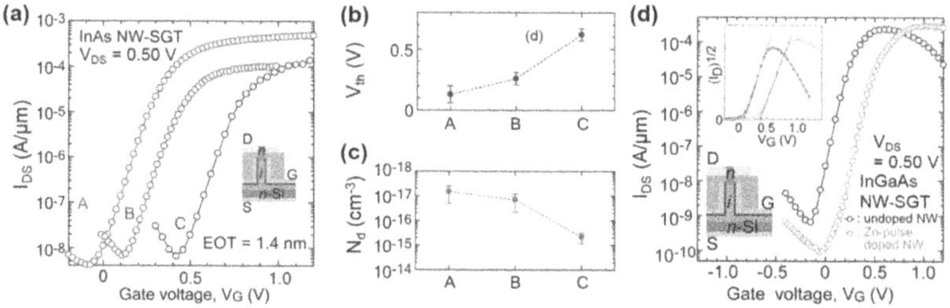

Figure 2.9 (a) Switching properties of InAs NW SGTs on Si. Curve A is for an InAs NW with an unintentional doped channel region. Curves B and C are for InAs NW channels made with the pulse doping technique. Curve B: Zn-pulse-doped channel (1 s Zn pulses at 29 s intervals). Curve C: Zn-pulse-doped channel (2 s Zn pulses at 28 s intervals). (b) V_{th} for InAs NW SGT for different channels (modified from Ref. [26]). (c) Carrier concentration (N_d) estimated from V_{th}. (d) I_D-V_G curves of InGaAs NW SGTs on Si of an undoped NW channel and Zn-pulse-doped NW channels.

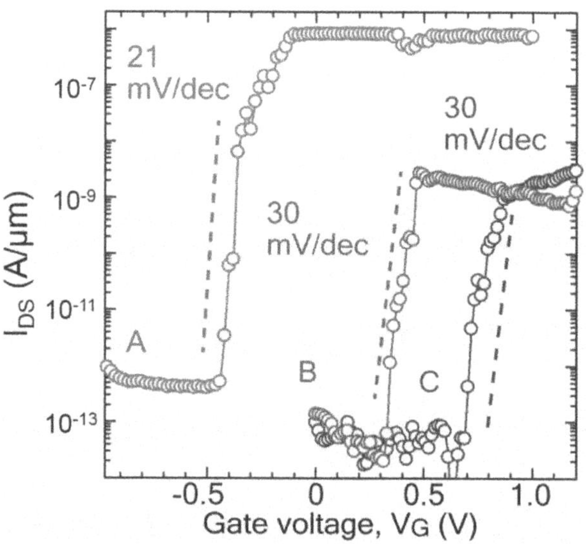

Figure 2.10 Doping effect on the steep-SS transistors using InAs NW/Si heterojunctions. Curve A is the transfer characteristic at $V_{DS} = 1.00$ V. Curves B and C show steep-SS characteristics at V_{DS} = 0.50 V obtained for InAs NW/Si heterojunctions. The NW growth conditions for the curves A, B, and C are the same as those for the corresponding curves in Fig. 2.9a.

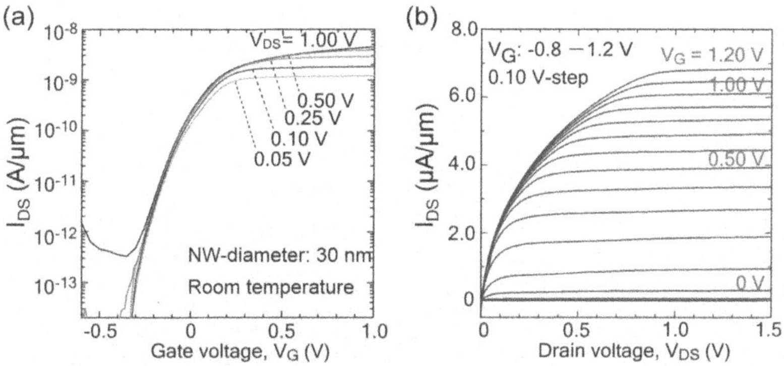

Figure 2.11 (a) I_D–V_G characteristic of the switching device using $In_{0.7}Ga_{0.3}As$ NW/Si heterojunctions with an EOT of 1.91 nm. L_G was 150 nm. The InGaAs NW diameter was 30 nm. The minimum SS was 23 mV/decade (b) The output characteristic of the TFET using a single $In_{0.7}Ga_{0.3}As$ NW/Si heterojunction with an EOT of 1.91 nm.

V_G. Even at V_{DS} = 0.50 V, the SS was only 30 mV/decade. This indicates that the carrier concentration of the NW-channel segment determines the threshold voltage of the TFET structure. Figure 2.11 shows the transfer curves of a TFET using the InGaAs NW/Si heterojunction with the Zn pulse-doped InGaAs NW-channel [27]. The NW was 30 nm in diameter. The channel length corresponds to the length of the Zn pulse region, 200 nm in this case. By using the pulse doping technique, we were able to improve the SS compared with that of an undoped InGaAs NW channel (Fig. 2.7). The advantage of the III-V/Si junction for a steep-SS switch is the suppression of parasitic leakage current thanks to band discontinuities across the III-V/Si heterojunction. In this InGaAs NW/Si heterojunction switch, the parasitic leakage current was suppressed effectively and the minimum SS became <60 mV/decade under a lower V_G. The minimum SS was 23 mV/decade, and the average SS over two decades (10^{-13} – 10^{-11} A/μm) was 40 mV/decade at V_{DS} = 0.25 V. However, the SS becomes larger as V_G increases. This is because tunneling transport involves pure BTBT and TAT since the InGaAs NW/Si heterojunction has misfit dislocations as shown in Fig. 2.3d. Therefore, reducing the InGaAs NW diameter and achieving coherent growth without misfit dislocations would enlarge the steep-SS region. An additional technique to boost I_{ON} is required for InGaAs NW/Si heterojunction tunnel FETs.

2.4.4.2 Heavy doping

Heavy doping is required to increase the on-state current and lower the contact resistance of a TFET. A conventional heavy doping method, such as ion implantation, usually forms an amorphous layer with large Urbach tails. Moreover, heavily doping bottom-up nanostructures basically changes their morphology. Especially in NW by selective-area growth, a huge number of dopant atoms disturb the normal surface diffusion process and result in the deterioration of a uniform NW growth. Therefore, a precise heavy doping maintaining the growth morphology is required for TFET

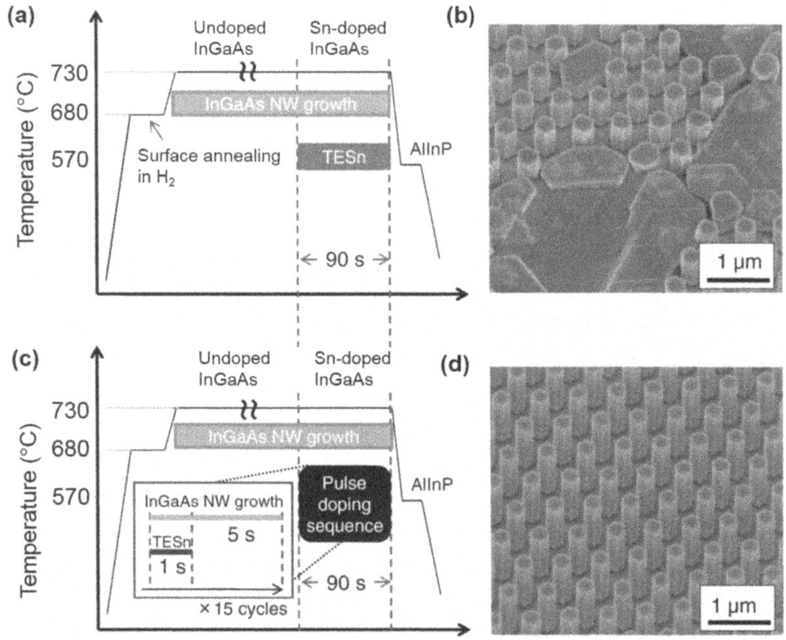

Figure 2.12 (a) Growth sequence of a conventionally doped InGaAs NW with tetra-ethyl tin (TESn) and (b) SEM image showing growth results. (c) Growth sequence of the NW array with pulse doping of TESn and (d) SEM image showing an as-grown NW array.

application. In this regard, pulse doping will be an effective doping method.

Figure 2.12 shows the growth sequences and morphologies of Sn-doped InGaAs NWs grown by selective-area growth [52]. The Sn atoms serve as n-type dopants in InGaAs. The morphologies of the heavily Sn-doped InGaAs NWs grown with a conventional tin-impurity flow show that the lateral overgrowth is enhanced and show the coalescence of the grown materials displayed in Fig. 2.12a. The morphologies of the InGaAs NWs grown with the pulse doping technique [26] shown in Fig. 2.12b demonstrated a uniform NW array. The pulse doping technique can suppress the lateral overgrowth and coalescing of the NWs. We confirmed that using this technique improved the performance of an InGaAs NW solar cell, which means that the heavily doped layer in the NW lowers the contact resistance. The n^+ contact layer can be formed inside a NW channel by the pulse doping technique regardless of the deteriorations.

2.5 Challenges in Increasing the On-State Current

The TFET works under the electrostatic gate-bias modulation of tunnel transport. The carrier transport depends on the tunneling probability. Tunneling is an energy filtering process resulting in a low conductance. And the on-state current of TFET is usually lower than that of conventional MOSFETs in which the carrier transport involves thermionic processes such as diffusion and drift processes. Thus, additional techniques of enhancing the inherent low conductance are required for practicing the TFET in integrated circuits.

2.5.1 Effect of Channel Length

Next, we investigated the effect of channel length (L_{ch}) on scaling [26]. Figure 2.13a shows the L_{ch} dependence of the transfer curve at a V_{DS} of 0.10 V. The L_G was fixed at 150 nm, and the growth condition and Zn pulse-doping sequence were the same for the

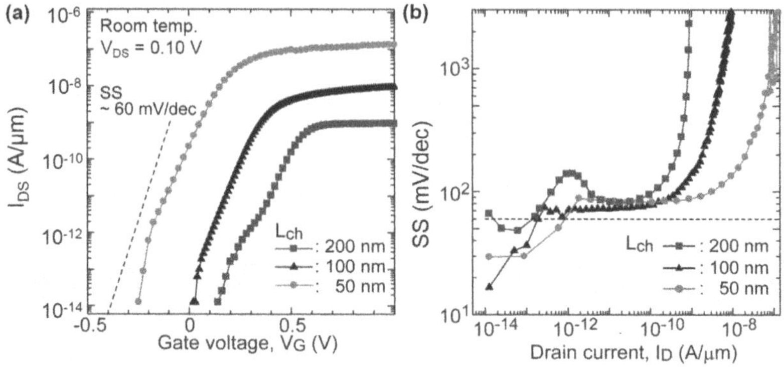

Figure 2.13 (a) Transfer curves for different values of L_{ch}. $V_{DS} = 0.10$ V at RT in the dark. The dashed line plots the theoretical limit of carrier thermal diffusion (SS = 60 mV/decade). (b) SS as a function of drain current. The dashed line shows the limit of MOSFETs (=60 mV/decade). The inset illustrates the NW channel.

different L_{ch}. We chose a low V_{DS} to characterize a steep turn-on behavior. Figure 2.13b shows SS as a function of drain current. The steep-SS region (< 60 mV/decade) in Fig. 2.13b tends to widen with a decreasing L_{ch}. The current at the boundary between sub- and super-60 mV/decade (I_{60}) [53] also increases for a shorter L_{ch}. The I_{60} was 10^{-12} A/μm for $L_{ch} = 50$ nm. Thus an internal electrical field was induced more effectively with a shorter L_{ch}. Another effect of L_{ch} on scaling is to enhance the I_D. The I_D in Fig. 2.13a achieved with the shortest L_{ch} reached 10^{-7} A/μm, which is 100 times that obtained with the longest L_{ch}. This indicates that a source edge closer to the tunnel junction would be better for enhancing the I_D because tunneling carriers are subject to influence from scattering processes in longer channels. In addition, the device with a longer L_{ch}, in this case, had a slight distance between the gated region with the heterojunction, which decreases the electrical field and degrades the SS. Scaling L_{ch} is, thus, predictably effective for $L_{ch} <$ 10 nm to increase the I_{ON}. This is because the tunnel distance is 7 nm. Thus, scaling L_{ch} has a synergetic effect, increasing both I_{60} and I_D.

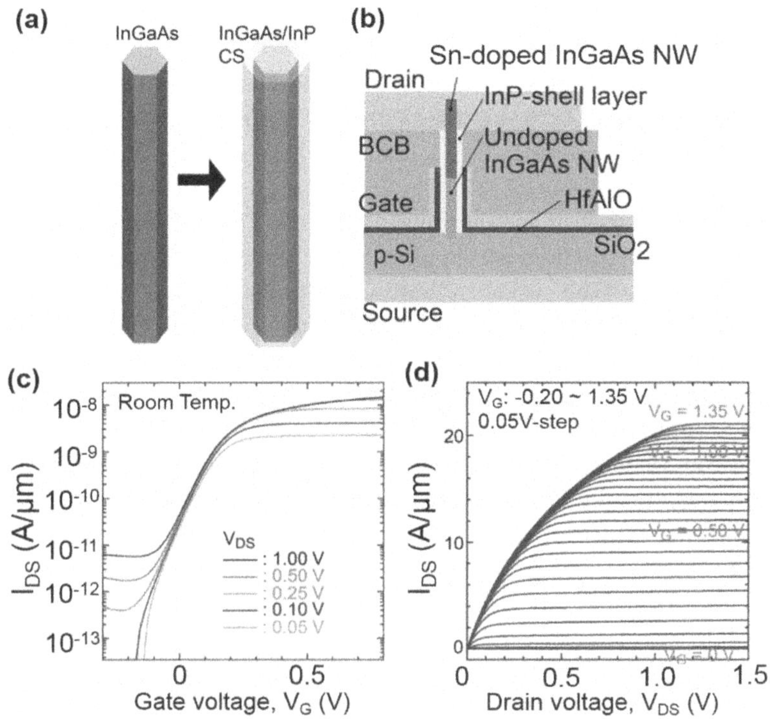

Figure 2.14 (a) Schematics of an InGaAs/InP core-shell (CS) NW. (b) Illustration of the device structure. (c) Switching properties of the vertical TFET using s InGaAs/InP CS NW. (d) Output properties of the TFET.

2.5.2 Effect on Strain

The core-shell structure shown in Fig. 2.14b can benefit the NW-based vertical TFET by inducing both a passivation and a strain effect. The direct integration can be used to make various types of core-shell NW structures. In this paragraph, we characterize the InGaAs NW/Si heterojunction TFET with a surrounding-gate structure and the effect of an InP shell layer on the TFET's properties.

Figure 2.14b shows the representative transfer properties of the InGaAs-InP core-shell NW/Si TFET at V_{DS} from 0.05 to 1.00 V. The L_G and L_{ch} were 200 nm. The core was 50 nm in diameter, and the thickness of the InP shell was 10 nm. The switching behavior due to the pure tunneling appeared at a V_{DS} as low as 10 mV. The I_{ON}/I_{OFF} ratio was approximately 10^5 at a V_{DS} of 0.10 V. The threshold voltage, V_T, of I_D was −0.10 V. The dark Zener current of the diode structure was modulated by V_G. The I_{ON} was about 10 nA/μm at $V_{DS} = V_G - V_T = 0.50$ V, which was 10 times higher than the bare InGaAs/Si [27]. The DIBL was 10 mV/V, which is much smaller than that of conventional MOSFETs. This indicates that the internal electrical field was applied at the tunneling junction despite the lowering of the surface potential due to V_{DS}.

The on-state current for the InGaAs-InP core-shell NW/Si TFET was three times higher than that for the bare InGaAs NW/Si TFET. The benefits of using the InP shell layer are an improvement in the SS and an enhancement of the I_{ON}. The InP shell has a good semiconductor/oxide interface [54], which decreases the interface-state density and thus results in an electrostatic improvement in the SS below 60 mV/decade. The strain effect of the InP shell layer also helps to improve the SS and the I_{ON}. The outer InP shell induces biaxial tensile strain against the whole core-InGaAs NW due to the lattice mismatch (−1.16%), and the uniaxial compressive strain is localized to near the tunnel junction (InGaAs/Si junction) and along the axial direction. The compressive strain possibly lowers the conduction band minimum of the core InGaAs NW, which has the effect to adjust the Fermi levels across the junction. It decreases the effective energy gap across the Si and InGaAs junction. Although further characterization of the strain is required, these results show that the strain of the core-shell structure can be used to increase the on-state current while maintaining a steep SS.

2.6 Conclusion

In this chapter, we described vertical surrounding-gate TFETs using III-V/Si NW heterojunctions. We discussed several steps for obtaining steep turn-on behavior due to a pure tunnel process and demonstrated that the exotic III-V/Si heterojunction was a moderate tunnel junction for the TFETs with a steep SS. Then, we introduced some approaches for enhancing the tunneling probability (on-state current) in TFETs while avoiding the insoluble dilemma in the quantum tunneling process.

Acknowledgment

The authors would like to thank Professor Emeritus Dr. Takashi Fukui. This work was financially supported by a grant-in-aid for scientific research from the Ministry of Education, Culture, Sports, Science, and Technology (MEXT) and the Japan Science and Technology Agency (JST) PRESTO program.

References

1. Ferrain, I., Colinge, C. A., and Colinge, J.-P. (2011). Multigate transistors as the future classical metal-oxide-semiconductor field-effect transistors, *Nature*, **479**, pp. 310–316.
2. Takato, H., Sunouchi, K., Okabe, N., Nitayama, A., Hieda, K., Horiguchi, F., and Masuoka, F. (1988). High performance CMOS surrounding gate transistor (SGT) for ultra high density LSIs, in *IEEE International Electron Devices Meeting (IEDM) Technical Digest*, pp. 222–225.
3. del Alamo, J. A. (2011). Nanometre-scale electronics with III-V compound semiconductors, *Nature*, **479**, pp. 317–323.
4. Seabaugh, A. C., and Zhang, Q. (2010). Low-voltage tunnel transistors for beyond CMOS logic, *Proc. IEEE*, **98**, pp. 2095–2110.
5. Takato, H., Sunouchi, K., Okabe, N., Nitayama, A., Hieda, K., Horiguchi, F., and Masuoka, F. (1991). Impact of surrounding gate transistor (SGT) for ultra-high-density LSI's, *IEEE Trans. Electron Devices*, **38**, pp. 573–578.
6. Nitayama, A., Takato, H., Okabe, N., Sunouchi, K., Hieda, K., Horiguchi, F., and Masuoka, F. (1991). Multi-pillar surrounding gate transistor (M-SGT) for compact and high-speed circuits, *IEEE Trans. Electron Devices*, **38**, pp. 579–583.
7. Yakimets, D., Eneman, G., Schuddinck, P., Bao, T. H., Bardon, M. G., Raghavan, P., Veloso, A., Collaert, N., Mercha, A., Verkest, D., Thean, A. V-Y., and Meyer, K. D. (2015). Vertical GAAFETs for the ultimate CMOS scaling, *IEEE Trans. Electron Devices*, **62**, pp. 1433–1439.

8. Tomioka, K., Ishizaka, F., and Fukui, T. (2015). Selective-area growth of InAs nanowires on Ge and transistor application, *Nano Lett.*, **15**, pp. 7253–7257.
9. Armstrong, Y., Davis, J. R., and Doyle, A. (1991). Characterization of bipolar snapback and breakdown voltage in thin-film SOI transistors by two-dimensional simulation, *IEEE Trans. Electron Devices*, **38**(2), pp. 328–336.
10. Pott, V., Kam, H., Nathanael, R., Jeon, J., Alon, E., and Liu, T-J. K. (2010). Mechanical computing redux: relays for integrated circuit applications, *Proc. IEEE.*, **98**, pp. 2076–2094.
11. Esaki, L. (1958). New phenomenon in narrow germanium p-n junctions, *Phys. Rev.*, **109**, pp. 603–604.
12. Zener, C. (1934). A theory of the electrical breakdown of solid dielectrics, *Proc. R. Soc. Lond.*, **145**(855), pp. 523–529.
13. Sedlmaier, S., Bhuwalka, K. K., Ludsteck, A., Schmidt, M., Hansch, W., and Eiseleless, I. (2004). Gate-controlled resonant interband tunneling in silicon, *Appl. Phys. Lett.*, **85**, pp. 1707–1709.
14. Choi, W. Y., Park, B. G., Lee, J. D., and Liu, T-J. K. (2007). Tunneling field-effect transistors (TFETs) with subthreshold swing (SS) less than 60 mV/dec, *IEEE Electron Device Lett.*, **28**(8), pp. 743–745.
15. Dewey, G. Chu-Kung, B., Boardman, J., Fastenau, J. M., Kavalieros, J., Kotlyar, R., Liu, W. K., Lubyshev, D., Metz, M., Mukherjee, N., Oakey, P., Pillarisetty, R., Radosavljevic, M., Then, H. W., and Chau, R. (2011). Fabrication, characterization, and physics of III-V heterojunction tunneling field effect transistors for steep sub-threshold swing, in *IEEE International Electron Devices Meeting (IEDM) Technical Digest*, pp. 785–788.
16. Iida, R., Kim, S.-H., Yokoyama, M., Taoka, N., Lee, S.-H., Takenaka, M., and Takagi, S. (2011). Planar-type In0.53Ga0.47As channel-band-to-band tunneling metal-oxide-semiconductor field-effect transistors, *J. Appl. Phys.*, **110**, p. 124505.
17. Mohata, D., Bijech, R., Mujumdar, S., Eaton, C., Engel-Herbert, R., Mayer, T., Narayanan, V., Fastenau, J. M., Loubychev, D., Liu, A. K., and Datta, S. (2011). Demonstration of MOSFET-like on-curent performance in arsenic/antimonide tunnel FETs with staggered heterojunctions for 300 mV logic applications, in *IEEE International Electron Devices Meeting (IEDM) Technical Digest*, pp. 781–784.
18. Fiori, G., and Iannaccone, G. (2009). Ultralow-voltage bilayer graphene tunnel FET, *IEEE Electron Device Lett.*, **30**(10), pp. 1096–1098.

19. Britnell, L., Gorbachev, R. V., Jalil, R., Belle, B. D., Schedin, F., Mishchenko, A., Georgiou, T., Katsnelson, M. I., Eaves, L., Morozov, S. V., Peres, N. M., Leist, J., Geim, A. K., Novoselov, K. S., and Ponomarenko, L. A. (2012). Field-effect tunneling transistor based on vertical graphene heterostructures, *Science*, **335**, pp. 947–950.
20. Tomioka, K., and Fukui, T. (2011). Tunnel field-effect transistor using InAs nanowire/Si heterojunction, *Appl. Phys. Lett.*, **98**, p. 083114.
21. Tomioka, K., Yoshimura, M., and Fukui, T. (2012). Steep-slope tunnel field-effect transistors using III–V nanowire/Si heterojunction, in *IEEE Symposium on VLSI Technology (VLSIT)*, pp. 47–48.
22. Sarkar, D., Xie, X., Liu, W., Cao, W., Kang, J., Gong, Y., Kraemer, S., Ajayan, P. M., and Banerjee, K. (2015). A subthermionic tunnel field-effect transistor with an atomically thin channel, *Nature*, **526**, pp. 91–95.
23. Memisevic, E., Svensson, J., Hellenbrand, M., Lind, E., and Wernersson, L.-E. (2016). Vertical InAs/GaAsSb/GaSb tunneling field-effect transistor on Si with S = 48 mV/decade and I_{on} = 10 μA/μm for I_{off} = 1 nA/μm at VDS = 0.3 V, in *IEEE International Electron Devices Meeting (IEDM) Technical Digest*, pp. 500–503.
24. Takagi, S., Ahn, D. H., Noguchi, M., Gotow, T., Nishi, T., Kim, M., and Takenaka, M. (2016). Tunneling MOSFET technologies using III-V/Ge materials, in *IEEE International Electron Devices Meeting (IEDM) Technical Digest*, pp. 516–519.
25. Tomioka, T., Yoshimura, M., and Fukui, T. (2013). Sub 60 mV/decade switch using an InAs nanowire-Si heterojunction, *Nano Lett.*, **13**, pp. 5822–5826.
26. Tomioka, K., and Fukui, T. (2014). Current increment of tunnel field-effect transistor using InGaAs nanowire/Si heterojunction by scaling of channel length, *Appl. Phys. Lett.*, **104**, p. 073507.
27. Tomioka, K., Yoshimura, M., Ishizaka, F., and Fukui, T. (2013). Integration of III-V nanowires on Si: From high-performance vertical FET to steep-slope switch, in *IEEE International Electron Devices Meeting (IEDM) Technical Digest*, pp. 88–91.
28. Gandhi, R., Chen, Z., Singh, N., Banerjee, K., and Lee, S. (2011). Vertical Si nanowire n-type tunneling FETs with low subthreshold swing (<500 mV/decade) at room temperature, *IEEE Electron Device Lett.*, **32**(4), pp. 437–439.
29. Gandhi, R., Chen, Z., Singh, N., Banerjee, K., and Lee, S. (2011). CMOS-compatible vertical-silicon-nanowire gate-all-around p-type tunneling

FETs with <50-mV/decade subthreshold slope, *IEEE Electron Device Lett.*, **32**(11), pp. 1504–1506.

30. Jeon, K., Loh, Y.-W., Patel, P., Kang, C. Y., Oh, J., Bowonder, A., Park, C., Park, C. S., Smith, C., Majhi, P., Tseng, H-H., Jammy, R., Liu, T.-J. K., and Hu, C. (2010). Si tunnel transistors with a novel silicide source and 46 mV/dec swing, in *IEEE Symposium on VLSI Technology (VLSIT)*, pp. 121–122.
31. Kim, S. H., Kam, H., Hu, C., and Liu, T. K. (2009). Germanium-source tunnel field effect transistors with record high I_{ON}/I_{OFF}, in *IEEE Symposium on VLSI Technology (VLSIT)*, pp. 121–122.
32. Mayer, F., Royer, C. L., Damlencourt, J.-F., Romanjek, K., Andrieu, F., Tabone, C., Previtali, B., and Deleonibus, S. (2008). Impact of SOI, Si1-xGexOI and GeOI substrates on CMOS compatible tunnel FET performance, in *IEEE International Electron Devices Meeting (IEDM) Technical Digest*, pp. 1–5.
33. Appenzeller, J., Lin, Y.-M., Knoh, J., and Avouris, P. (2004). Band-to-band tunneling in carbon nanotube field-effect transistors, *Phys. Rev. Lett.*, **93**, p. 196805.
34. Knoll, L., Zhao, Q.-T., Nichau, A., Trellenkamp, S., Richter, S., Schafer, A., Esseni, D., and Selmi, L., Bourdelle, K. K., and Mantl, S. (2013). Inverters with strained Si nanowire complementary tunnel field-effect transistors, *IEEE Electron Device Lett.*, **34**, pp. 813–815.
35. Lrishnamohan, T., Kim, D., Raghunathan, S., and Saraswat, K. (2008). Double-gate strained-Ge heterostructure tuuneling FET (TFET) with record high drive currents and <60 mV/dec subthreshold slope, in *IEEE International Electron Devices Meeting (IEDM) Technical Digest*, pp. 1–3.
36. Huang, Q., Huang, R., Wu, C., Zhu, H., Chen, C., Wang, J., Guo, L., Wang, R., Ye, L., and Wang, Y. (2014). Comprehensive performance reassessment of TFETs with a novel design by gate and source engineering from device/circuit perspective, in *IEEE International Electron Devices Meeting (IEDM) Technical Digest*, pp. 335–338.
37. Villalon, A. Royer, C. L., Casse, M., Cooper, D., Previtali, B., Tabone, C., Hartmann, J.-M., Perreau, P., Rivallin, P., Damlencourt, J.-F., Allain, F., Andrrieu, F., Weber, O., Faynot, O., and Poiroux, T. (2012). Strained tunnel FETs with record I_{ON}: first demonstration of ETSOI TFETs with SiGe channel and RSD, in *IEEE Symposium on VLSI Technology (VLSIT)*, pp. 49–50.

38. Ganjipour, B., Wallentin, J., Börgstri, M. T., Samuelson, L., and Thelander, C. (2012). Tunnel field-effect transistors based on InP-GaAs heterostructure nanowires, *ACS Nano*, pp. 3109–3113.
39. Kim, M., Wakabayashi, Y., Nakane, R., Yokoyama, M., Takenaka, M., and Takagi, S. (2014). High I_{ON}/I_{OFF} Ge-source ultrathin body strained-SOI tunnel FETs – impact of channel strain, MOS interfaces and back gate on the electrical properties, in *IEEE International Electron Devices Meeting (IEDM) Technical Digest*, pp. 331–334.
40. Lu, W., Guo, A., Vardi, A., and del Alamo, J. A. (2014). A test structure to characterize nano-scale Ohmic contact in III-V MOSFETs, *IEEE Electron Device Lett.*, **35**(2), pp. 181–183.
41. Bessire, C. D., Björk, M. T., Schmidt, H., Schenk, A., Reuter, K. B., and Riel, H. (2011). Trap-assisted tunneling in Si-InAs nanowire heterojunction tuunel diodes, *Nano Lett.*, **11**, pp. 4195–4199.
42. Fang, Z. M., Ma, K. Y., Cohen, R. M., and Stringfellow, G. B. (1991). Effect of growth temperature on photoluminescence of InAs grownby-organometallic vapor phase epitaxy, *Appl. Phys. Lett.*, **59**, p. 1446.
43. Bijesh, R., Lie, H., Madan, H., Mohata, D., Li, W.,Nguyen, N. V., Gundlach, D., Richter, C. A., Maier, J., Wang, K., Clarke, T., Fastenau, J. M., Loubychev, D., Liu, W. K., Narayanan, V., and Datta, S. (2013). Demonstration of $In_{0.9}Ga_{0.1}As/GaAs_{0.18}Sb_{0.82}$ near broken-gap tunnel FET with $I_{ON} = 740$ μA/μm, $G_M = 700$ μS/μm and gigahertz switching performance at $V_{DS} = 0.5$ V, in *IEEE International Electron Devices Meeting (IEDM) Technical Digest*, pp. 28.2.1–28.2.4.
44. Fukui, T., Ando, S., Tokura, Y., and Toriyama, T. (1991). GaAs tetrahedral quantum dot structures fabricated using selective area metalorganic chemical vapor deposition, *Appl. Phys. Lett.*, **58**, p. 2018.
45. Yoshida, A., Tomioka, K., Ishizaka, F., Chiba, K., and Motohisa, J. (2016). Selective-area growth of vertical InGaAs nanowires on Ge for transistor application, *ECS Trans.*, **75**, pp. 265–270.
46. Yoshida, A., Tomioka, K., Ishizaka, F., and Motohisa, J. (2017). Growth of InGaAs nanowires on Ge(111) by selective-area metal-organic vapor phase epitaxy, *J. Cryst. Growth*, **464**, pp. 75–79.
47. Tomioka, K., Tanaka, T., Hara, S., Hiruma, K., and Fukui, T. (2011). III-V nanowires on Si substrate: selective-area growth and device applications, *IEEE J. Sel. Top. Quantum Electron.*, **17**, p. 1112–1129.
48. Tomioka, K., Yoshimura, M., and Fukui, T. (2012). A III-V nanowire channel on silicon for high-performance vertical transistors, *Nature*, **488**, pp. 189–192.

49. Tomioka, K., and Fukui, T. (2013). Gate-first process and EOT-scaling of III-V nanowire-based vertical transistros on Si, in *IEEE 71st Device Research Conference*, pp. 15–16.
50. Fischetti, M. V., Jin, S., Tang, T.-W., Asbeck, P., Taur, Y., Laux, S. E., Rodwell, M., and Sano, N. (2009). Scaling MOSFETs to 10 nm: Coulomb effectsm source starvation, and virtual source model, *J. Comput. Electron.*, **8**(2), pp. 60–77.
51. Zamit, U., Madhusoodanan, K. N., Marinelli, M., and Wesch, L. W. (1994). Optical-absorption studies of ion-implantation damage in Si on sapphire, *Phys. Rev. B*, **49**(20), pp. 14322–14330.
52. Nakai, E., Yoshimura, M., Tomioka, K., and Fukui, T. (2013). GaAs/InGaAs core-multishell nanowire-array-based solar cells, *Jpn. J. Appl. Phys.*, **52**(5), p. UNSP 055002.
53. Vandenberge, W. G., Verhulst, A. S., Soree, B., Magnus, W., Groeseneken, G., Smets, Q., Heyns, M., and Fischetti, M. V. (2013). Figure of merit for and identification of sub-60mV/decade devices, *Appl. Phys. Lett.*, **102**, p. 013510.
54. Radsavljevic, M., Dewey, G., Fastenau, J. M., Kavalieros, J., Kotlyar, R., Chu-Kung, B., Liu, W. K., Lubychev, D., Metz, M., Millard, K., Mukherjee, N., Pan, L., Pillarisetty, R., Rachmady, W., Shah, U., and Chau, R. (2010). Non-planar, multi-gate InGaAs quantum well field effect transistors with high-k gate dielectric and ultra-scaled gate-to-drain/gate-to-source separation for low power logic applications, in *IEEE International Electron Devices Meeting (IEDM) Technical Digest*, pp. 126–129.

Chapter 3

Nanoelectromechanical Switches

Chuang Qian and Tsu-Jae King Liu
Department of Electrical Engineering and Computer Sciences, University of California at Berkeley, Berkeley, CA 94720, USA
cqian@berkeley.edu, tking@eecs.berkeley.edu

Digital logic integrated circuits (ICs) implemented with complementary metal-oxide-semiconductor (CMOS) transistors have a fundamental lower limit in energy efficiency due to off-state leakage current. Nanoelectromechanical (NEM) switches are promising alternatives to CMOS transistors for ultra-low-power digital IC applications due to their virtually zero off-state leakage current and abrupt switching characteristics. Among various actuation mechanisms, electrostatic actuation is the most widely used, thanks to its scalability and compatibility with CMOS integration. This chapter firstly introduces a NEM switch design (relay) suitable for digital computing and then discusses body biasing schemes for minimizing the operating voltage and switching energy. Relay-based IC design for optimal energy-delay performance and novel computing architectures enabled by hybrid CMOS+NEMS technology are also discussed.

Emerging Devices for Low-Power and High-Performance Nanosystems: Physics, Novel Functions, and Data Processing
Edited by Simon Deleonibus

ISBN 978-981-4800-11-2 (Hardcover), 978-0-429-45873-6 (eBook)
www.panstanford.com

3.1 Introduction

3.1.1 Background: CMOS Energy Efficiency Limit

Over the past few decades, complementary metal-oxide-semiconductor (CMOS) field-effect transistor technology has been the backbone of growth for the semiconductor industry. However, due to the inherently nonideal switching behavior of a transistor, digital logic integrated circuits (ICs) implemented with CMOS transistors have a fundamental lower limit in energy efficiency.

Figure 3.1a illustrates the current-versus-voltage (I–V) transfer characteristics of a metal-oxide-semiconductor field-effect transistor (MOSFET), where the drain-to-source current I_{DS} is plotted on a logarithmic scale while the gate voltage V_{GS} (referenced to the source voltage) is plotted on a linear scale. As can be seen from this plot, a nonzero current flows for V_{GS} equal to zero; in other words, a MOSFET cannot be turned off completely. The leakage current I_{OFF}, that is, the current I_{DS} when $V_{GS} = 0$ V and V_{DS} is equal to the power-supply voltage V_{DD}, is given by

$$I_{OFF} = I_{TH} 10^{-\frac{V_T}{SS}} \tag{3.1}$$

where V_T is the threshold voltage, I_{TH} is the current I_{DS} at $V_{GS} = V_T$, and SS is called the subthreshold swing and defined as

$$SS \equiv \frac{1}{\text{slope}} = \frac{1}{d(\log I_{DS})/dV_{GS}} \tag{3.2}$$

in the subthreshold region of operation.

Clearly, for a given I_{TH} and V_T, I_{OFF} is limited by SS: the smaller the SS, the smaller is the I_{OFF}. For a conventional MOSFET, it can be shown that

$$SS = \ln 10 * \frac{kT}{q} \left(1 + \frac{C_{dep}}{C_{ox}}\right) \tag{3.3}$$

where k is the Boltzmann constant, T is the absolute temperature, q is the electronic charge, C_{ox} is the gate-to-channel capacitance per unit area, and C_{dep} is the depletion region capacitance per unit area. $\frac{kT}{q}$ is called the thermal voltage and has a value of about 26 mV at room temperature (300 K). Since the factor $(1+\frac{C_{dep}}{C_{ox}})$ is larger than 1, SS is always larger than 60 mV/decade at room temperature.

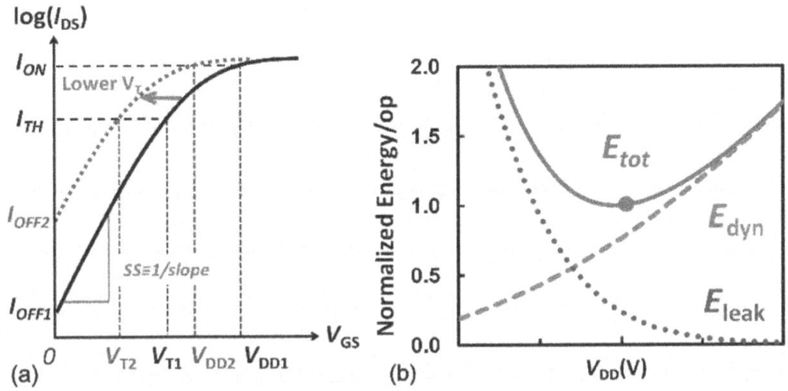

Figure 3.1 (a) Typical I–V transfer characteristic of a MOSFET, $V_{DS} = V_{DD}$. (b) Trade-off between dynamic energy consumption and leakage energy consumption of CMOS digital ICs. There is a lower limit in energy efficiency for any CMOS-based digital IC due to the nonideal switching behavior of MOSFETs.

The total energy consumption E_{tot} per clock cycle of a CMOS IC comprises two parts, the dynamic energy E_{dyn}, which dissipates during the charging and discharging of capacitances, and the static energy E_{leak}, which is the result of transistor off-state leakage current:

$$E_{tot} = E_{dyn} + E_{leak}, \tag{3.4}$$

where

$$E_{dyn} = C_{eff} V_{DD}^2; \tag{3.5}$$

$$E_{leak} = \alpha I_{OFF} V_{DD} L_{dep} t_{delay}; \tag{3.6}$$

C_{eff} is the effective capacitance, which gets charged during the clock cycle; L_{dep} is the logic depth; t_{delay} is the average propagation delay per logic gate; and α is a fitting parameter. Further, t_{delay} can be written as

$$t_{delay} \approx \frac{C_L V_{DD}}{2 I_{ON}} \tag{3.7}$$

where C_L is the load capacitance of a logic gate and I_{ON} is the on-state current of the driving transistor.

Equation 3.5 indicates that decreasing V_{DD} is an effective way to reduce E_{tot}; however, this also decreases I_{ON} (cf the solid curve in

Fig. 3.1a), which in turn increases t_{delay} and E_{leak} (Eq. 3.6 and Eq. 3.7). To maintain I_{ON}, the gate-overdrive voltage (i.e., $V_{DD} - V_T$) should be maintained at a constant level, which requires V_T to decrease with V_{DD} (cf the dashed curve in Fig. 3.1a). However, a linear decrease in V_T results in an exponential increase in I_{OFF}, which in turn also increases E_{leak}. In either case, then, a decreasing E_{dyn} results in an increasing E_{leak}, as shown in Fig. 3.1b. Therefore, there is always a lower limit in total energy dissipated per operation for any CMOS-based digital IC.

3.1.2 Why Nanoscale Electromechanical Switches

The energy efficiency limit of a CMOS-based digital IC stems from the nonzero leakage current and nonabrupt switching behavior of the MOSFET. To break this limit, new electronic switching devices are needed. A nanoscale electromechanical (NEM) switch is one of the most promising candidates. Unlike a MOSFET, which utilizes (electron or hole) energy barrier modulation as the switching mechanism, a NEM switch controls current flow by making/breaking physical contact between two conductive electrodes so that it can achieve zero off-state leakage current and abrupt switching behavior. Typical I–V characteristics of a NEM switch are shown in Fig. 3.2. It has virtually zero leakage current and abrupt switching behavior (although there is a hysteresis voltage between the turn-on voltage V_{ON} and the turn-off voltage V_{OFF}, which will be discussed later). In principle, a NEM switch can be more energy efficient than a MOSFET. Therefore, there has been a resurgence of interest in mechanical computing for ultra-low-power applications, in recent years [1–9].

3.2 Actuation Mechanisms

A NEM switch comprises mechanical-actuation and current-conduction portions. Various transduction mechanisms can be employed for mechanical actuation, among which electromagnetic, electrothermal, piezoelectric, and electrostatic mechanisms have been widely explored.

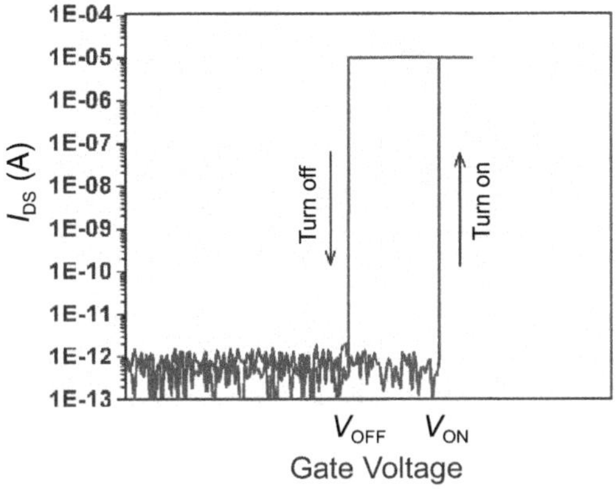

Figure 3.2 Typical *I*–*V* transfer characteristic of a NEM switch. The current compliance limit was set to 10 μA to avoid contact welding during the DC sweep.

3.2.1 Electromagnetic

An electromagnetic switch uses a magnetic force to actuate a movable element to make a contact. It usually involves the magnetization of the movable element, which subsequently moves in response to a magnetic field. Once the movable element is demagnetized (or reverses its magnetic polarity), it returns to its original position by elastic restoring force (or repulsive magnetic force).

Figure 3.3 shows a magnetic microelectromechanical (MEM) switch whose cantilever beam is a bilayer composite consisting of a soft magnetic material (e.g., NiFe permalloy) on the top side and a conductive material (e.g., Au) at the bottom side. The temporary magnetic field generated by a current pulse in the buried microcoils can change the polarity of the magnetic moment of the permalloy such that the cantilever has either a clockwise or a counterclockwise torque exerted by the external magnetic field H_0. By changing the current pulse direction, the cantilever can be switched between the contact position (on state) and the detach position (off state).

Similar to the design described above, most magnetic MEM switches reported in the literature [10–14] employ external permanent magnets, which is a major disadvantage since this is not

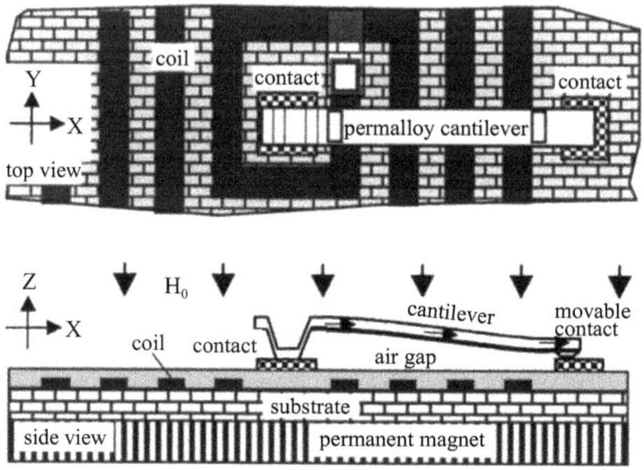

Figure 3.3 Structure of a magnetic MEM switch. (a) Top view; (b) cross-section view. Reprinted from Ref. [10], Copyright (2001), with permission from Elsevier.

compatible with conventional CMOS ICs. There are a few other designs [15, 16] that don't require an external magnetic field but generally require static power dissipation (i.e., constant current in microcoils) to maintain the switch in the on or off state and therefore are not suitable for low-power IC applications.

3.2.2 Electrothermal

An electrothermal switch relies on thermal expansion/contraction to actuate the movable element. A thermal actuator usually comprises either stacked layers of materials having different thermal expansion coefficients so that it bends upon heating (Fig. 3.4a) [17] or a single material with a specific geometry favoring elongation (e.g., the V shape in Fig. 3.4b) [18, 19] or rotation (e.g., the Vernier shape in Fig. 3.4c) [20, 21].

The advantage of thermal actuation is that it can provide a large contact force (thus possibly a smaller contact resistance) with a low voltage; however, the disadvantages are obvious: it tends to consume more power and have a slower switching speed than other types of actuators.

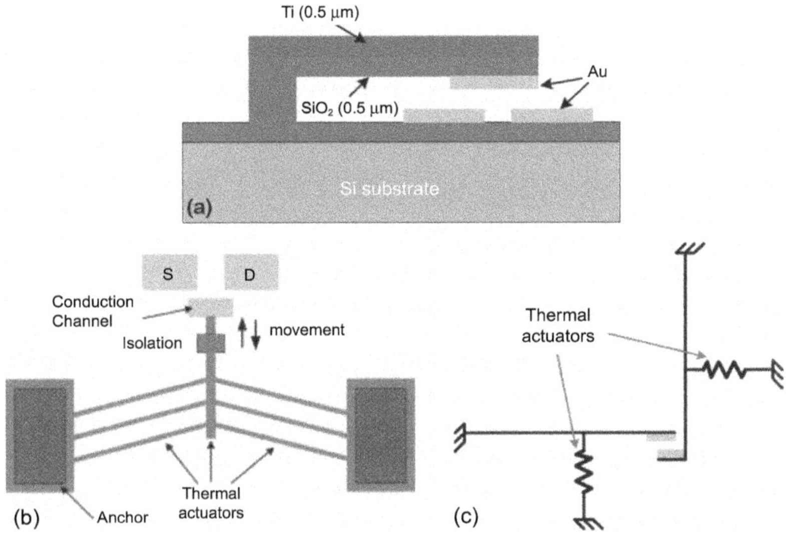

Figure 3.4 Examples of thermally actuated mechanical switch designs. (a) Cross-section view of a bilayer cantilever beam switch [17]; (b) top view of a V-shape switch [19]; (c) schematic of a Vernier-shape switch [20].

3.2.3 Piezoelectric

The piezoelectric effect refers to the generation of charge within a material in response to mechanical stress. It is a reversible process, in that a piezoelectric material also generates a mechanical stress when an electric field is applied. Piezoelectric NEM switches rely on the latter transduction mechanism to open or close air gaps between electrical contacts. The most common piezoelectric materials used for NEM/MEM switch applications include ZnO, AlN, and lead zirconate titanate (PZT) [22, 23].

Figure 3.5 shows the structure of an exemplary piezoelectric NEM switch. Details of the fabrication process are reported in Ref. [24]. A 10-nm-thick piezoelectric film (AlN) is sandwiched between the gate and body metal layers and bonded on an elastic layer (also AlN). Upon the application of a sufficiently large voltage between the gate and the body, the generated lateral strain in the piezoelectric film causes an up-bending motion of the cantilever beam, opening the normally closed gaps between the tip

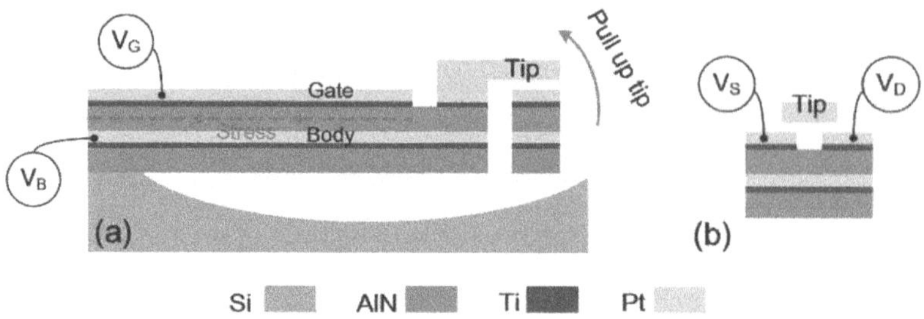

Figure 3.5 Schematics of a normally on piezoelectric NEM switch [24]. (a) Cross-section view along the cantilever beam; (b) cross-section view transverse to the length of the beam, in the contact region.

and source/drain electrodes. The switch has a sub-1-V threshold voltage V_T, which is the smallest among piezoelectric NEM switches reported to date. However, the device footprint is still very large; and the level of gate-to-body leakage current has not been disclosed.

Compared to electromagnetic and electrothermal NEM switches, piezoelectric NEM switches show more promise for mechanical computing. However, to make them compelling for IC application, the following challenges must be addressed: (i) scaling down the lateral dimension to increase the device density for improved functionality per unit area, (ii) scaling down the piezoelectric film thickness (which is essential to achieve a low voltage) while maintaining low leakage current through the piezoelectric film, and (iii) achieving a uniform stress profile across many devices, especially in terms of the stress gradient across the multilayer film stack of the piezoelectric actuator, so as to achieve a uniform V_T (as it becomes more challenging to control stress uniformity when scaling down film thickness).

3.2.4 Electrostatic

An electrostatic NEM switch utilizes the electrostatic force between charges of opposite polarity on two opposing electrodes that form a capacitor; usually one electrode is fixed and the other is movable [25]. By charging or discharging the capacitor, the movable electrode can be attracted to (by electrostatic force) or move away from (due

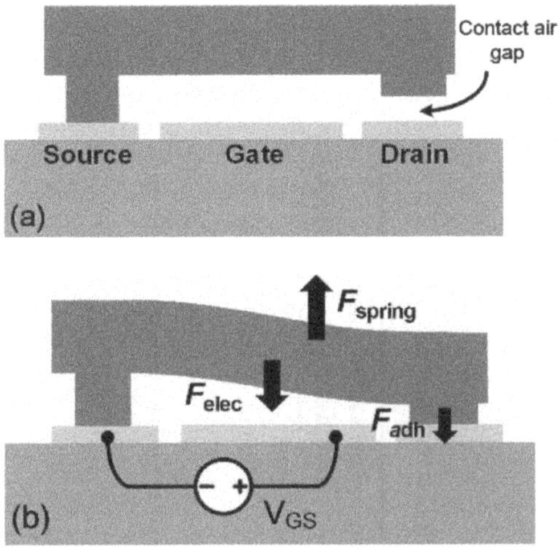

Figure 3.6 Schematics illustrating an electrostatically actuated cantilever NEM switch. (a) Off state; (b) on state.

to elastic force) the fixed electrode so as to close or open a contact air gap, respectively, as illustrated in Fig. 3.6.

In the off state (Fig. 3.6a), the source electrode is physically separated from the drain electrode by an air gap so no current can flow. When a voltage V_{GS} is applied between the gate and the source, the electrostatic force F_{elec} actuates the cantilever beam downward whereas the spring restoring force F_{spring} counteracts this movement such that a balance in forces is reached. If V_{GS} is larger than a certain value (called the turn-on voltage V_{ON}), F_{elec} is sufficient (i.e., $F_{elec} > F_{spring}$) to pull the cantilever beam down to make contact with the drain electrode such that a current I_{DS} can suddenly flow between the drain and the source, that is, the switch is turned on abruptly (Fig. 3.6b). When V_{GS} decreases to below the turn-off voltage V_{OFF} such that F_{spring} can overcome F_{elec} and the contact adhesive force F_{adh} (i.e., $F_{spring} > F_{elec} + F_{adh}$), the contact between the drain and the source breaks and I_{DS} drops to zero suddenly. A typical I–V characteristic of an electrostatic NEM switch is shown in Fig. 3.2.

Different switching mechanisms are compared in Table 3.1. Electrostatic actuation is the most widely used mechanism for

Table 3.1 Comparison of different actuation mechanisms for a mechanical switch

Actuation mechanism	Size	Switching voltage	Power consumption	Turn-on speed	Fabrication complexity	CMOS compatibility
Electromagnetic	Large	Low	High	Medium	High	Poor
Electrothermal	Medium	Low	High	Slow	Medium	Medium
Piezoelectric	Medium	Medium	Medium	Fast	High	Medium
Electromechanical	Small	Medium	Low	Fast	Low	Good

mechanical switches because electrostatic switches are easy to fabricate over CMOS circuitry and more scalable than the other types of mechanical switches mentioned previously. Therefore, the remainder of this chapter focuses on electrostatic NEM switches.

3.3 Electrostatic Relay Design for Digital Computing

A basic electrostatic relay has three terminals, namely drain, source, and gate, as illustrated in Fig. 3.6. Note that the state of the relay depends on both the gate voltage V_G and the source voltage V_S. For IC applications, it is often required to connect relays in series (i.e., the drain of relay in the previous stage is connected to the source of a relay in the next stage). In such a configuration, variation in V_S (hence variation in V_{GS}) may lead to undesirable switching of the relay. To solve this shortcoming, a separate reference terminal is needed to isolate the switching-control signal path from the current conducting path.

Figure 3.7 shows cross-section views of a 4-terminal (4-T) relay. In the off state, the channel (a metal strip attached to the underside of the movable body electrode via a dielectric insulation layer) is separated from the source/drain electrodes by air gaps; in the on state, the air gaps are closed, so current can flow between the source and the drain through the channel.

3.3.1 Modeling and Analysis

3.3.1.1 Parallel-plate capacitor model

The 4-T relay illustrated in Fig. 3.7 can be simply modeled as a parallel-plate capacitor, as shown in Fig. 3.8. The top plate (body) is electrically grounded and mechanically suspended by a spring with stiffness k_{eff}. The fixed bottom plate (gate) has almost the same area as the body since the drain/source electrodes are much smaller than the gate (cf Fig. 3.7). The actuation gap (i.e., the distance between the two plates) and the contact gap sizes are g and g_{CONT}, respectively. As fabricated (with no applied voltage), $g = g_0$ and $g_{\text{CONT}} = g_d$. When a

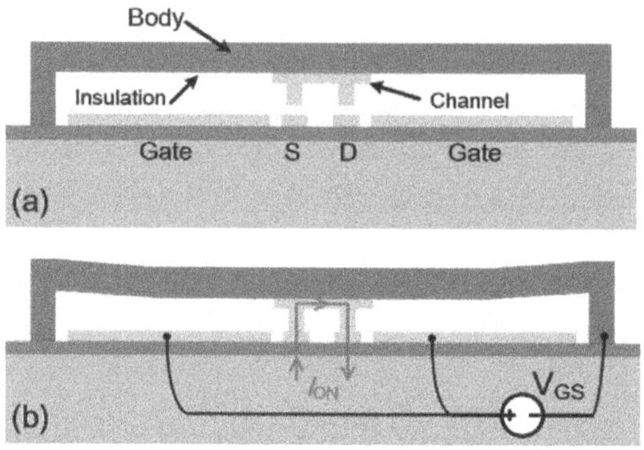

Figure 3.7 Schematics of a 4-T relay. (a) Off state; (b) on state.

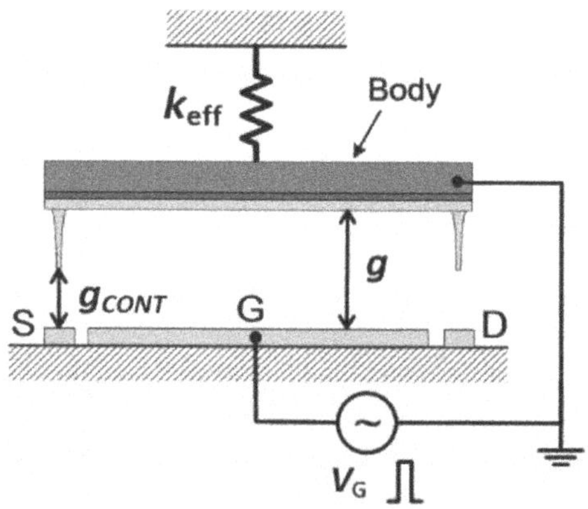

Figure 3.8 Parallel-plate capacitor model of the electrostatic NEM switch in Fig. 3.7.

gate voltage V_G is applied, F_{spring} and F_{elec} in the steady state can be written as

$$F_{spring} = k_{eff}(g_0 - g), \tag{3.8}$$

and

$$F_{elec} = \frac{\varepsilon_0 A_{ACT} V_G^2}{2g^2} \tag{3.9}$$

respectively, where A_{ACT} is the actuation area (i.e., the gate or body area) and ε_0 is the vacuum permittivity.

3.3.1.2 Operation modes

As can be seen from Eq. 3.8 and Eq. 3.9, F_{spring} increases linearly whereas F_{elec} increases superlinearly when g decreases. It can be easily shown that there is a critical body displacement (equal to $g_0/3$), beyond which F_{elec} not only surpasses F_{spring} but also increases at a faster rate as g decreases. So the net force on the movable body increases with a positive feedback, which pulls the body down to the gate quickly. This phenomenon is called the pull-in (PI) effect. Relays working in this fashion are referred to as PI-mode relays. The minimum gate voltage for the PI effect to happen is

$$V_{\text{PI}} = \sqrt{\frac{8k_{\text{eff}}g_0^3}{27\varepsilon_0 A_{\text{ACT}}}}. \tag{3.10}$$

The switching energy of a PI-mode relay is

$$E_{\text{PI}} = C_{\text{on}}V_{\text{PI}}^2 = \frac{\varepsilon_0 A_{\text{ACT}}}{g_0 - g_{\text{d}}}\frac{8k_{\text{eff}}g_0^3}{27\varepsilon_0 A_{\text{ACT}}}, \tag{3.11}$$

where C_{on} is the parallel-plate capacitance in the on state. However, if the as-fabricated contact gap size g_{d} is less than $g_0/3$, the relay would turn on before the PI effect could occur. Such a relay is referred to as a non-pull-in (NPI)-mode relay. The turn-on voltage of a NPI-mode relay can be easily derived from Eq. 3.8 and Eq. 3.9:

$$V_{\text{on-NPI}} = \sqrt{\frac{2k_{\text{eff}}g_{\text{d}}(g_0 - g_{\text{d}})^2}{\varepsilon_0 A_{\text{ACT}}}}, \tag{3.12}$$

where $g_{\text{d}} < g_0/3$ for NPI-mode operation. The switching energy of a NPI-mode relay is

$$E_{\text{NPI}} = C_{\text{on}}V_{\text{on-NPI}}^2 = 2k_{\text{eff}}g_{\text{d}}(g_0 - g_{\text{d}}). \tag{3.13}$$

3.3.1.3 Hysteresis voltage

To turn off a relay, V_{GB} is lowered to be smaller than the release voltage V_{RL} so that F_{spring} overcomes $F_{\text{elec}} + F_{\text{adh}}$ and the movable structure moves away from the contacting position. From Eq. 3.8 and Eq. 3.9 V_{RL} can be derived to be

$$V_{\text{RL}} = \sqrt{\frac{2(k_{\text{eff}}g_{\text{d}} - F_{\text{adh}})(g_0 - g_{\text{d}})^2}{\varepsilon_0 A_{\text{ACT}}}}. \tag{3.14}$$

When a PI-mode relay is turned on (i.e., $V_{GB} > V_{ON} = V_{PI}$), F_{elec} is much larger than F_{spring} due to the PI effect. To turn it off, V_{RL} must be smaller than V_{PI} to a certain extent, that is there is hysteresis in the switching characteristic due to PI-mode operation. The contact adhesive force F_{adh} increases the magnitude of the hysteresis voltage $V_H \stackrel{def}{=} |V_{ON} - V_{RL}|$. For NPI-mode relays, F_{adh} is the only factor contributing to V_H.

3.3.1.4 Mechanical delay

After a high V_G ($> V_{ON}$) is applied to the gate, it takes time for the movable structure (body) to move from its suspended position in the off state to its (channel) contacting position in the on state. This is referred to as the mechanical delay, which is proportional to $\sqrt{m_{eff}/k_{eff}}$, where m_{eff} and k_{eff} are effective mass and effective spring constant, respectively, of the spring-mass oscillation system (Fig. 3.8). The mechanical delay usually is orders of magnitude larger than the capacitive charging time (i.e., the "RC" delay) of the gate electrode. For example, the mechanical delay of a typical micro-electromechanical logic relay reported in literature [26] is about 1 μs whereas the RC delay is less than 1 ns. Thus the operating speed of a relay is limited by a mechanical delay.

3.3.2 Challenges for Reducing Operating Voltage

Reducing the operating voltage is an effective way to decrease the energy consumption of relay-based ICs. A sufficiently large V_{GB} is necessary to generate a strong enough F_{elec} so as to turn on the relay (i.e., $F_{elec} > F_{spring}$). According to Eq. 3.9, increasing A_{ACT} or decreasing g_0 can provide for a larger F_{elec} to reduce the voltage required to turn on the relay. However, A_{ACT} is constrained from being very large by the requirement of a high device density and g_0 is constrained from being very small by process technology limitations. Therefore, reducing F_{spring} (by using a spring with lower stiffness) has been the preferred approach for reducing operating voltage. The challenge is that a relay with k_{eff} that is too small is

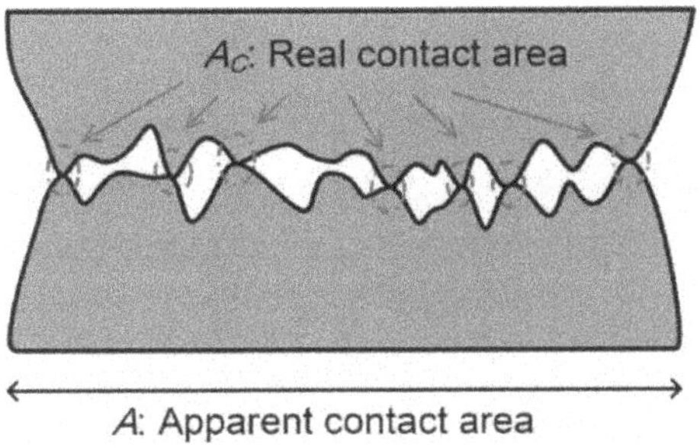

Figure 3.9 Illustration of a microscale interface. A_C is much smaller than A due to surface roughness.

prone to stuck-on failure if $F_{spring} = k_{eff} \times g_d < F_{adh}$. Clearly, the switching energy of a relay is fundamentally limited by F_{adh}.

Figure 3.9 illustrates a microscale contact. The real contact area A_C is much smaller than the apparent contact area A since the two surfaces make contact with each other only at a few asperities due to surface roughness. The exact value of A_C depends on both surface morphology and contact force and thus is hardly known. In addition, A_C can change over many on/off switching cycles due to surface wear and tear. In addition, the contact adhesive force can be substantially affected by surface oxidation, interface contamination, etc. These render the precise control of F_{adh} very challenging, if not impossible. To avoid potential stuck-on failure, the effective spring constant k_{eff} should be sufficiently large to ensure that F_{spring} is always greater than F_{adh}. This would seem to limit the down-scaling of relay operating voltage.

3.4 Body-Biased Relay Technology

By applying a DC bias voltage V_B to the body electrode, the magnitude of the gate driving voltage V_G required to switch the relay can be adjusted. If a negative V_B is applied, V_G needs only to switch between 0 V and $V_{DD} = V_{ON} + V_B$ to turn on/off the relay. The overall energy consumption of a relay-based IC can be greatly reduced by decreasing V_{DD}.

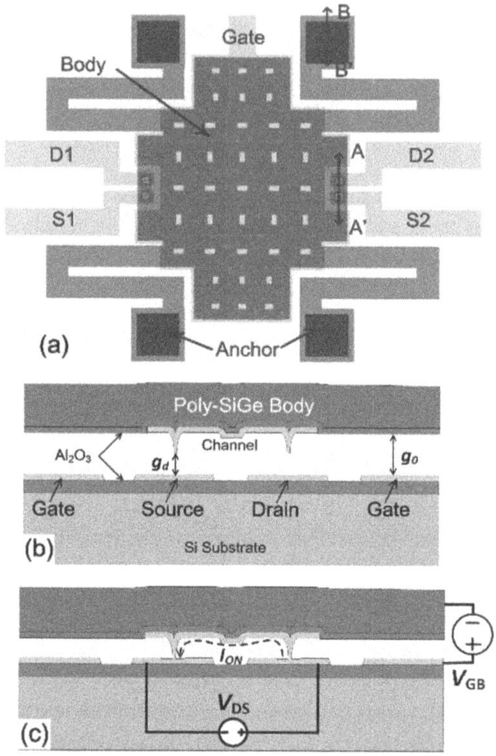

Figure 3.10 Structure of a 6-T relay. (a) Layout; (b) cross-section view along the cutline AA' in the off state; (c) cross-section view along the cutline AA' in the on state.

3.4.1 Relay Design and Fabrication Process

Figure 3.10 illustrates the structure of a 6-terminal (6-T) electrostatic MEM relay developed for digital logic applications, which comprises a movable poly-$Si_{0.4}Ge_{0.6}$ body structure suspended by four folded-flexure beams over a fixed tungsten (W) gate electrode and two pairs of W source/drain electrodes coplanar with the gate electrode. The W channel (a narrow W strip) is attached to the underside of the body via an Al_2O_3 insulation layer. The 6-T relay design has two switches incorporated into the same footprint, which effectively doubles the device density. The cross-section views along cutline AA' in the off and on states are shown in Fig. 3.10b and

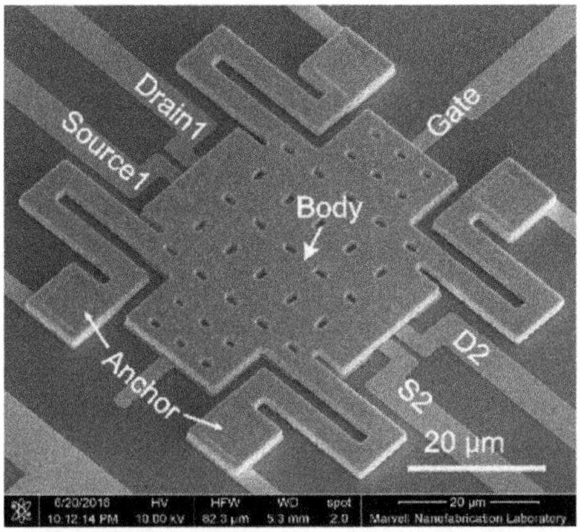

Figure 3.11 SEM image of a fabricated 6-T relay.

Fig. 3.10c, respectively. Note that the body electrode overlaps with the gate electrode over most of its area except in the regions of the four flexure beams and the channels, thereby forming a parallel-plate capacitor. The scanning electron microscope (SEM) image of a fabricated 6-T is shown in Fig. 3.11.

One of the main advantages of electrostatic actuation over other actuation mechanisms is its relatively simple actuator structure (an air-gap capacitor). The fabrication process of the 6-T logic relay in Fig. 3.11 involves neither exotic materials nor high-temperature processing and thus is fully compatible with CMOS circuitry. As illustrated in Fig. 3.12, starting from a silicon substrate, an isolation layer of 80 nm Al_2O_3 was coated by atomic layer deposition (ALD). A 50 nm tungsten layer was deposited by DC-sputtering deposition, patterned by stepper lithography and etched by reactive ion etching (RIE) to form the source/drain and gate electrodes (Fig. 3.12a). Next, a first sacrificial layer of low-temperature silicon oxide (LTO-1) was deposited by low-pressure chemical vapor deposition (LPCVD) at 400°C with SiH_4 and O_2 precursor gases. The contact holes on top of the source/drain electrodes were formed by patterning and etching of LTO-1 (Fig. 3.12b). Afterward, LPCVD was used again to deposit a second layer of sacrificial SiO_2 (LTO-2), whose thickness defined the nominal contact gap size g_d. The total thickness of LTO-1 and LTO-2 determined the nominal actuation gap size g_0. Then a thin (60 nm)

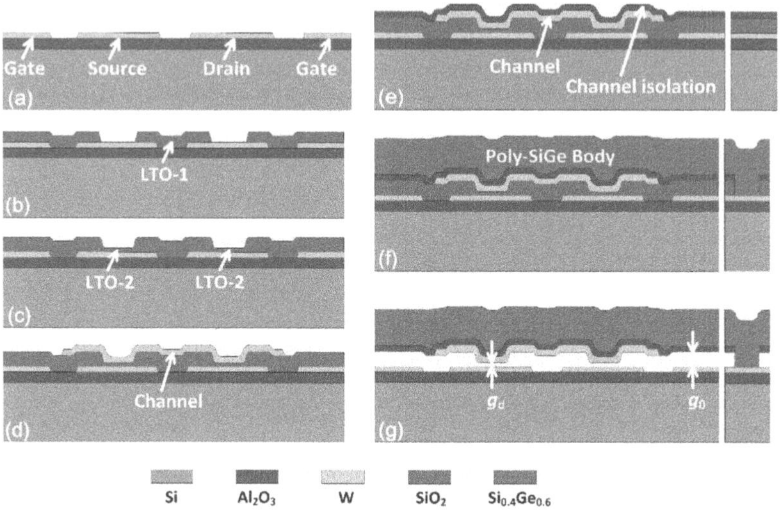

Figure 3.12 Cross-sectional schematics illustrating key steps in the fabrication process of a 6-terminal logic relay. (a–d) and the left part of (e–g) are images along the cutline AA′ in Fig. 3.10a; the right part of (e–g) are images along the cutline BB′ in Fig. 3.10a.

layer of tungsten was sputter-deposited and patterned to form the current-conducting channels above the source/drain electrodes (Fig. 3.12d). Next the entire surface was coated conformally with a second isolation Al_2O_3 layer (50 nm) by ALD (Fig. 3.12e). After that, four anchor vias (right part of Fig. 3.12f) were opened by patterning and etching through the whole stack (the top Al_2O_3 isolation layer, LTO-2, and LTO-1) until the bottom tungsten anchor electrode was exposed. The main structural layer (boron-doped polycrystalline $Si_{0.4}Ge_{0.6}$) was then deposited with an in situ two-step LPCVD process. The first step was to deposit a Si seed layer using Si_2H_6 as the precursor gas at 410°C for 5 min. This Si seed layer affects the microcrystalline structure and hence the stress gradient of the subsequently deposited $Si_{0.4}Ge_{0.6}$ film. After the seed layer growth, 1.75 μm of boron-doped polycrystalline $Si_{0.4}Ge_{0.6}$ was deposited using the precursor gases SiH_4, $GeHe_4$, and BCl_3 at 410°C for about 5 h, in the same furnace without breaking vacuum (Fig. 3.12f). Then, 400 nm of LTO was deposited as a hard mask layer by LPCVD. In the end, the top three layers, that is the LTO hard mask, the poly-

$Si_{0.4}Ge_{0.6}$ structural layer, and the channel isolation Al_2O_3 layer, were patterned by lithography and RIE, leaving the sacrificial LTO layer exposed. To avoid stiction, ethanol-assisted anhydrous HF vapor was used to selectively etch away the sacrificial LTO-1 and LTO-2. The final released structure is shown in Fig. 3.12g.

3.4.2 Scaling Theory

Conventional wisdom suggests that the structural stiffness of the NEM relay should be minimized to achieve minimum switching voltage and hence operating energy. But this does not take into account the possibility of body biasing to lower the operating voltage [26]. As briefly mentioned above, a body-biased relay can operate with a lower gate switching voltage V_{DD}. The minimum possible V_{DD} is equal to $V_H = V_{ON} - V_{RL}$ in the case that a maximum possible $|V_B|$(i.e., $V_B = -V_{RL}$) is applied. Note that V_B cannot be smaller than $-V_{RL}$; otherwise the relay would fail to turn off at $V_G = 0$ V. To gain insight into the scaling of V_{DD}, a quantitative analysis is necessary.

According to Eqs. 3.10, 3.12, and 3.14, the minimum V_{DD} of body-biased PI-mode relay is

$$V_{DD} = V_{H\text{-}PI} = \sqrt{\frac{8k_{eff}g_0^3}{27\varepsilon_0 A_{ACT}}} - \sqrt{\frac{2(k_{eff}g_d - F_{adh})(g_0 - g_d)^2}{\varepsilon_0 A_{ACT}}}, \quad (g_d < g_0 < 3g_d), \tag{3.15}$$

and of NPI-mode relay is

$$V_{DD} = V_{H\text{-}NPI} = \sqrt{\frac{2k_{eff}g_d(g_0 - g_d)^2}{\varepsilon_0 A_{ACT}}} - \sqrt{\frac{2(k_{eff}g_d - F_{adh})(g_0 - g_d)^2}{\varepsilon_0 A_{ACT}}}, \quad (g_0 \geq 3g_d). \tag{3.16}$$

If there is no body bias (i.e., $V_B = 0$), it is easy to show that the minimum V_{DD} is

$$V_{min} = \sqrt{\frac{8F_{adh}g_d^2}{27\varepsilon_0 A_{ACT}}}, \tag{3.17}$$

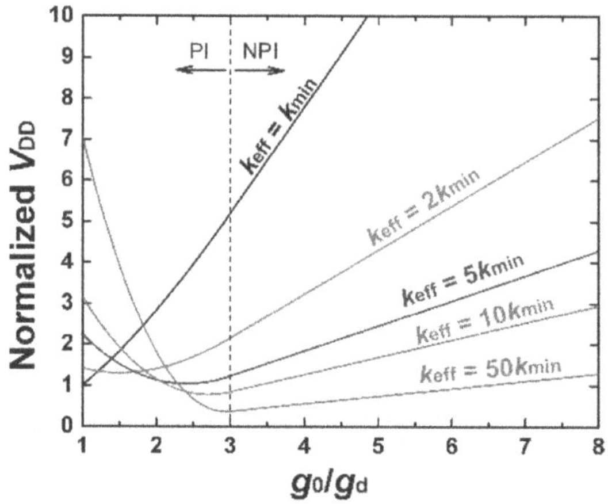

Figure 3.13 Dependence of $V_{\rm DD}$ on $k_{\rm eff}$ and $g_0/g_{\rm d}$. $V_{\rm DD}$ is minimized when $k_{\rm eff}$ increases and $g_0/g_{\rm d}$ approaches 3.

which can be achieved under the following idealistic conditions: $g_0 = g_{\rm d}$, and $k_{\rm eff} = \frac{F_{\rm adh}}{g_{\rm d}} (\stackrel{\rm def}{=} k_{\rm min})$. If a body bias is applied, $V_{\rm DD}$ is a function of $k_{\rm eff}$ and g_0 according to Eqs. 3.15 and 3.16. This relation is visualized in Fig. 3.13, where the voltage is normalized to $V_{\rm min}$ (i.e., $\bar{V_{\rm DD}} \stackrel{\rm def}{=} V_{\rm DD}/V_{\rm min}$). For small $k_{\rm eff}$, relays designed for PI-mode operation have smaller $V_{\rm DD}$; however, as $k_{\rm eff}$ increases, the minimum $V_{\rm DD}$ design point shifts toward NPI-mode operation and the value of minimum $V_{\rm DD}$ decreases. It can be shown that the ratio $g_0/g_{\rm d}$ for minimum $V_{\rm DD}$ always falls within the PI-mode design region (i.e., $g_0 < 3g_{\rm d}$) but approaches the boundary (i.e., $g_0 = 3g_{\rm d}$) as $k_{\rm eff}$ approaches infinity; the minimum $V_{\rm DD}$ is

$$\begin{aligned}\lim_{k_{\rm eff}\to\infty} V_{\rm DD\_min} &= \lim_{\substack{k_{\rm eff}\to\infty \\ g_0\to 3g_d}} V_{\rm DD} \\ &= \lim_{\substack{k_{\rm eff}\to\infty \\ g_0=3g_d}} V_{\rm min}\sqrt{\frac{27}{4}\left(\frac{g_0}{g_{\rm d}}-1\right)^2}\left(\sqrt{\frac{k_{\rm eff}}{k_0}}-\sqrt{\frac{k_{\rm eff}}{k_0}-1}\right) \\ &= \lim_{k_{\rm eff}\to\infty} \sqrt{27}V_{\rm min}\left(\sqrt{\frac{k_{\rm eff}}{k_0}}-\sqrt{\frac{k_{\rm eff}}{k_0}-1}\right) = 0. \quad (3.18)\end{aligned}$$

Equation 3.18 dictates that the minimum $V_{\rm DD}$ can approach 0 V as $k_{\rm eff}$ increases to infinity, so there is no fundamental limit for scaling down $V_{\rm DD}$ of body-biased relays. However, a large $k_{\rm eff}$ results in large

V_{RL} (Eq. 3.14), which sets a practical upper limit for k_{eff} and thus a practical lower limit for V_{DD}. Note also that V_{DD} increases quickly with decreasing g_0/g_d if $g_0/g_d < 3$ and k_{eff} is large (Fig. 3.13). To ensure low V_{DD} in practice, it is usually better to design g_0/g_d to be slightly larger than 3, namely for the relay to operate in the NPI-mode region. In addition, both k_{eff} and V_B affect the mechanical delay. Interested readers are referred to Ref. [27] for details. With this method (i.e., to design a relay in NPI mode with relatively strong k_{eff} and body biasing), a sub-100-mV electrostatic MEM relay was recently demonstrated (cf the next section).

3.4.3 Recent Progress

Thanks to their ability to operate at a reduced voltage, body-biased relays have attracted interest and recently been demonstrated to operate reliably at sub-100 mV [28].

The device structure and the fabrication process are similar to those described in Fig. 3.10 and Fig. 3.12, respectively. A thick poly-$Si_{0.4}Ge_{0.6}$ layer (1.75 μm) was used to provide a large k_{eff}. The thickness of LTO-1 and LTO-2 is 160 nm and 60 nm, respectively, and the nominal gap size of g_0 and g_d as fabricated should be 220 nm and 60 nm, respectively. (Due to a nonzero stress gradient, the poly-$Si_{0.4}Ge_{0.6}$ structure may buckle up slightly after release, causing both g_0 and g_d to increase. The nominal value of g_0 was intentionally designed to be larger than $3g_d$ to take into account this gap-size-increasing effect, to ensure relay operation in NPI mode.) Measured relay *I*–*V* characteristics are shown in Fig. 3.14. Clearly, the drain-to-source current I_{DS} can switch between the off state (immeasurably low current) and the on state (current subjected to a preset compliance limit of 10 μA) abruptly. The hysteresis voltage is about 70 mV. With $V_B = -11$ V, the relay can be turned on/off with sub-100-mV gate swing voltage (inset of Fig. 3.14a). A relay-based inverter that uses a body-biased relay as a pull-down device and a resistive load as a pull-up device (Fig. 3.14b) was successfully operated with an 80 mV gate swing voltage (Fig. 3.14c).

Such a small gate swing voltage can be explained by the model in the last section, as follows: assuming the as-fabricated g_0 and g_d are 240 nm and 80 nm, respectively (i.e., a 20 nm increase caused by the

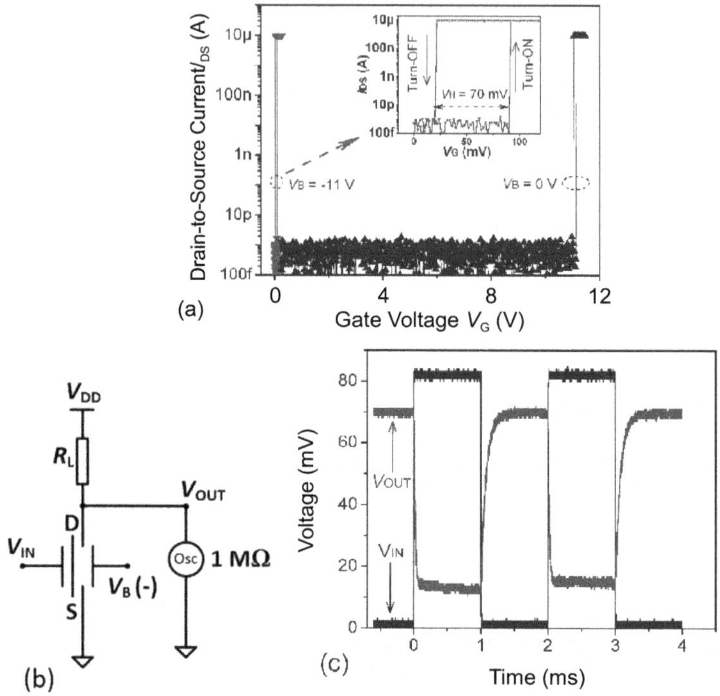

Figure 3.14 (a) Measured *I–V* characteristics of a 6-T relay with sub-100-mV hysteresis voltage; (b) a schematic of a relay-based inverter that uses a body-biased relay as a pull-down device and a resistor as a pull-up device. The output voltage is monitored by an oscilloscope with 1 MΩ input resistance; (c) measured voltage waveforms of input/output of the inverter, demonstrating sub-100-mV operation.

upward buckling of the poly-$Si_{0.6}Ge_{0.4}$ structure after release), and the contact adhesive force $F_{adh} = 1$ μN [29], it can be calculated that $k_{min} = 12.5$ N/m and $V_{min} = 0.38$ V according to Eq. 3.17, where $A_{ACT} = 1460$ μm^2 is used [28]. Given $k_{eff} = 486$ N/m [28], the hysteresis voltage V_H is predicted to be about 0.16 V according to Eq. 3.16. Considering the error in the estimation of the above parameters, the predicted V_H matches the experimental result well.

To further reduce V_H, the contact adhesive force F_{adh} needs to be reduced. In this regard, materials with lower contact adhesive force yet good conductivity and robustness (i.e., resistance to welding, mechanical wear and tear, chemical oxidation, etc.) are desirable.

W electrodes have the advantage of superior hardness and melting point. However, metallic contacts usually have much larger F_{adh} than metal-oxide contacts due to their large Hamaker constant and metallic bonds [30]. The formation of native WO_x is beneficial for reducing F_{adh} but dramatically increases the contact resistance (which in fact is the main cause of failure of W-contact relay circuits). Ruthenium (Ru) is a better alternative since both Ru and its native oxide (RuO_2) are mechanically robust and electrically conductive. The integration of Ru electrodes into a logic relay fabrication process was recently reported [31]. Another way of reducing F_{adh} is to coat the metal contact surfaces with antistiction material. Relays with sub-50-mV V_H achieved with this method have been recently reported [32].

3.5 Relay-Based Integrated Circuit Design

3.5.1 Relay-Based Logic Gates

A relay that has terminals similar what a MOSFET has, namely drain, source, gate, and body, can be configured in a way similar to a MOSFET for digital logic circuit application. To support complementary logic circuits, a p-type relay, which turns off when V_G is high and turns on when V_G is low (similarly to a p-type MOSFET), is needed. Thanks to the ambipolarity of electrostatic force (i.e., always attractive regardless of the polarity of applied voltage), this p-type device behavior can be easily achieved by applying a large positive voltage to the body. Figure 3.15 shows measured I–V characteristics of a p-type relay that is positively body-biased and has a $V_H = 62$ mV ($V_B = 11.2$ V).

Two relays can be wired together as shown in Fig. 3.16a to form a complementary inverter [9, 33]. By tuning the body bias V_{Bp} (+) and V_{Bn} (−), the pull-up relay is turned on when V_{IN} is low whereas the pull-down relay is turned on when V_{IN} is high. Only one relay is on at a time, to avoid crowbar current [34]. Dynamic inverter operation at 50 Hz with $V_{DD} = 2$ V is shown in Fig. 3.16b [9]. Figure 3.16c shows a schematic of a relay-based NAND gate. Other logic gates can be formed similarly.

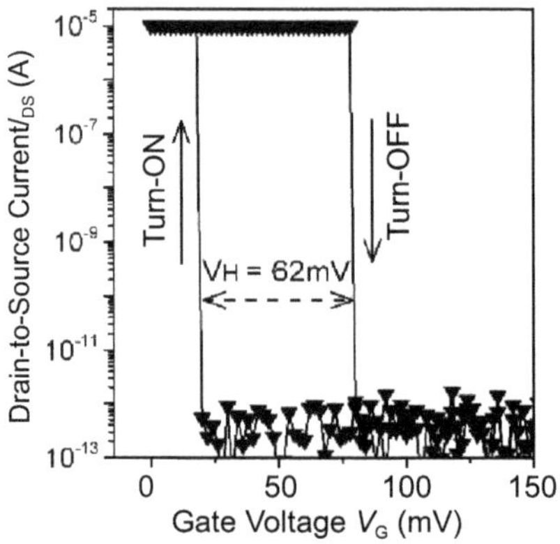

Figure 3.15 Measured I–V characteristics of a positively body-biased relay with 62 mV V_H.

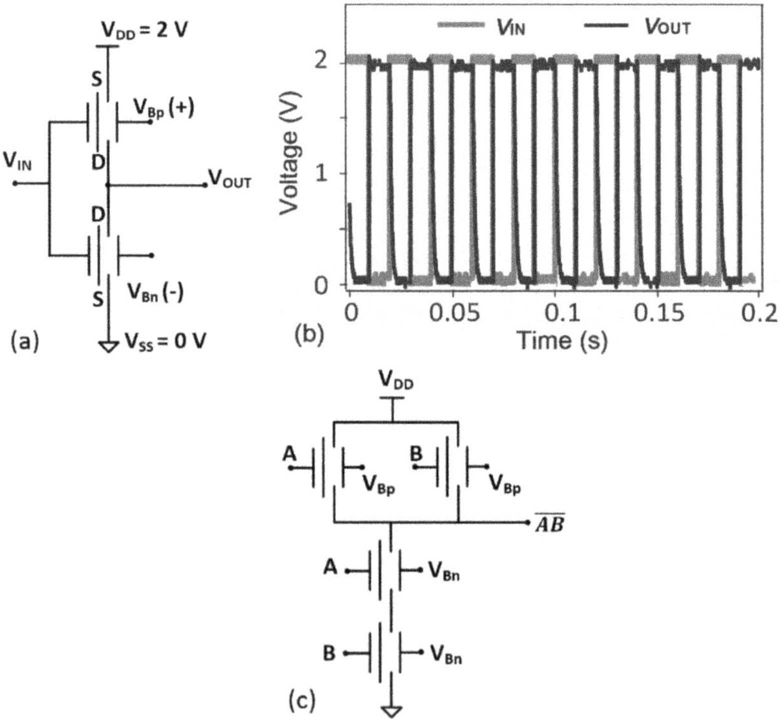

Figure 3.16 (a) Schematic of a complementary relay-based inverter; (b) input/output waveforms of the inverter in (a); (c) schematic of a relay-based NAND gate.

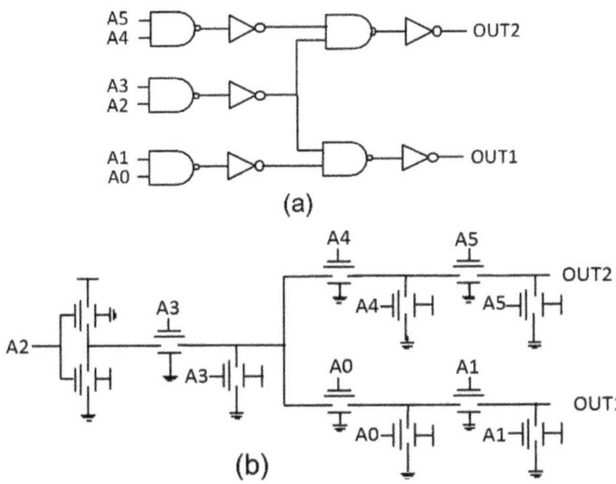

Figure 3.17 (a) AND logic implemented by cascading CMOS logic gates; (b) implementation of the same AND logic function with relay-based pass gates. All mechanical movement happens within one mechanical delay.

3.5.2 Pass-Gate Logic

The electrical delay of a CMOS circuit is largely set by the time it takes to charge or discharge the output capacitance, that is the RC delay. In contrast, the delay of a mechanical relay circuit is dominated by the mechanical delay (remember that the RC delay of a relay is orders of magnitude smaller than the mechanical delay). Cascading relay logic gates serially in a CMOS circuit fashion (e.g., Fig. 3.17a) would lead to a long critical path and thus a low operation frequency. To remedy this drawback, an optimized relay-based circuit should arrange relays in a way that all mechanical movements on a signal path happen simultaneously within one mechanical delay, as exemplified in Fig. 3.17b [35]. Of course, this increases the total electrical delay of charging or discharging the output node (e.g., Out1 and Out2) drastically since the on-resistances of all the relays on the signal path add together. Given the large ratio of mechanical delay to electrical delay, many relays can be serially connected in a single complex pass-gate-style logic block until the total RC delay dominates the mechanical delay (in which case it might be beneficial to partition the logic block into two stages and to rebuffer the signal).

3.6 Hybrid CMOS+NEMS Technology

Despite their superior energy efficiency, purely relay-based ICs will always be slower than their CMOS counterparts, even if the pass-gate logic style described in the last section is used [35]. In addition, the endurance of relays is inferior to CMOS transistors [36]. Another challenge of relay technology is a large device footprint; it is difficult for relays to be miniaturized to be as compact as CMOS transistors. As discussed in Section 3.3.2, given a spring stiffness value and actuation air gap size, a sufficiently large actuation area is necessary to generate a strong enough electrostatic force to turn on a relay. This limits the device density, cost, and functionality of purely relay-based ICs. By integrating relays with CMOS transistors, however, the advantages of both technologies can be realized simultaneously.

3.6.1 Low-Cost Integration Approach

NEM switches can be monolithically integrated over CMOS circuitry in a cost-effective manner by leveraging the back-end-of-line (BEOL) process [37–39]. Over the years, the number of metal wiring layers in CMOS ICs has steadily increased with the increasing density of transistors. The metal wire pitch size (both line width and line-to-line space) shrinks as IC technology advances. Recently, air gaps have been incorporated in-between metal wires to reduce undesirable parasitic capacitance [40]. These trends (increasing number of metal wiring layers, decreasing wire pitch, and incorporation of air gaps) can be leveraged to implement NEM switches in the BEOL layers with a relatively low incremental process cost. The monolithic integration of NEM switches and CMOS transistors allows for minimal interconnect parasitic resistance and capacitance, resulting in low energy consumption and a better system performance.

Some preliminary demonstrations of NEMS+CMOS integration have been reported recently [39, 41]. Figure 3.18 illustrates

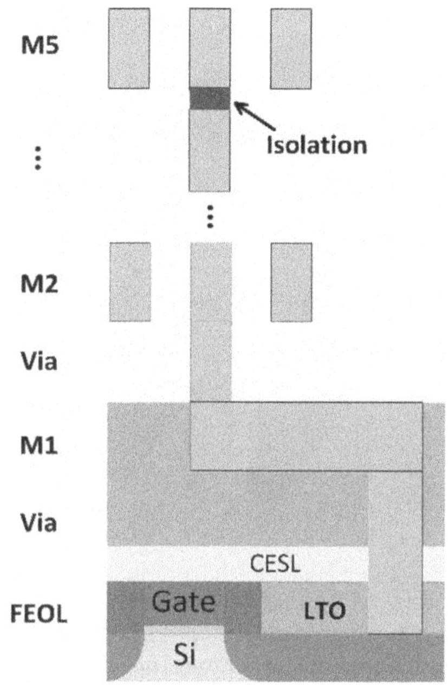

Figure 3.18 Schematic illustration of integrating a vertically oriented relay on top of CMOS circuits with BEOL process.

schematically a cross-sectional view of a relay/MOSFET integrated structure where a vertically orientated relay is implemented using BEOL layers of a standard CMOS technology.

The 3-D structure of the vertical relay is illustrated in Fig. 3.19 [42], which shows a single-pole/double-throw (SPDT) switch design. Metal layers 2-4 (M2-4) are used to form the actuation portions of the movable structure while metal layer 5 (M5) is used to form the contact portion. Note that a portion of the movable structure can be formed with an inter-metal dielectric to electrically insulate the input from the output. All the key design parameter values of a relay (e.g., actuation area, spring stiffness, contact, and actuation air gap sizes) can be well controlled. The effective actuation area can be increased by using longer beams and more metal layers. The spring stiffness can be tuned by adjusting the total number of metal layers and the number of metal vias in-between adjacent metal layers. As the metal line pitch decreases

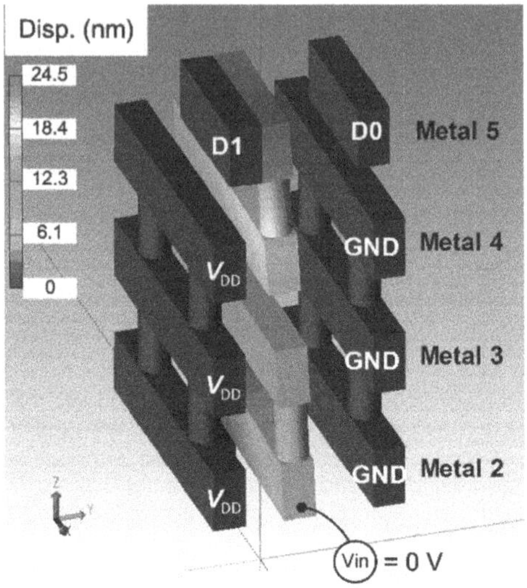

Figure 3.19 Structure of a SPDT NEM switch implemented with BEOL metal and via layers in a conventional CMOS process. The middle movable element, which is anchored at the bottom (not shown), consists of several layers of metal lines and vias. The two actuation electrodes as well as the drain electrodes on each side are fixed. By applying voltage to an actuation electrode on either side, the movable electrode can be attracted to make contact with the corresponding drain (i.e., D0 or D1). The color scale indicates physical displacement.

with manufacturing technology advancement (thus both actuation and contact air gap sizes decrease), the operation voltage scales down. By taking advantage of an advanced BEOL process, this vertically oriented NEM switch design can have a relatively compact footprint.

3.6.2 Application Examples

In addition to functioning as logic switches, NEM switches can also serve as nonvolatile memory (NVM) devices, which can be leveraged to implement novel circuit architectures for ultrafast and energy-efficient computing.

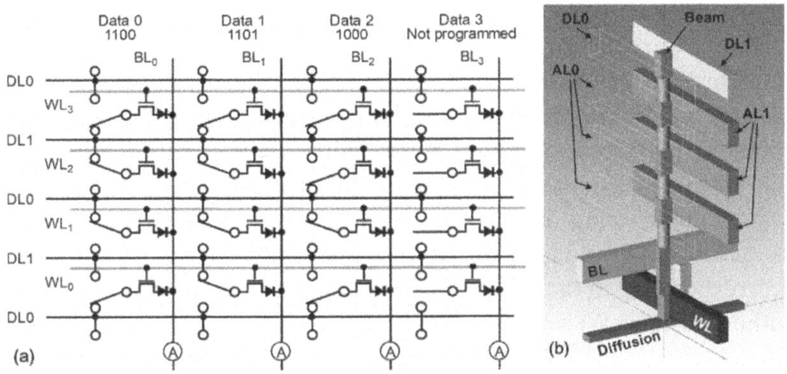

Figure 3.20 (a) Circuit architecture of a CMOS+NV-NEMS array suitable for fast data searching; (b) 3D structure of a memory cell comprising an access transistor and a nonvolatile NEM switch memory element. To set the state of NVM cell, the beam is grounded (by driving the WL high, thereby electrically connecting the beam through the transistor to the bit-line, which is driven low during the programming operation), and a voltage pulse is applied to either actuation electrode (AL0 or AL1, which is shared between all memory cells of the same row) to attract the beam to make contact with the corresponding data line (DL0 or DL1, which is also shared between all cells of the same row).

3.6.2.1 Data searching

Data searching is an essential task in the processing of large and complex data sets. Conventionally, a data searching operation involves frequent data transfer between a memory chip (e.g., nonvolatile flash or dynamic random-access memory [DRAM]) and a central processing unit (CPU), with significant associated energy consumption and wire delay.

Recently, a novel memory architecture based on NEM switches that has lower latency and more energy-efficient data searching capability has been proposed [43]. As shown in Fig. 3.20a, each cell of the memory array consists of an access transistor and a nonvolatile nanoelectromechanical (NV-NEM) switch. The 3D structure of an integrated CMOS+NV-NEM memory cell is illustrated in Fig. 3.20b. The state ("0" or "1") of each cell can be programmed by applying a high voltage on either actuation electrode (i.e., AL0 or AL1), with the two data lines (DLs) (i.e., DL0 and DL1) floating, and word-line (WL) = V_{DD} and bit-line (BL) = 0 V applied to the access transistor. To find the column in the array that stores data matching an input string, two read operations need to be performed across all the programmed columns in parallel (see Ref. [43] for details): (i) Match "0" to find all columns whose data strings match all "0" bits of the input string and (ii) Match "1" to find the columns whose data string matches all "1" bits of the input string. In the Match "0" operation, all of the DL1 lines and only the WLs of cells corresponding to the "0" bits of the input string are driven to V_{DD}; all other DLs and WLs are floating. A high current in a BL (monitored by a sensing amplifier) indicates at least one of the accessed cells has "1" stored, which means the string stored in that column doesn't match the input string. The Match "1" operation works analogously. With such a search scheme, the location of the desired string can be found very efficiently with only two read operations. It should be noted that a nonlinear selector (indicated by the diode symbol in Fig. 3.20a) should be incorporated in series with each NEM switch to prevent sneak leakage current during a Match operation. This can be achieved by leaving an insulating layer on the sidewalls of the contacting electrodes so that a metal/insulator/metal diode is formed upon contact.

The CMOS+NV-NEM cell has a compact layout (the footprint of each cell is only $8F^2$, where F is the minimum half pitch of the metal interconnect layer), which is suitable for high-density storage. Circuit simulations indicate that this hybrid circuit architecture can achieve orders of magnitude improvement in latency and energy consumption for data searching applications, compared to the conventional CPU+DRAM approach [43].

3.6.2.2 Look-up table

A look-up table (LUT) operation is more energy efficient than conventional Von Neumann computation because it avoids the need for data transfer between a CPU and off-chip memory. In a LUT, run-time computation is replaced by a simpler memory-array indexing operation. Figure 3.21a shows the architecture of a traditional CMOS-based LUT. A LUT can also be implemented

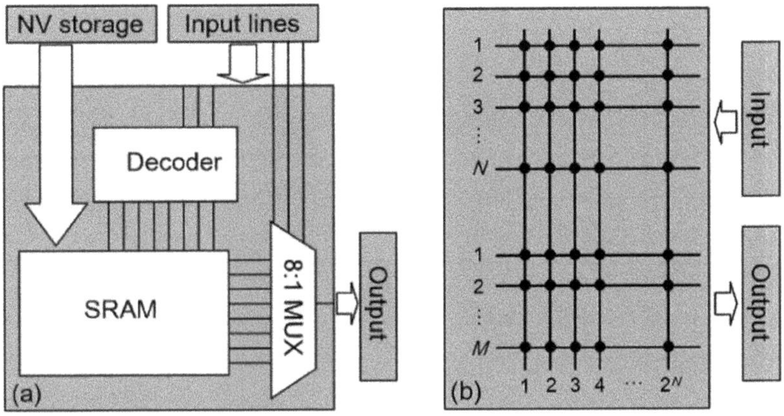

Figure 3.21 (a) CMOS-based LUT; (b) cross-point NEMory-based LUT.

with a more compact cross-point nonvolatile resistive random-access memory (ReRAM) array [44]. Recently, a nonvolatile NEM memory (NV-NEMory)-based LUT architecture has also been proposed (Fig. 3.21b) [45], since NV-NEMory is more energy efficient than ReRAM.

Figure 3.22a illustrates a 5-input/2-output NV-NEMory based LUT that has an input (address) portion and an output (result) portion; the BLs of the input portion (IBLs) are connected to the BLs of the output portion (OBLs) through CMOS-based tristate gates (TSGs). The total number of rows (i.e., number of I/OBLs) equals 2^N, where N is the number of input columns (i.e., input bits). Each memory cell within the NV-NEM array (the left shadowed box in Fig. 3.21b) comprises a SPDT NEM switch as shown in Fig. 3.22b and highlighted in red in Fig. 3.22a, where the movable electrode anchored on the bottom BL can make contact with either DL (D0 or D1) upon a program voltage V_{prog} pulse applied between the BL and the actuation electrode (P0 or P1). Note that P0 and P1 are shared across all cells within a single column. To write the LUT, each IBL/OBL is grounded one at a time, during which V_{prog} is firstly applied to P0 electrodes of all "0"-bit columns and then applied to P1 electrodes of all "1"-bit columns. Note that D0/D1 are electrically floating during programming so no direct current flows;

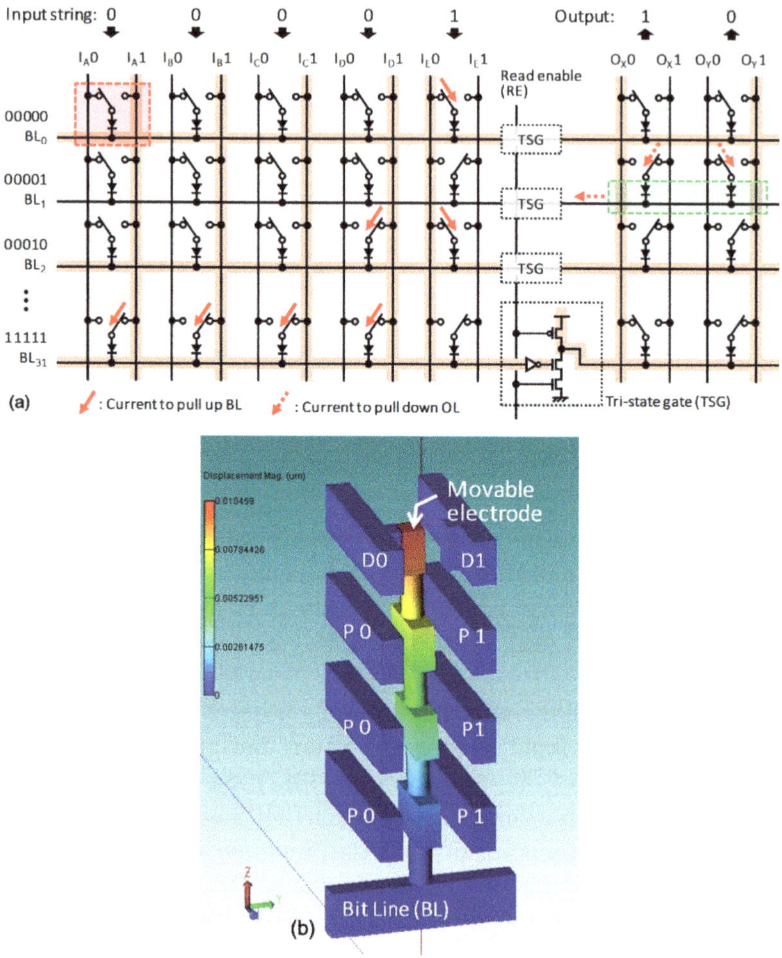

Figure 3.22 (a) Circuit diagram of a 5-input/2-ouput NV-NEMory-based LUT; (b) structure of a NV-NEMory cell.

this "cold-switching" operation contributes to not only ultra-low-power consumption but also improved endurance.

A look-up operation of such a NV-NEMory LUT has three steps: (i) with the read-enable signal (RE) grounded, all IBLs and input DLs (I0 and I1) are precharged low while all OBLs and output DLs (O0 and O1) are precharged high; (ii) the input DLs are then driven such that

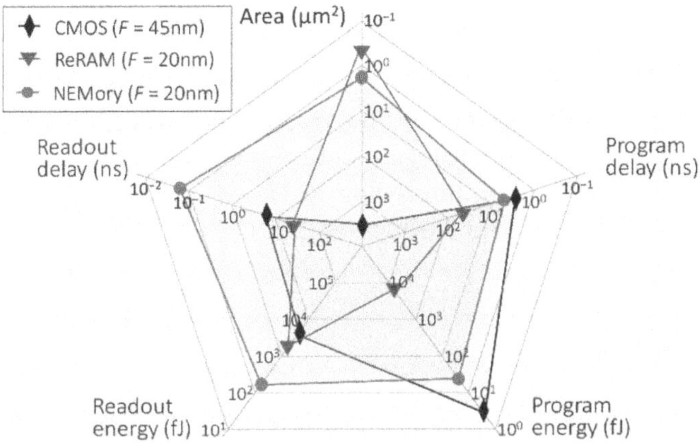

Figure 3.23 Radar plot comparing the projected performance of a NEMory-based LUT against a conventional CMOS-based LUT and ReRAM-based in-memory computing [45].

all IBLs but one—the one corresponding to the input bit string—are charged high, as illustrated in Fig. 3.22a; and (iii) the RE is pulsed high such that the OBL corresponding to the low IBL is discharged low via a TSG so that the states of the OBL-connected memory cells (highlighted in green in Fig. 3.22a) can be read out.

The key performance characteristics of the NV-NEMory LUT are benchmarked against those of ReRAM-based and CMOS-based LUTs in Fig. 3.23 [45]. Clearly, the NV-NEMory-based LUT has remarkable advantages in readout energy and delay. However, practical challenges in implementation, such as contact resistance and potential reliability issue, remain to be experimentally investigated.

3.7 Conclusion

Electrostatically actuated NEM switches are promising alternatives to MOSFETs for ultra-low-power digital logic applications due to their virtually zero off-state leakage current and abrupt switching characteristics. Body-biased NEM relays have been demonstrated to operate reliably at sub-100 mV, which is a major milestone toward the goal of realizing ultra-low-power mechanical computing. NEM switches also can be utilized as nonvolatile memory elements in future energy-efficient data-centric computing systems. Monolithic integration of NEM switches with CMOS transistors can leverage the advantages of both technologies. Future research and development efforts should focus on improving NEM switch reliability and the demonstration of hybrid CMOS+NEMS ICs.

References

1. Lee, T. H., Bhunia, S., and Mehregany, M. (2010). Electromechanical computing at 500°C with silicon carbide, *Science*, **329**, pp. 1316–1318.
2. Lee, J. O., Song, Y. H., Kim, M. W., Kang, M. H., Oh, J. S., Yang, H. H., and Yoon, J. B. (2013). A sub-1-volt nanoelectromechanical switching device, *Nat. Nanotechnol.*, **8**, pp. 36–40.
3. Jang, J. E., Cha, S. N., Choi, Y. J., Kang, D. J., Butler, T. P., Hasko, D. G., Jun, J. E., Kim, J. M., and Amaratunga, G. (2008). Nanoscale memory cell based on a nanoelectromechanical switched capacitor, *Nat. Nanotechnol.*, **3**, pp. 26–30.
4. Lee, S. W., Park, S. J., Campbell, E. E. B., and Park, Y. W. (2011). A fast and low-power microelectromechanical system-based non-volatile memory device, *Nat. Commun.*, **2**, pp. 1–6.
5. Sinha, N., Jones, T. S., Guo, Z., and Piazza, G. (2010). Demonstration of low voltage and functionally complete logic operations using body-biased complementary and ultra-thin ALN piezoelectric mechanical switches, in *IEEE 23rd International Conference on Micro Electro Mechanical Systems (MEMS)*, pp. 751–754.
6. Spencer, M., Chen, F., Wang, C. C., Nathanael, R., Fariborzi, H., Gupta, A., Kam, H., Pott, V., Jeon, J., Liu, T. J. K., Markovic, D., Alon, E., and Stojanovic, V. (2011). Demonstration of integrated micro-electro-mechanical relay circuits for VLSI applications, *IEEE J. Solid-State Circuits*, **46**(1), pp. 146–148.
7. Venkatasubramanian, R., Manohar, S. K., and Balsara, P. T. (2013). NEM relay-based sequential logic circuits for low-power design, *IEEE Trans. Nanotechnol.*, **12**, pp. 386–398.
8. Sinha, N., Jones, T. S., Guo, Z., and Piazza, G. (2012). Body-biased complementary logic implemented using AlN piezoelectric MEMS switches, *IEEE J. Microelectromech. Syst.*, **21**, pp. 484–496.

9. Nathanael, R., Pott, V., Kam, H., Jeon, J., and Liu, T. J. K. (2009). 4-terminal relay technology for complementary logic, in *IEEE International Electron Devices Meeting (IEDM) Technical Digest*, pp. 223–226.
10. Ruan, M., Shen, J., and Wheeler, C. B. (2001). Latching microelectromagnetic relays, *Sens. Actuators A*, **91**, pp. 346–350.
11. Cho, I., Song, T., Baek, S. H., and Yoon, E. (2005). A low-voltage and low-power RF MEMS series and shunt switches actuated by combination of electromagnetic and electrostatic forces, *IEEE Trans. Microwave Theory Tech.*, **53**, pp. 2450–2457.
12. Ruan, M., Shen, J., and Wheeler, C. B. (2001). Latching micromagnetic relays, *IEEE J. Microelectromech. Syst.*, **10**(4), pp. 511–517.
13. Cho, I., and Yoon, E. (2010). Design and fabrication of a single membrane push-pull SPDT RF MEMS switch operated by electromagnetic actuation and electrostatic hold, *J. Micromech. Microeng.*, **20**, pp. 1–7.
14. Gray, G. D. Jr., and Kohl, P. A. (2005). Magnetically bistable actuator: Part 1. Ultra-low switching energy and modeling, *Sens. Actuators A*, **119**, pp. 489–501.
15. Taylor, W. P., Brand, O., and Allen, M. G. (1998). Fully integrated magnetically actuated micromachined relays, *IEEE J. Microelectromech. Syst.*, **7**, pp. 181–191.
16. Glickman, M., Tseng, P., Harrison, J., Niblock, T., Goldberg, I. B., and Judy, J. W. (2011). High-performance lateral-actuating magnetic MEMS switch, *IEEE J. Microelectromech. Syst.*, **20**, pp. 842–851.
17. Raychowdhury, A., Kim, J. I., Peroulis, D., and Roy, K. (2006). Integrated MEMS switches for leakage control of battery operated systems, in *IEEE Custom Integrated Circuits Conference*, pp. 457–460.
18. Wang, Y., Li, Z., McCormick, D. T., and Tien, N. C. (2003). A micromachined RF microrelay with electrothermal actuation, *Sens. Actuators A*, **103**, pp. 231–236.
19. Li, X., Lang, L., Liu, J., Xia, Y., Yin, L., Hu, J. B., Fang, D., and Zhang, H. X. (2010). Electro-thermally actuated RF MEMS switch for wireless communication, in *IEEE 5th International Conference on Nano/Micro Engineered and Molecular Systems*, pp. 497–500.
20. Dellaert, D., and Doutreoigne, J. (2015). Compact thermally actuated latching MEMS switch with large contact force, *Electron. Lett.*, **51**(1), pp. 80–81.
21. Daneshmand, M., Fouladi, S., Mansour, R. R., Lisi, M., and Stajcer, T. (2009). Thermally actuated latching RF MEMS switch and its

characteristics, *IEEE Trans. Microwave Theory Tech.*, **57**(12), pp. 3229–3238.

22. Eom, C. B., and Trolier-McKinstry, S. (2012). Thin-film piezoelecetric MEMS, *Mater. Res. Soc. Bull.*, **37**, pp. 1007–1017.
23. Zou, J., Lin, C. M., Chen, Y. Y., and Pisano, A. P. (2014). Theoretical study of thermally stable SiO2/AlN/SiO2 Lamb wave resonators at high temperatures, *J. Appl. Phys.*, **115**(9), p. 094510.
24. Zaghloul, U., and Piazza, G. (2014). Sub-1-volt piezoelectric nanoelectromechanical relays with millivolt switching capability, *IEEE Electron Device Lett.*, **35**(6), pp. 669–671.
25. Liu, T. J. K., Xu, N., Chen, I. R., Qian, C., and Fujiki, J. (2014). NEM relay design for compact, ultra-low-power digital logic circuits, in *IEEE International Electron Devices Meeting*, pp. 13.1.1–13.1.4.
26. Qian, C., Peschot, A., Connelly, D. J., and Liu, T. J. K. (2015). Energy-delay performance optimization of NEM logic relay, in *IEEE International Electron Devices Meeting*, pp. 18.1.1–18.1.4.
27. Qian, C., Peschot, A., Chen, I. R., Chen, Y., Xu, N., and Liu, T. J. K. (2015). Effect of body biasing on the energy-delay performance of logic relays, *Electron Device Lett.*, **36**(8), pp. 862–864.
28. Qian, C., Peschot, A., Osoba, B., Ye, Z. A., and Liu, T. J. K. (2017). Sub-100 mV computing with electro-mechanical elays, *IEEE Trans. Electron Devices*, **64**(3), pp. 1323–1329.
29. Yaung, J., Hultin, L., Jeon, J., and Liu, T. J. K. (2014). Adhesive force characterization for MEM logic relays with sub-micron contacting regions, *J. Microelectromech. Syst.*, **23**(1), pp. 198–203.
30. Israelachvili, J. N. (2011). Van der Waals forces between particles and surfaces, in *Intermolecular and Surface Forces*, Academic Press, Elsevier, MA, USA, pp. 253–290.
31. Chen, I. R., Chen, U., Hutin, L., Pott, V., Nathanael, R., and Liu, T. J. K. (2013). Stable ruthenium-contact relay technology for low-power logic, in *17th IEEE Solid-State Sensors, Actuators and Microsystems (Transducers'2017)*, pp. 896–899.
32. Osoba, B., Saha, B., Dougherty, L., Edgington, J., Qian, C., Niroui, F., Lang, J. H., Bulovic, V., Wu J., and Liu, T. J. K. (2016). Sub-50 mV NEM relay operation enabled by self-assembled molecular coating, in *IEEE International Electron Devices Meeting (IEDM)*, pp. 655–658.
33. Nathanael, R., Pott, V., Kam, H., Jeon, J., Alon, E., and Liu, T. J. K. (2010). Four-terminal-relay body-biasing schemes for complementary logic circuits, *IEEE Electron Device Lett.*, **31**(8), pp. 890–892.

34. Fujiki, J., Xu, N., Hutin, L., Chen, I. R., Qian, C., and Liu, T. J. K. (2014). Microelectromechanical relay and logic circuit design for zero crowbar current, *IEEE Trans. Electron Devices*, **61**(9), pp. 3296–3302.

35. Chen, F., Kam, H., Markovic, D., Liu, T. J. K., Stojanovic, V., and Alon, E. (2008). Integrated circuit design with NEM relays, in *IEEE/ACM International Conference on Computer-Aided Design*, pp. 750–757.

36. Peschot, A., Qian, C., and Liu, T. J. K. (2015). Nanoelectromechanical switches for low-power digital computing, *Micromachines*, **6**(8), pp. 1046–1065.

37. Fedder, G. K., Howe, R. T., Liu, T. J. K., and Quevy, E. P. (2008). Technologies for cofabricating MEMS and electronics, *Proc. IEEE*, **96**(2), pp. 306–322.

38. Pott, V., Kam, H., Nathanael, R., Jeon, J., Alon, E., and Liu, T. J. K. (2010). Mechanical computing redux: relays for integrated circuit applications, *Proc. IEEE*, **98**(12), pp. 2076–2094.

39. Xu, N., Sun, J., Chen, I. R., Hutin, L., Chen, Y., Fujiki, J., Qian, C., and Liu, T. J. K. (2014). Hybrid CMOS/BEOL-NEMS technology for ultra-low-power IC applications, in *IEEE International Electron Devices Meeting*, pp. 28.8.1–28.8.4.

40. Natarajan, S., et al. (2014). A 14nm logic technology featuring 2nd-generation FinFET, air-gapped interconnects, self-aligned double patterning and a 0.0588 μm^2 SRAM cell size, in *IEEE International Electron Devices Meeting*, pp. 71–73.

41. Riverola, M., Vidal-Alvarez, G., Torres, F., and Barniol, N. (2014). CMOS-NEM relay based on tungsten VIA layer, in *IEEE SENSORS 2014 Proceedings*, pp. 162–165.

42. Liu, T. J. K., Sikder, U., Kato, K., and Stojanovic, V. (2017). There's plentry of room at the top, in *IEEE 30th International Conference on Micro Electro Mechanical Systems (MEMS)*, pp. 1–4.

43. Kato, K., Stojanovic, V., and Liu, T. J. K. (2016). Non-volatile nano-electro-mechanical memory for energy-efficient data searching, *IEEE Electron Devices Lett.*, **37**(1), pp. 31–34.

44. Chen, B., Cai, F., Zhou, J., Ma, W., Sheridan, P., and Lu, W. D. (2015). Efficient in-memory computing architecture based on crossbar arrays, in *IEEE International Electron Devices Meeting (IEDM)*, pp. 17.5.1–17.5.4.

45. Kato, K., Stojanović, V., and Liu, T. J. K. (2016). Embedded nano-electro-mechanical memory for energy-efficient reconfigurable look-up tables, *IEEE Electron Device Lett.*, **37**(12), pp. 1563–1565.

Chapter 4

Adiabatic Solutions for Ultra-Low-Power Electronics

Hervé Fanet
CEA-LETI MINATEC Campus, 38000 Grenoble, France
herve.fanet@cea.fr

This chapter is a description of adiabatic logic introduced 40 years ago but not yet applied in the industry. First we will demonstrate the theoretical interest of this principle and also calculate the limits of use of CMOS technology. The adiabatic principle of charge and discharge of a capacitance is identified as the optimal solution for ultra-low-power electronics. Nevertheless, the use of CMOS for adiabatic logic introduces nonadiabatic dissipation due to threshold voltages, and finally the gain is far from the theoretical limit. That explains the interest in investigating other technologies. The nano-electrical mechanical systems (NEMS) technology appears to be very interesting because the leakage dissipation is reduced compared to CMOS and because threshold effects can be reduced in adiabatic implementation. A detailed analysis and comparison of NEMS and CMOS technologies for low-power adiabatic logic implementation is presented. The advantage of combining NEMS technology and adiabatic logic is demonstrated. Furthermore, NEMS-based adiabatic gates architectures are described. The chapter concludes with a

Emerging Devices for Low-Power and High-Performance Nanosystems: Physics, Novel Functions, and Data Processing
Edited by Simon Deleonibus

ISBN 978-981-4800-11-2 (Hardcover), 978-0-429-45873-6 (eBook)
www.panstanford.com

detailed comparison of the energy-performance properties of the different explored technologies and with an introduction of a new paradigm: the capacitance-based adiabatic logic.

4.1 Introduction

The management of power dissipation in electronic systems represents the main concern in new component development and architecture evolution. This is especially true for portable systems and Internet of Things electronics, where energy consumption is limited by battery performance or by energy harvesting constraints. This is also true for high-performance computing or data center electronic systems, where heat dissipation and energy consumption have to be limited. In the current complementary metal-oxide-semiconductor (CMOS) technology, the two main sources of dissipation are leakage (subthreshold and tunnel currents) and active power dissipation (current due to charge and discharge of capacitances when logic states are changing), which increases with the operating frequency and the square of the supply voltage. To reduce active power dissipation, standard solutions for current CMOS technologies consist in developing components working at low voltage and/or to design low-frequency architectures. However, such simplified approaches are not able to solve the trade-off between energy dissipation and performance if ultra-low-power constraints need to be applied.

Therefore, various solutions have been suggested to drastically improve the performance of low-power circuits. These may fundamentally be divided into two categories: a device-based approach and a circuit-based approach. On a device level, the use of advanced device technologies, such as Fin field-effect transistor (FinFET) and fully depleted silicon-on-insulator (FDSOI) metal-oxide-semiconductor field-effect transistor (MOSFET), improves electrostatic control over the conducting channel. Other emerging devices, such as tunnel field-effect transistor (TFET), promise to improve on the 60 mV/decade maximum allowed subthreshold slope in current MOSFETs. Another promising emerging solution is the use of nanoelectromechanical (NEM) relays, which eliminates the leakage current and therefore the static power loss. Another approach consists in changing circuit-level architectures and power management solutions in order to reduce power consumption. Such solutions include subthreshold circuits, power gating, asynchronous architectures, energy recovery supplies, and adiabatic logic.

4.2 Sources of Dissipation in CMOS Circuits

In this paragraph we will identify the main causes of heat dissipation. The physical origins of thermal dissipation are based on rather complex models that are detailed in solid-state physics. In general, Joule's law is applied; it links dissipation to the current flowing through a resistive device resistance. Dissipated power, which is to say energy transformed into heat by time units, is written in the classical way:

$$P(t) = RI(t)^2 \tag{4.1}$$

In this relation, R is the resistance in question and $I(t)$ is the value of the instantaneous current assumed to be flowing through the device. As the difference in potential at the terminal is $U = RI$, it can also be written in the following form:

$$P(t) = U(t)I(t) \tag{4.2}$$

This equation is more general and is also applied to devices where Ohm's law is not valid, such as metal-semiconductor junctions or semiconductor-semiconductor junctions. In reality, the physics of this phenomenon is described in a much more complex way when it is modeled by the carrier's wave function interactions with phonons. This result can be applied to pure or intrinsic resistances but also to devices of the following types: metal-semiconductor, metal-metal, and semiconductor-semiconductor.

Let us now take the example of a series resistance and a capacitor like the one shown in Fig. 4.1. This simple schematic is fundamental because it is the electrical representation of a change of logic state in a gate. There is no heat dissipation in the capacitor. The energy E_c is brought about by the following relationship:

$$E_c = 1/2 \cdot CV^2 \tag{4.3}$$

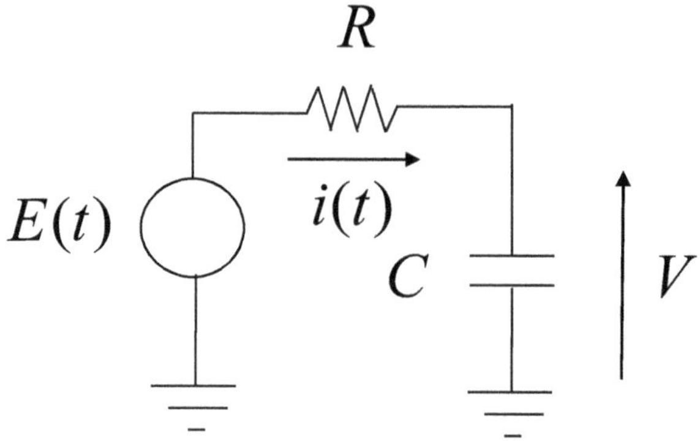

Figure 4.1 Charge and discharge of a capacitor.

The voltage V is the voltage applied at the capacitor terminals.

The two elements are linked to a voltage generator, which is itself linked to the ground. The source of this generator output voltage may be of different natures: chemical, mechanical, or electrical. The power delivered by this generator is $E(t)i(t)$. To simplify the problem, let us assume that the voltage $E(t)$ form has a constant value of V_{DD}. Then we calculate the voltage at the capacitor terminals. The system is represented with a very simple differential equation:

$$E(t) = Ri(t) + V(t) \tag{4.4}$$

$$i(t) = C\frac{dV}{dt} \tag{4.5}$$

Initially, the voltage at the capacitor terminals, as well as the electrode charges, is null. Equations 4.4 and 4.5 can be summarized by one pair of equations:

$$E(t) = RC\frac{dV}{dt} + V(t) \tag{4.6}$$

$$V(0+) = 0 \tag{4.7}$$

This initial condition is a consequence of the capacitors' behavior; starting from 0, another voltage value immediately after applying the generator voltage would lead to an infinite current, which is physically impossible. This remark leads to the following well-known equation:

$$V(t) = V_{\mathrm{DD}}\left(1 - e^{-\frac{t}{RC}}\right) \quad \text{for } t \geq 0 \tag{4.8}$$

The voltage at the capacitor terminals starts from the value of 0 and moves toward the value V_{DD} following an exponential law of time constant *RC*. From this, we deduct the value of the current in resistance and then the value of dissipated energy in resistance throughout the infinite duration of the capacitor charge.

$$E_{\mathrm{R}} = R\frac{V_{\mathrm{DD}}^2}{R^2}\int_0^{\infty}\left(e^{-\frac{t}{RC}}\right)^2 \mathrm{d}t \tag{4.9}$$

Calculating this integral allows us to obtain the following well-known value:

$$E_{\mathrm{R}} = \frac{1}{2}C V_{\mathrm{DD}}^2 \tag{4.10}$$

Therefore, the dissipated energy does not depend on the resistance value but on the capacitance value. This value is equal to the energy stored in the capacitor at the end of charging, which is in theory at the end of an infinite time. Both of these results may come as a surprise, but nevertheless, they remain to be true if the resistance varies with the terminal voltage.

To conclude, we will give a few points on static consumption. In a circuit containing a conduction path between the supply line and the ground, a continuous current circulates permanently and generates rather bad dissipation as it doesn't connect with any logic operation. Transistors, even cut-off ones, meaning those that are controlled below the threshold, do not have a strong enough resistance value and so a current called a subthreshold current circulates. This leakage current can be written more precisely as:

$$I = I_0 e^{-\frac{1}{n_0}\cdot\frac{V_{\mathrm{T}}}{k_{\mathrm{B}}T/q}} \tag{4.11}$$

In this expression, V_{T} is the threshold voltage that corresponds to the chosen technology, k_{B} is the Boltzmann constant, n_0 is a coefficient depending on the transistor electrostatics (the subthreshold slope), T is the temperature, and q is the charge of an electron. The threshold voltages are different for both n-type metal-oxide semiconductor (NMOS) and p-type metal-oxide semiconductor (PMOS) and values of 300 mV are the usual ones for current technologies. The exponential dependence of the subthreshold current shows how acutely sensitive static dissipation is to the choice of the threshold voltage. This phenomenon is amplified by the effects of variability. When the transistor size diminishes, the threshold voltage static dispersions increase. This happens even though a non-negligible number of transistors may have threshold voltage values that are far lower than the average value and significantly contribute to the subthreshold current increase. This is a major difficulty in CMOS technology and explains how many current technologies and architectures have evolved the way they have. The fundamental problems facing circuit designers are how to restrict threshold currents and the variation in average values for threshold voltages. Increasing the average value of the threshold voltage is not a viable solution in this case because the supply voltage would also need to be increased, leading to an important active power. Breakthrough technologies may be invented to radically treat this problem by designing devices with negligible leakage current.

Now, we have to study the dissipation in logic gates. A digital system can be described as a set of subsystems as depicted in the example of Fig. 4.2. The combinational block implements a Boolean function. The different inputs are either latch outputs or external signals. Latch outputs are synchronized by the clock signal. The clock period is T. During this period, only one logical transition is possible. It is a property of synchronous systems, and the probability α of this transition is called the activity factor.

To calculate the dissipation during a period T it is necessary to estimate the dissipation of all the combinational gates they are commuting and after to add the dissipation in the latches. In this chapter we only detail the dissipation in the combinational gates and we choose the inverter because all the other gates have the same architecture and consequently the results transpose very easily to more complex logic gates (AND, OR, NAND, NOR, etc.). This calculation-illustrated Fig. 4.3 looks at the gate transition from the off state to the on state at the time t_1.

Before t_1 the PMOS is conductive and the NMOS is "off." The leakage current of the NMOS is flowing between supply and ground,

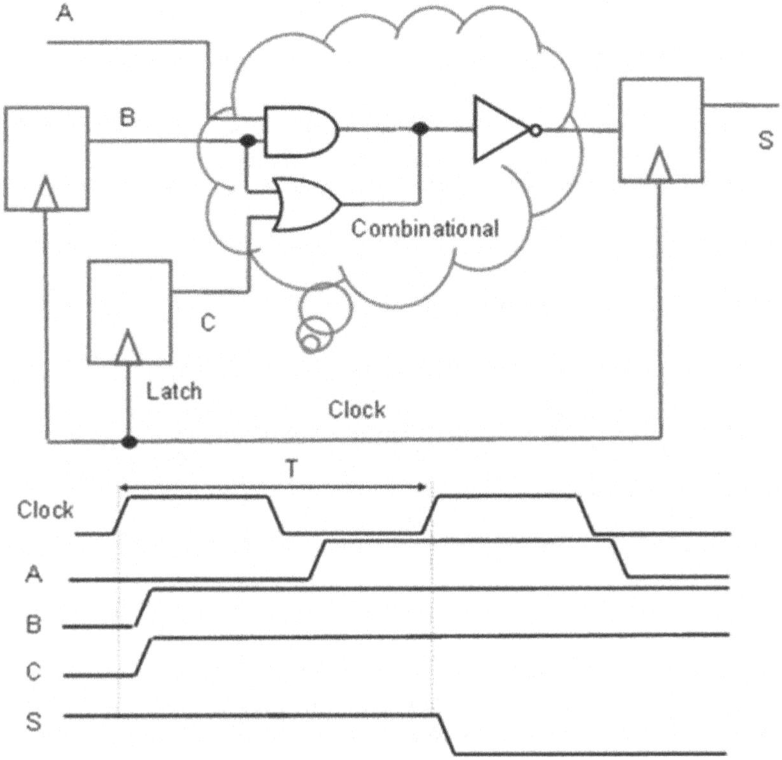

Figure 4.2 A simple example of synchronous logic.

and the output voltage is close to V_{DD}. At t_1 the NMOS becomes conductive and the PMOS is "off." The output capacitance C_L is discharged at the ground. At the time of logical transition, the two inverter transistors can both simultaneously be conductors for a brief time. The associated dissipation is weak though and will not be taken into account in this calculation. After this transition the leakage current of the PMOS is flowing into the gate. To simplify the calculation, we assume both leakage currents have the same value. In summary, two terms will be taken into account: the NMOS discharge current and the subthreshold current passing through the MOSFET during the period T considered.

The dynamic consumption is $1/2\alpha \cdot C_L V_{DD}^2$, an expression in which α is called the gate activity factor and expresses the probability

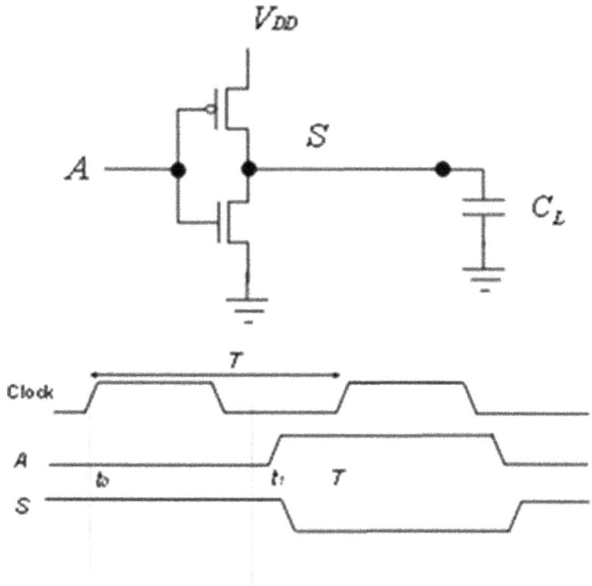

Figure 4.3 The CMOS inverter.

that a logic transition will take place during the period T. This value can vary widely according to the considered logic operator. In the pipeline of a calculating operator, values between 0.1 and 1 are possible. To simplify this somewhat, we can consider that these two subthreshold currents are approximately the same during the period T, creating thermal dissipation equal to:

$$E_{\text{stat}} = I_0 e^{-\frac{V_T}{n_0 \varphi_t}} \left(1 - e^{-\frac{V_{DD}}{\varphi_t}}\right) V_{DD} T \tag{4.12}$$

Before the transition, the values of V_{GS} for the NMOS or V_{SG} for the PMOS are close to zero and V_{DD} and the values of V_{DS} and V_{SD} are close to V_{DD} and zero. The value I_0 depends on the chosen model for the subthreshold current. In the first step we will neglect the short-channel effects. The parameters to be optimized are the threshold voltage and the supply voltage, assuming that the activity factor as well as the period T are given. We will also assume that the speed constraints impose a relation between the supply voltage and the threshold voltage.

$$V_{DD} = kV_T \tag{4.13}$$

Conforming with a popular factor of merit, we can choose k equal to 3. Although we keep the factor k here as a parameter that is adjustable according to the application, we can still determine the speed constraints. Using Eq. 4.13, the total dissipated energy is:

$$E = I_0 e^{-\frac{V_T}{n_0 \varphi_t}} \left(1 - e^{-\frac{kV_T}{\varphi_t}}\right) kV_T T + \frac{1}{2}\alpha C_L k^2 V_T^2 \qquad (4.14)$$

Let us define the normalized threshold:

$$y = \frac{V_T}{n_0 \phi_t} \qquad (4.15)$$

$$\phi_t = \frac{k_B T}{q} \qquad (4.16)$$

And the dissipated energy becomes:

$$E = I_0 e^{-y} \left(1 - e^{-kn_0 y}\right) kn_0 y \phi_t T + \frac{\alpha}{2} C_L \phi_t^2 k^2 n_0^2 y^2 \qquad (4.17)$$

Neglecting the term $e^{-kn_0 y}$, we obtain:

$$E = \frac{\alpha}{2} C_L \phi_t^2 k^2 n_0^2 \left(y^2 + mye^{-y}\right) \qquad (4.18)$$

$$m = \frac{2 I_0 T}{\alpha C_L k n_0 \phi_t} \qquad (4.19)$$

This factor m depends on some technological factors, such as I_0 and n_0, but also on architectural parameters (α, T, and k). The vast majority of digital applications are characterized by values of m in the range 10 to 1000. For example, m is equal to 70 for a standard inverter in CMOS 45 nm technology. The transistor size is included in the term I_0.

Studying the function $f(y) = y^2 + mye^{-y}$ allows one to easily find the optimum, as shown Fig. 4.4. This curve is very generic and, therefore, fundamental. The case ($m = 1000$) corresponding to very advanced technologies shows clearly that there is an optimum value of y near 6 that corresponds to an optimal threshold:

$$V_{Topt} = 6 n_0 \phi_t \qquad (4.20)$$

With $n_0 = 1.3$, a typical value in CMOS, we obtain:

$$V_{Topt} = 195 \text{ mV}$$

This value is quite close to the values effectively chosen in the applications, but it is weaker. This difference can be explained by taking the drain-induced barrier lowering (DIBL) into account.

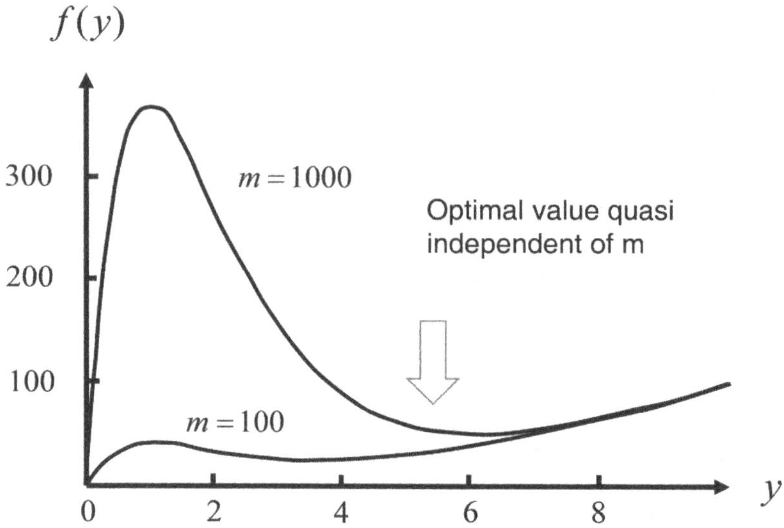

Figure 4.4 Minimization of the dissipated energy.

Now let us introduce the short-channel effects and the DIBL. The dissipation becomes:

$$E = I_0 e^{\frac{k\eta V_T}{n_0\phi_t}} e^{-\frac{V_T}{n_0\phi_t}} \left(1 - e^{-\frac{kV_T}{\phi_t}}\right) kn_0 y\phi_t T + \frac{\alpha}{2} C_L k^2 V_T^2 \tag{4.21}$$

The parameter η measures the DIBL effect.

$$y = \frac{V_T}{n_0\phi_t}(1 - k\eta) \tag{4.22}$$

We can easily obtain:

$$E = \frac{\alpha}{2} C_L \frac{\phi_t^2}{(1 - k\eta)^2} k^2 n_0^2 \left(y^2 + mye^{-y}\right) \tag{4.23}$$

The optimal value of y is the same, but the corresponding threshold voltage and the corresponding dissipation increase. For the high values of m, we obtain an optimal value of the threshold voltage that is:

$$V_{\mathrm{Topt}} = 6n_0\phi_t \frac{1}{1 - k\eta} \tag{4.24}$$

For advanced technologies, we obtain optimal values around 300–400 mV. As the equation gives the dissipated energy value to be

Table 4.1 Active and leakage power

Parameter m	Optimal parameter y	Active energy	Leakage energy	Ratio
100	4	16	7.3	45%
500	5.5	30	11	36%

optimum, the ratio between static energy and dynamic energy can also easily be determined according to the parameter m. Two values are calculated in Table 4.1.

Finally, the conclusion is that it is impossible using CMOS to achieve ultra-low power dissipation because the optimal value of the threshold is not very small. Use of advanced devices (FinFET, FDSOI FET, nanowireFET, graphene FET, etc.) only modifies the subthreshold coefficient and does not change significantly the optimization. A major reduction of dissipation needs a major change of gate architecture and use of a zero-leakage technology. In a first step we have to consider a new paradigm for gate implementation: the adiabatic principle.

4.3 The Adiabatic Principle

It consists of optimizing the charge of a constant or variable value capacitance through a constant or variable value resistance. The aim is to minimize the energy dissipated in heat form.

4.3.1 Charging a Constant Capacitance through a Constant Resistance

In the first instance, we will assume that the value of the capacitance is constant, as well as that of the resistance. Later, it will be shown that it is not necessary to assume that the resistance is constant, but this very simple case is a good introduction to the problem. This example is not just an academic exercise; it is the basic pattern for explaining dissipation in electronic circuits.

The second paragraph showed that a change of a logical state transmitted from one gate to another corresponds to the charging

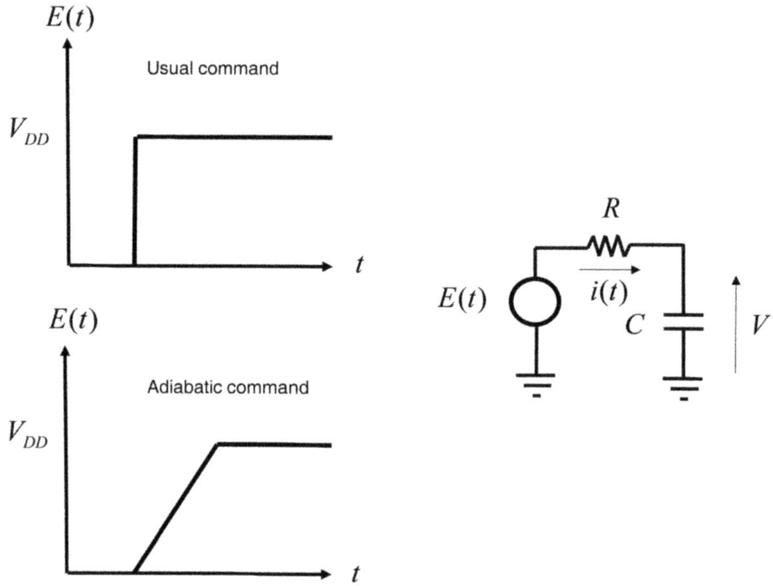

Figure 4.5 Capacitor charge.

or discharging of a capacitor. This is the case for all logic families. The considered capacitance is the sum of the input capacitance of the connected gates, output capacitance of the gate itself, and interconnect capacitances.

The very simple electric pattern shown in Fig. 4.5 leads to the following equations:

$$E(t) = Ri(t) + V(t) \tag{4.25}$$

$$i(t) = \frac{\mathrm{d}Q}{\mathrm{d}t} = C\frac{\mathrm{d}V}{\mathrm{d}t} \tag{4.26}$$

The equation above assumes that there is no capacitance variation in the considered time. The generator provides a constant voltage V_{DD} that starts from zero time.

This differential equation of the first order with constant coefficients is very easily solved. The output voltage is written as:

$$V(t) = V_{\mathrm{DD}}\left(1 - e^{-\frac{t}{RC}}\right) \tag{4.27}$$

It is zero at $t = 0$ and exponentially tends toward V_{DD} when the time is infinite or at least a lot greater than the time constant RC. The current circulating in the circuit is then:

$$i(t) = \frac{V_{\mathrm{DD}}}{R} e^{-\frac{t}{RC}} \tag{4.28}$$

The current moves from V_{DD}/R to zero. Unlike for the voltage, this discontinuity is not forbidden by physics. It extends toward zero when the capacitor is fully charged. The dissipated energy in the resistance is calculated by applying Joule's law.

$$E_{\mathrm{R}} = \int_0^{\infty} (V_{\mathrm{DD}} - V) \frac{V_{\mathrm{DD}}}{R} e^{-\frac{t}{RC}} \, \mathrm{d}t \tag{4.29}$$

A simple calculation gives the following:

$$E_{\mathrm{R}} = \int_0^{\infty} \frac{V_{\mathrm{DD}}^2}{R} e^{-\frac{t}{RC}} \, \mathrm{d}t - \int_0^{\infty} \frac{V_{\mathrm{DD}}^2}{R} e^{-\frac{2t}{RC}} \, \mathrm{d}t = C V_{\mathrm{DD}}^2 - \frac{1}{2} C V_{\mathrm{DD}}^2 = \frac{1}{2} C V_{\mathrm{DD}}^2 \tag{4.30}$$

Notice that the first term is the energy supplied by the generator and the second is the energy stored in the capacitor. To put it definitively, the energy converted into heat is half of the energy supplied by the generator. It is also equal to the energy stored in the capacitor.

We can ask ourselves whether these results remain the same when the resistance and the capacitor are no longer constant. We also notice that the value of energy dissipated in the resistance does not depend on the resistance value itself in this case. This may seem to be a paradox, but it can be easily explained when one considers that the resistance value is acting on the maximum range and on the current form.

4.3.2 Charging a Constant Capacitor through a Variable Resistance

The schematics for this is the same as it was in the first case, but now the resistance may vary with time. This is the case for the transistor conduction resistance.

$$E(t) = R(t)i(t) + V(t) \tag{4.31}$$

$$i(t) = \frac{\mathrm{d}Q}{\mathrm{d}t} = C\frac{\mathrm{d}V}{\mathrm{d}t} \tag{4.32}$$

It is no longer so simple to solve the differential equation, as one must first specify the relationship between the resistance and the voltage at the terminals. The result is more easily obtained by writing:

$$E_{\mathrm{R}} = \int_0^{V_{\mathrm{DD}}} V_{\mathrm{DD}} C dV - \int_0^{V_{\mathrm{DD}}} V C dV \tag{4.33}$$

With the final voltage being V_{DD} and the capacitance value being constant, Eq. 4.34 is obtained:

$$E_{\mathrm{R}} = C V_{\mathrm{DD}}^2 - \frac{1}{2} C V_{\mathrm{DD}}^2 = \frac{1}{2} C V_{\mathrm{DD}}^2 \tag{4.34}$$

The result is thus identical to the simplified case, and the dissipated energy is half the energy supplied by the generator.

4.3.3 Charging a Variable Capacitor Using a Variable Resistance

Although this is a more general case, it is necessary to know the relation between the capacitance and the voltage at the terminals. If the capacitance is the input capacitance of a MOSFET, this capacitance between gate and source is considered constant, but correction terms may be introduced. If the input capacitance is that of a nanorelay, it can vary in a more significant way.

In the first instance let us use a simple relation between capacitance and voltage, such as a linear law:

$$C = C_0 + a_1 V \tag{4.35}$$

The basic equations are written as:

$$E(t) = R(t) i(t) + V(t) \tag{4.36}$$

$$i(t) = \frac{\mathrm{d}Q}{\mathrm{d}t} = C\frac{\mathrm{d}V}{\mathrm{d}t} + \frac{\mathrm{d}C}{\mathrm{d}t} V \tag{4.37}$$

In the linear case, one obtains:

$$i(t) = \frac{\mathrm{d}Q}{\mathrm{d}t} = C\frac{\mathrm{d}V}{\mathrm{d}t} + \frac{\mathrm{d}C}{\mathrm{d}t} V = (C_0 + a_1 V)\frac{\mathrm{d}V}{\mathrm{d}t} + a_1 \frac{\mathrm{d}V}{\mathrm{d}t} V$$
$$= (C_0 + 2a_1 V)\frac{\mathrm{d}V}{\mathrm{d}t} \tag{4.38}$$

A fairly simple calculation leads to:

$$E_R = C_0 V_{DD}^2 + a_1 V_{DD}^3 - \frac{1}{2} C_0 V_{DD}^2 - \frac{2}{3} a_1 V_{DD}^3$$
$$= \frac{1}{2} C_0 V_{DD}^2 + \frac{1}{3} a_1 V_{DD}^3 \tag{4.39}$$

The final capacitance value is $C_F = C_0 + a_1 V_{DD}$. The dissipated energy is therefore not equal to $1/2 C_F V_{DD}^2$.

What does this result become when applied to a more general case? Let us assume that the dependence between capacitance and voltage is a polynomial function in the following form:

$$C = C_0 + a_1 V + a_2 V^2 + \cdots + a_i V^i + \cdots . \tag{4.40}$$

The same calculation (left up to the reader) leads to the following relationship:

$$E_R = \frac{1}{2} C_0 V_{DD}^2 + \frac{1}{3} a_1 V_{DD}^3 + \frac{1}{4} a_2 V_{DD}^4 + \cdots + \frac{1}{i+2} a_i V_{DD}^{i+2} + \cdots . \tag{4.41}$$

This relationship is the general expression when the variable capacitance is charged.

We have to note that the energy necessary for this operation is not only the electrostatic energy $1/2 C_F V_{DD}^2$. Additional energy has to be considered. In the case of an electromechanical device, this additional term is the work of the electrical force between the electrodes.

A rigorous approach based on a thermodynamic argument [1] allows one to write the power supplied by the generator, if we assume there is no loss inherent to the physical mechanism leading to the capacitance variation, as follows:

$$P_e(t) = \frac{d}{dt}\left[\frac{CV^2}{2}\right] + \frac{1}{2}\frac{dC}{dt} V^2 \tag{4.42}$$

4.3.4 Charging a Constant Capacitor in Adiabatic Mode

"Adiabatic" means without heat. The question addressed in this section is the following: Is it possible to charge a constant capacitance while dissipating less than the value established in the first example, namely $1/2 \cdot C V_{DD}^2$?

To answer to the above question, one must refer to one of the gate model hypotheses: the supply voltage is constant. If the voltage supplied by the generator is not constant, but increases in a linear way from zero to the final voltage V_{DD} with a ramp time T that is greater than the circuit time constant *RC*, the voltage may be established with no delay at the capacitor terminals. The voltage at the resistor terminals is almost zero, although the current flowing through it is very weak, which leads to very weak dissipation in the resistor. A fairly basic calculation enables one to quantify the gain obtained.

$$V_{\mathrm{DD}}\frac{t}{T} = Ri(t) + V(t) \qquad \text{for} \qquad t \geq 0 \tag{4.43}$$

$$i(t) = C\frac{\mathrm{d}V}{\mathrm{d}t} \tag{4.44}$$

The solution to this differential equation with constant coefficients is:

$$V = V_{\mathrm{DD}}\frac{t}{T} - \left(\frac{RC}{T}\right) V_{\mathrm{DD}}\left[1 - e^{-\frac{t}{RC}}\right] \text{ for } t \geq 0 \text{ and } t \leq T \tag{4.45}$$

$$V = V_{\mathrm{DD}} - \left(\frac{RC}{T}\right) V_{\mathrm{DD}}\left[1 - e^{-\frac{T}{RC}}\right] e^{-\frac{t-T}{RC}} \text{ for } t \geq T \tag{4.46}$$

The dissipating energy is calculated in the same way. The intermediate calculations are not detailed in this section, but they are very simple.

$$E_{\mathrm{R}} = \int_0^{\infty} \left(V_{\mathrm{DD}}\frac{t}{T} - V\right) i(t)\mathrm{d}t \tag{4.47}$$

When the ramping time T is a lot greater than the time constant *RC*, the expression of dissipated energy has the following value:

$$E_{\mathrm{R}} = \left(\frac{RC}{T}\right) C V_{\mathrm{DD}}^2 \tag{4.48}$$

This expression is the fundamental formula of adiabatic logic, and it will be used at various times throughout this chapter. The adiabatic charge of capacitors is an efficient method when the time constant *RC* has a very weak value, which allows one to choose an equally weak value for the ramping time T and thus to ensure a convenient operating frequency. The reader should keep in mind the difference between the normal regime and the adiabatic one. In the normal regime, the circuit supply voltage is constant, whereas in the adiabatic regime, it is no longer constant and varies in a linear manner.

Let us take a few examples from the microelectronics industry. A gate charge capacitance has a typical value of 1 fF in an advanced technology. A contact resistance of 10 kΩ leads to a time constant of 10 picoseconds. A ramping time of 10 nanoseconds ensures that the dissipation will be reduced by a factor of 1000, in comparison to a classical method. The operating frequency is around 100 MHz. This type of frequency will be completely impossible if one used other techniques to ensure very low consumption, such as subthreshold CMOS [2].

Also note that the simplified formula of adiabatic dissipation corresponds to heat creation in charging a capacitor when the generator can be considered a source of constant current. In fact, when the time constant is weak compared to the ramping time of voltage supplied by the generator, the load can be considered at the first order as a capacitance C. The current is $i = CV_{\mathrm{DD}}/T$ throughout the time period T, and then the value is reduced. The dissipated energy is:

$$E_{\mathrm{R}} = R\left(\frac{CV_{\mathrm{DD}}}{T}\right)^2 T \text{ or } \left(\frac{RC}{T}\right) CV_{\mathrm{DD}}^2 \tag{4.49}$$

When the resistance value is equal to R when the voltage is greater than a threshold voltage noted V_{T} the current does not flow instantaneously in the circuit and the results are slightly modified. Although the calculations are longer, it is possible to show that the energy dissipated during the charging process can be expressed as:

$$E_{\mathrm{R}} = \left(\frac{RC}{T}\right) CV_{\mathrm{DD}}^2 + \frac{1}{2} CV_{\mathrm{T}}^2 \tag{4.50}$$

This calculation shows the first effect of the threshold voltage on silicon technology.

4.3.5 Optimizing a Capacitor Charge: General Case

The previous section showed that in a simple case where resistance and capacitance are constant, a linear ramp applied during a time T followed by a constant value V_{DD} leads to a significant reduction in energy loss in heat form. There is nothing to prove that this form is optimum and likewise, there is nothing to prove that this technique remains valid when the resistance and capacitance values are no longer constant. The aim of this section is to respond to these questions. The analysis that follows is based on an article published in July 2000 by Paul [3].

$$E(t) = R(t)i(t) + V(t) \tag{4.51}$$

$$i(t) = \frac{\mathrm{d}q}{\mathrm{d}t} = C\frac{\mathrm{d}V}{\mathrm{d}t} + V\frac{\mathrm{d}C}{\mathrm{d}t} \tag{4.52}$$

Thus, the problem to solve is how to determine the input function (voltage supplied by the generator) that minimizes thermal dissipation in the resistor. To do this, we will employ the Pontryagin method, which is presented in works dealing with control-command systems.

In the first instance, we will assume that the resistance is constant in the given time. The first stage is to write the state equation for the system. The charge stored in the capacitor is chosen as the state variable. The voltage E is the command variable, and it is unknown.

$$\frac{\mathrm{d}q}{\mathrm{d}t} = \frac{E}{R} - \frac{q}{RC} \tag{4.53}$$

The second step consists in writing Hamilton's function:

$$H(\lambda, q, E, t) = \frac{1}{R}\left(E - \frac{q}{C}\right)^2 + \lambda(t)\left(\frac{E}{R} - \frac{q}{RC}\right) \tag{4.54}$$

It corresponds to minimizing the power dissipated in the resistor during a charging operation of duration T. The function $\lambda(t)$ is an additional unknown variable. The initial and final conditions allow one to completely solve the problem. The capacitor voltage is zero. The initial voltage at the terminal is zero. After a given time T, it is equal to a given value V_{DD}. Therefore, it can be proven that the optimal input voltage is the solution to the following three valid equations between 0 and T:

$$\frac{\partial H}{\partial E} = 0 \tag{4.55}$$

$$\frac{\mathrm{d}\lambda}{\mathrm{d}t} = -\frac{\partial H}{\partial q} \tag{4.56}$$

$$\frac{\partial H}{\partial \lambda} = \frac{\mathrm{d}q}{\mathrm{d}t} \tag{4.57}$$

Equation 4.55 allows one to write:

$$\frac{\partial H}{\partial E} = \frac{2}{R}\left(E - \frac{q}{C}\right) + \lambda(t)\left(\frac{1}{RC}\right) = 0 \tag{4.58}$$

Equation 4.56 can be written as:

$$\frac{\mathrm{d}\lambda}{\mathrm{d}t} = -\frac{\partial H}{\partial q} = \frac{2}{R}\left(E - \frac{q}{C}\right) + \lambda(t)\left(\frac{1}{RC}\right) \tag{4.59}$$

This latter expression is zero according to Eq. 4.58. This result enables the following calculations to be considerably simplified. The parameter λ is constant in this case.

$$\left(E - \frac{q}{C}\right) = -\frac{1}{2C}\lambda_0 \tag{4.60}$$

$$E - V = RC\frac{\mathrm{d}V}{\mathrm{d}t} \tag{4.61}$$

It can be deduced that:

$$V(t) = \frac{-\lambda_0}{2RC^2}t + A \quad \text{for } t \geq 0 \tag{4.62}$$

As the initial voltage is 0 and the voltage is V_{DD} at the time T, one can deduce the following:

$$-\lambda_0 = -\frac{2RC^2 V_{\mathrm{DD}}}{T} \tag{4.63}$$

A remarkable property of the solution is established: the potential difference at the resistor terminal is constant, meaning that the operation is performed at a constant current.

This result is only identical to that found in the previous section when the ratio between T and the time constant *RC* is far greater than 1. This means that applying a linear ramp is not the optimal solution but a quasi-optimal one. The optimal solution is slightly different to the linear ramp and is written with a positive value of t:

$$E(t) = \frac{V_{\mathrm{DD}}t}{T} + \frac{RCV_{\mathrm{DD}}}{T} \tag{4.64}$$

It is also interesting to compare the dissipated energies in both cases: optimal and quasi optimal. In the optimal solution, the

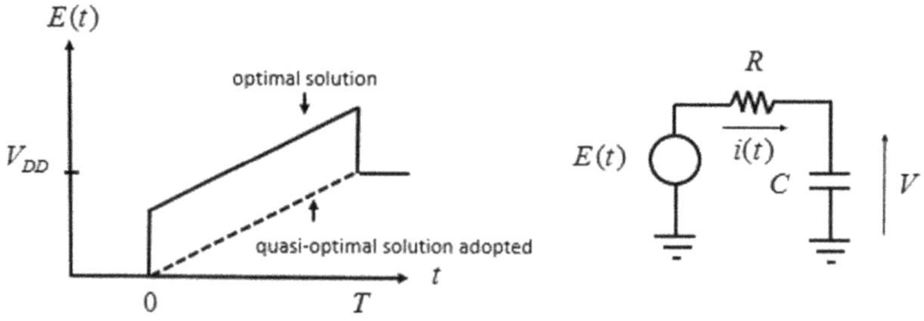

Figure 4.6 Optimal and quasi-optimal solutions in a capacitor charge.

dissipated energy is $E_R = RC/T.CV_{DD}^2$. In the quasi-optimal solution (used in practice), it is slightly different. The two values can become confused when the *T/RC* ratio is high.

After a time T, the input signal can revert to V_{DD}, which is the optimal solution. It can also retain the value it reached in T, which is another quasi-optimal solution that introduces a minimal dissipation increase when T is far greater than *RC*. Both solutions are presented in Fig. 4.6. In practice, the generators used in these low-consumption solutions deliver voltages that are close enough to those in the ramp shown in Fig. 4.6 but are nonetheless different. Very often, this form is approximated by a sinusoid shifted by its average value so as to remain positive for all the time values. All these solutions are however close enough to the optimal solution when the *RC/T* factor is far less than 1 and lead to a real reduction in heat dissipation.

The ratio η between the stored energy and the sum (stored energy and dissipated energy) is shown in Table 4.2 according to the *RC/T* ratio. The considered duration is T. This ratio is the charge energy efficiency.

Table 4.2 Energy efficiency of the optimal solution

RC/T	**0.1**	**0.2**	**0.3**	**0.4**	**0.5**	**1**
η optimal	0.83	0.71	0.62	0.55	0.5	0.33
η quasi optimal	0.82	0.69	0.60	0.52	0.46	0.28

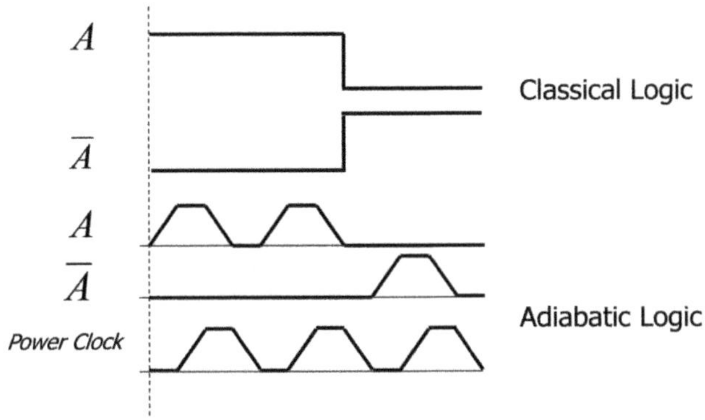

Figure 4.7 Coding the logical states in CMOS and in adiabatic logic.

4.4 Adiabatic Principle and Reversibility: A Circuit Approach

The relations between reversibility and adiabatic principle have been studied from the angle of thermodynamics and information theory [4]. They are complex and still spark many debates. A more original approach may be made by starting from the architectural constraints imposed by the research on minimal dissipation.

4.4.1 Implementation of the Adiabatic Principle

The approach is as follows. Starting with the capacitance's optimal charge and discharge principle presented in the previous section, it consists of creating the circuit architectures, meaning to cascade the logic gates. The previous sections showed that it is necessary to perform capacitor charge with a constant current or similarly, applying a linear ramp so as to produce something close to the energetic optimum. A simple solution to satisfy this requirement is to replace the constant supplies with variable supplies. The signals generated are in trapeze form. This technique is the only one used today and is shown Fig. 4.7. The consequence of the use of this solution is fundamental: logical states are not coded with electrical levels (the case of CMOS) but with the presence or the absence of electrical pulses.

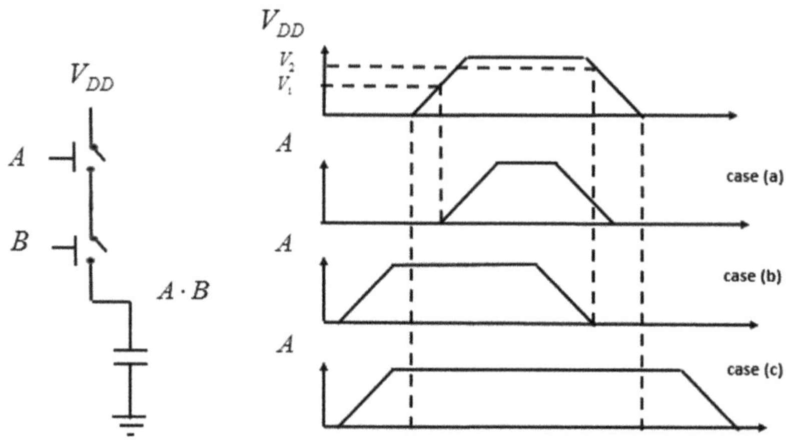

Figure 4.8 A logic gate with an adiabatic command.

Let us assume in the AND gate depicted Fig. 4.8 that the input B is in the 1 state. The capacitor shown represents the effects of interconnect and the other connected gates. Applying a linear variation of the supply voltage is equivalent to provide a constant current. In fact, assuming the switches' resistances are very weak, performing the logical function goes back to electrically linking the supply output to the charge capacitor. The current is then:

$$i = C\frac{\mathrm{d}V_{\mathrm{DD}}}{\mathrm{d}t} \tag{4.65}$$

It is constant because the voltage variation is linear in the time and because the capacitance is assumed to be constant. The adiabatic principle is implemented. Certain precise requirements must be fulfilled with regard to synchronizing the signals.

Let us take the first case (a). The supply signal starts to increase while the state A is at 0. The supply is not linked to the capacitor. When the signal A moves to 1, voltage V_1 is abruptly applied to the capacitor terminals. In this example, we assume that the very weak voltage is enough to ensure that switching occurs. It is easy to show that this operation induces a nonadiabatic dissipation $1/2CV_1^2$.

In the example (b), the capacitor is charged adiabatically because the input signal A does not change during this period. Discharging the capacitor, on the other hand, is not done adiabatically because the voltage applied to the capacitor terminals remains in the constant value V_2 when the input signal reverts to zero. It can perhaps be connected to the supply zero value in the following event, which brings about a dissipation $1/2C V_2^2$.

There is only one case (c), which ensures that the capacitor is charged and discharged adiabatically. The duration of the input signal is therefore higher than that of the supply signal. This mode is called "Benett clocking," paying homage to Charles Benett, IBM researcher. In this example, we assumed that the command signal's rise and fall times were identical to those of the supply voltage. This is a logical hypothesis if we imagine a system with cascaded logic gates. As one gate input is the output of the previous gate, it is necessary that the rise and fall times are conserved in the chain. Figure 4.9 shows an adiabatic architecture based on Benett clocking.

Each of the supply voltage rise and fall times is carried out only when the inputs are stable, either in the on state or in the off state. This structure does, however, have some significant drawbacks. Each gate corresponds to a different form of supply voltage, which imposes a very complex logic gate supply system. The frequency of these input state changes is very weak by principle and decreases with the number of gates in a series.

This architecture indirectly introduces a link between the adiabatic control and reversibility. In fact, if we significantly increase the duration of input states, it is possible to consider that no gate is put back to zero and no information is lost. This is the fundamental property of a reversible gate [5]. The ramping time of the supply ramps can be chosen to be far greater than the electric time constants, which allows the heat dissipation to tend toward zero. This architecture can be considered to be asymptomatically reversible with zero dissipation.

A more realistic solution for implementing the adiabatic principle is to carry out a pipeline [6]. In this instance, we will make an effort to make the signals have the same shape (rise and fall time) as is the case in classical dynamic logic. The main idea is to systematically delay the supply signal in comparison to the input signals. In general, the value of the ramping time is chosen. This time delay forbids any input variation during the capacitor charge and

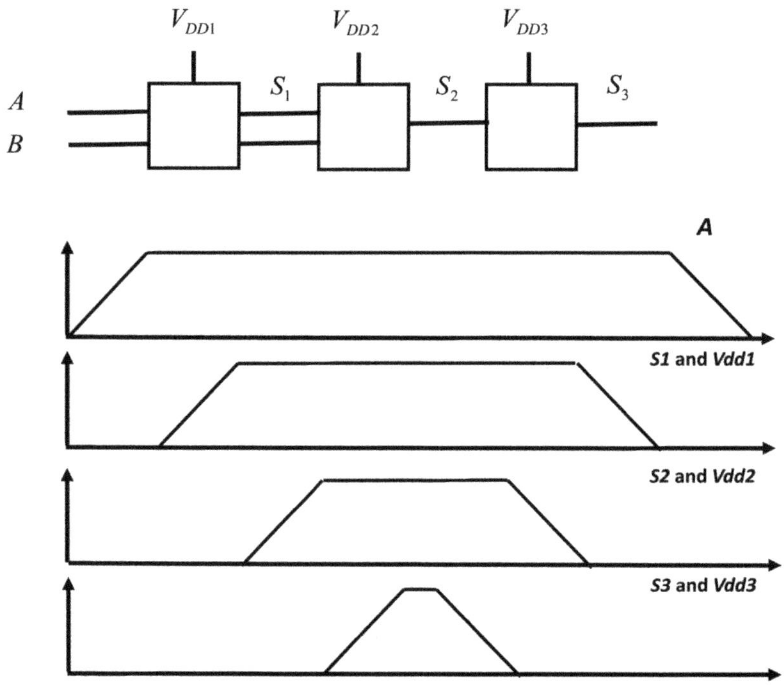

Figure 4.9 The Benett clocking principle.

therefore also ensures that the charging process will be adiabatic. Figure 4.10 describes a stage of pipeline logic. This schematic is incomplete, as the following analysis will show.

We assume that the input B is in the on state and that the switches have no threshold, meaning that a very small positive voltage applied is enough to make them change state. The input voltage increases and reaches the value V_{DD}. In the second phase, the input voltage stays in the on state and the supply voltage increases. The output voltage that is directly connected follows with the same slope. When the input voltages go back to zero in the third phase, the output is isolated from the supply. The other gates' inputs have high impedance values and the output node is floating, retaining its value V_{DD}. In the fourth phase, the supply voltage goes back to zero, but the two input switches are open and the floating node retains its value. When the input voltages go back to the on state, the floating

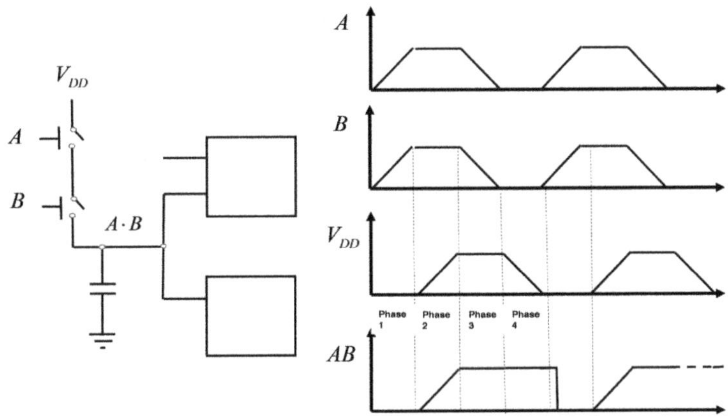

Figure 4.10 An incomplete pipeline.

node is again connected to the supply. As it is at zero potential, there is a sudden nonadiabatic discharge and a dissipation $1/2C\,V_{\mathrm{DD}}^2$.

In summary, thermal dissipation is significant and the signal shape is modified. The desired properties are not verified. Tor them to be, the signal must return in a linear manner to zero at the end of phase 3. Another conduction path must be introduced, as shown Fig. 4.11. The device doing that must have a certain intelligence. In fact, when the output voltage is at 1, it is necessary to connect the output to the supply voltage. If, on the other hand, the output is at 0, it is crucial not to establish a connection because the output has to stay at zero volts. The question that remains to be solved is how to define the command signal for this intelligent resetting to zero. Two solutions have been proposed: the quasi-adiabatic solution and the adiabatic solution.

4.4.2 Quasi-Adiabatic Solution

This solution is the one that is most frequently used. It simply consists of using the output signal as a command signal. In fact, when the output is in the on state, it is necessary to establish a second conduction path so that the output returns to zero following the decrease in the supply voltage. The signal can also come from a complementary logic part.

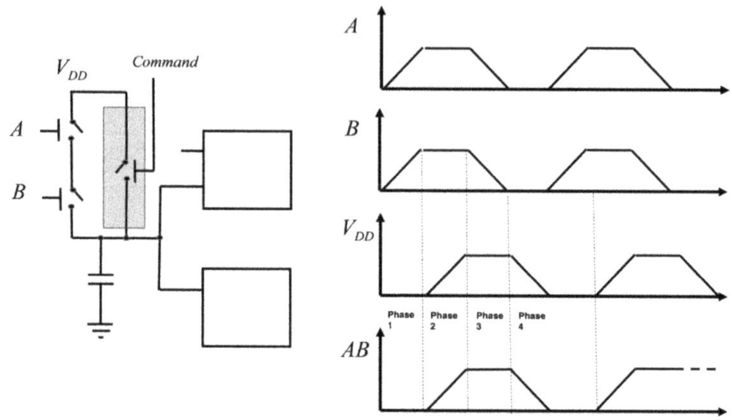

Figure 4.11 An operational adiabatic pipeline.

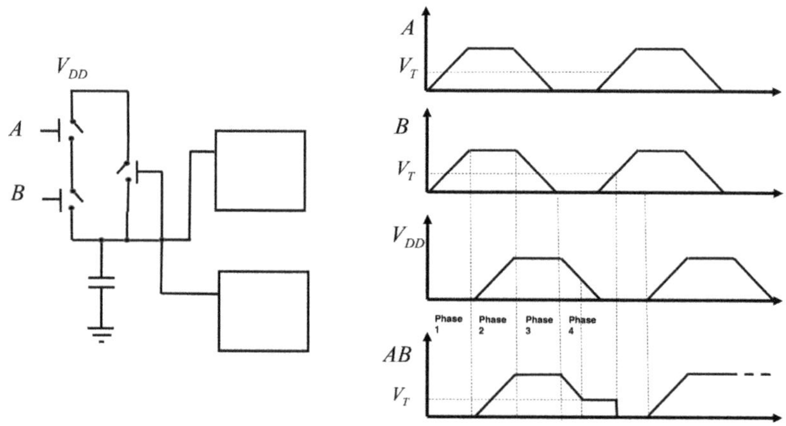

Figure 4.12 A quasi-adiabatic gate.

The only drawback of this technique comes from the existence of thresholds. In practice, the switches are not perfect as they only start to be conductors from a given value of *VT*, rather than from zero. As the output voltage is greater than the threshold voltage, the reset path to zero is possible, but when the output voltage is weaker than the threshold voltage, this reset path to zero is open and the output voltage remains at the threshold voltage. This function is illustrated Fig. 4.12. When the input signals move back to the on state, a

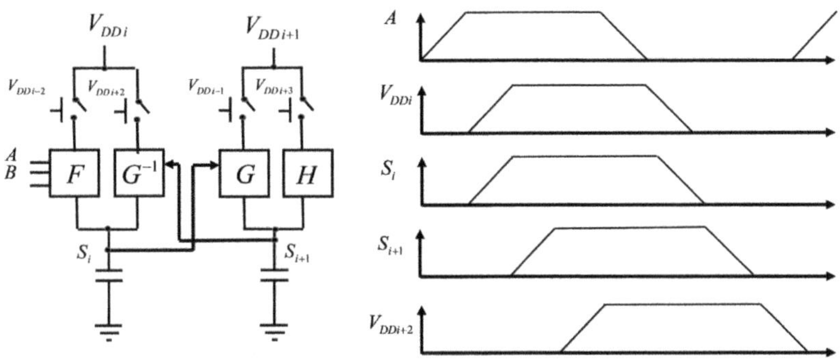

Figure 4.13 The reversible pipeline.

conduction path is opened from the output threshold and connected to the zero potential. There is therefore a nonadiabatic dissipation equal to $1/2CV_T^2$, which explains the term "quasi-adiabatic." This dissipation is weaker than that of the classic logic $1/2CV_{DD}^2$ but represents a serious limitation, especially when one uses advanced CMOS technology.

4.4.3 Full Adiabatic Solution and Reversible Pipeline

This solution is closer to the theoretical optimum but causes great complexity when it is being implemented, as it makes use of reversible logic. The principle is, when carrying out the resetting-to-zero process, to use a delayed output signal that is an exact copy of the output signal. Figure 4.13 illustrates the principle of this architecture.

This diagram is made simple because a single output is represented by a gate. In reality, the gates used in this structure are reversible and there are as many outputs as there are inputs. The necessary supply signals are each shifted from the other ones and noted V_{DDi+j}; i is the index of the gate in question linked to its supply, and j is the index measuring the time delay. The architecture is formed by putting the gates into a series, using the functions *F*, *G*, *H*, etc. The inverse functions are also performed, which means that the gates must be reversible. The output signal propagation through the *G* gate G^{-1} allows the logical signal to be delayed so as to create a conduction path at the correct moment. The switches figuring in the diagram all equally contribute to the set functioning. It is said to be an eight-phase system as it is only after eight ramping times that the next logical operation can be performed. This architecture is used infrequently due to the complexity of performing it.

The reversible gates allow an adiabatic pipeline to be produced without nonadiabatic dissipation. There is nothing to prove that this is the only way to do it, but it is the only one known today. Note however that the reversible functions are only used in this structure to perform a simple delay. In contrast, it is possible to prove (in a rather absurd way) that following the Landauer principle [7], reversible logic must be used to achieve logic that has dissipation tending toward zero when the operating frequency also tends toward zero. Given the fact that it was possible to achieve adiabatic logic with classic gates, it would also be possible to have dissipation by a gate as small as possible, particularly weaker than the Landauer limit, by choosing a very high ramping time value. Classical logic, in fact, loses information bits. This result would be incompatible with Landauer's principle as it affects each bit with an information loss with a minimal value of $k_{\mathrm{B}}T\ln 2$.

These different considerations feed debates on whether or not it is necessary to move from classical logic to reversible logic. The problem is still very theoretical because heat creation by processed bit $1/2CV_{\mathrm{DD}}^2$ is approximately 10^{-15} joules while $k_{\mathrm{B}}T$ is equal to $4 \cdot 10^{-21}$ joules at room temperature.

4.5 The CMOS Quasi-Adiabatic Solutions

4.5.1 ECRL Architecture

Let us choose the CMOS inverter shown Fig. 4.14 and replace the constant supply with a power clock supply [8]. The input is established in phase 1. The NMOS is then a conductor, and the output is at zero. When the supply voltage rises in phase 2, the PMOS source-gate voltage remains less than the threshold. In phase 3, the voltage A diminishes and when the source-gate voltage becomes greater than the threshold, the PMOS becomes conductive. The

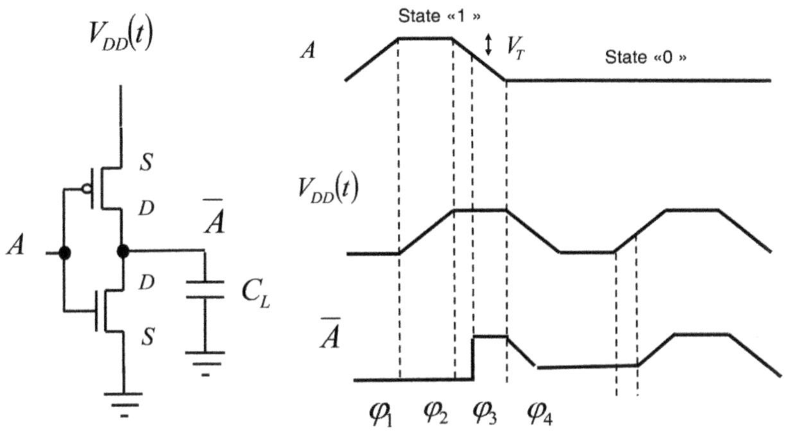

Figure 4.14 CMOS patterns' incompatibility with the adiabatic principle.

output voltage then takes on the value of the supply voltage. It then sinks back down toward zero in phase 4 following the decrease in the supply voltage. However, it cannot actually reach zero, as when the supply voltage is less than the threshold, the PMOS is no longer conductive. It blocks this value, while waiting for the following event.

In summary, the output is very different to the signal representing the complementary input function A. The inverter function cannot be transposed into adiabatic. Different schemes need to be performed.

As shown in Fig. 4.15, efficient charge recovery logic (ECRL)-type logic is a first solution to this. For the sake of simplicity, the rationale is based on the buffer/inverter gate, but it can be transposed to any gate. In the remainder of this chapter, the supply voltage will be written as $\phi(t)$ rather than $V_{DD}(t)$. One must keep the following elements in mind. An NMOS transistor is conductive when the voltage between the gate and the source is greater than the threshold voltage. A PMOS transistor is conductive when the voltage between the source and the gate is greater than the absolute value of the threshold voltage (negative for a PMOS). The current provided by the supply circulates from the source toward the drain in a PMOS transistor and from the drain toward the source in an NMOS one. As a first step, let us use the example of an inverter/buffer. Note that this gate architecture processes not only the logic signals but

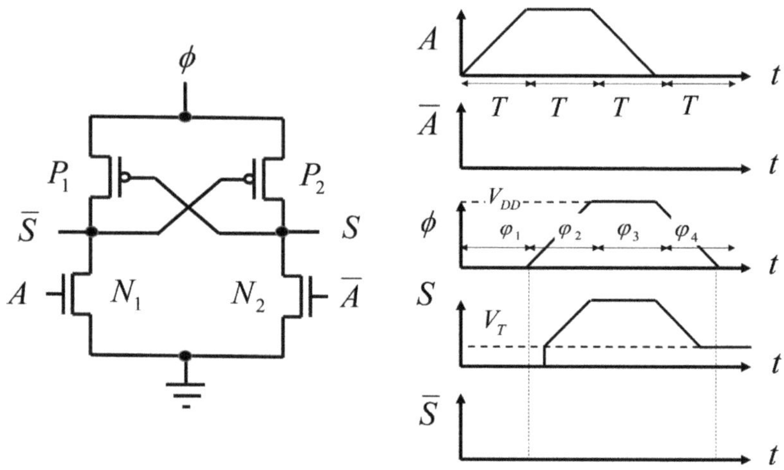

Figure 4.15 ECRL buffer/inverter.

also the complementary signals. For this reason, it is called dual-rail logic, because a state is coded by two electrical signals. Remember that the logic states are not coded by high and low voltage, as in CMOS, but instead by the presence or absence of the trapezoidal signal.

Phase 1: Input establishment phase

Before the signal A is applied, all the voltages are at zero. The signal A is then applied and turns the NMOS N_1 into the conductive state. The NMOS N_2 remains nonconductive because a zero voltage is applied to the gate defined by the complementary input state. At the beginning of the cycle, the PMOS gates are at zero, which enables P_2 to be the conductor when the voltage ϕ becomes greater than the threshold voltage V_{TP}.

Phase 2: Evaluation phase

The signal A retains its value. The supply voltage rises. The gate output capacitance at the output S is then charged through the transistor P_2. This transistor gate is maintained at zero potential by the NMOS N_1 conductor because the grid voltage at V_{DD} is greater than the threshold. Note that the output voltage jump with a value equal to the threshold voltage V_{TP} is the cause of nonadiabatic dissipation equal to $1/2C_L V_{TP}^2$.

Phase 3: Hold phase

The signal A returns to the zero value, and the power clock voltage remains at V_{DD}. The transistor P_2 is still active because the grid remains at zero. The output then follows the supply voltage and remains at the value V_{DD}. The transistor P_1 is blocked because the voltage between the source and the grid is higher than the threshold. The complementary output is at zero potential.

Phase 4: Resetting to zero

The supply voltage begins decreasing linearly toward zero. The transistor P_2 is conductive as long as the supply voltage is greater than the threshold. Then it is blocked although the output node is electrically isolated, because the transistor N_2 is itself also blocked, with the grid at zero potential. The output voltage is then also at the potential V_{TP}. The output node remains isolated and keeps the value V_{TP}. This residue would not be a drawback if the output returned to the on state in the following event as the output signal moves from V_T to V_{DD}, instead of moving from 0 to V_{DD}. On the other hand, if the output is in the off state in the following event, the output quickly discharges to the ground through the transistor N_2 because in this case the input is at zero. The consequence is nonadiabatic dissipation equal to $1/2C_L V_{TP}^2$. In summary, adiabatic functioning is ensured but nonadiabatic residue appears. This residue is linked to the presence of a threshold voltage and can be far greater than the adiabatic dissipation itself. This effect represents a serious limitation to CMOS technology.

This architecture is simple, and it uses a limited number of transistors. It can be implemented by performing any Boolean function, as shown Fig. 4.16.

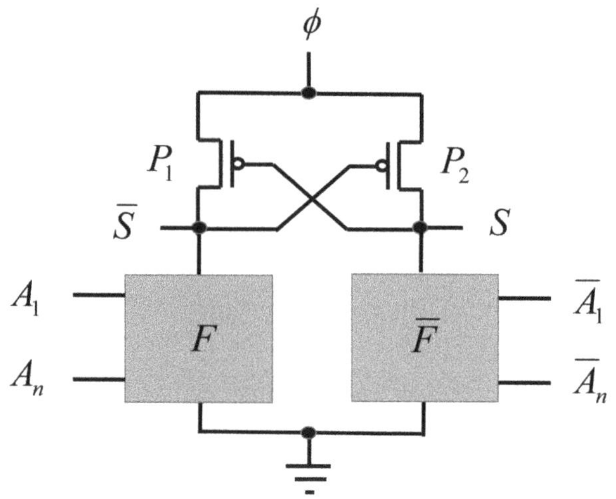

Figure 4.16 A generic ECRL gate.

4.5.2 PFAL Architecture

Positive feedback adiabatic logic (PFAL) is an improvement on ECRL, and its style is the one that is most commonly retained [8]. The improvement comes from replacing the two transistors P_1 and P_2 by the transmission gates, meaning PMOS sets in parallel with NMOS transistors commanded by complementary signals. The transmission resistance, meaning the ratio $\Delta V_{\text{DS}}/\Delta I_{\text{DS}}$, in the triode-mode transistor model is more or less constant, depending on the current, and has a fairly weak value in comparison to those of NMOS and PMOS taken separately. The value reached by this resistance is written as follows, assuming that the threshold voltages are equal.

$$R = \frac{L}{W}\frac{1}{\mu C'_{\text{ox}}(V_{\text{DD}} - 2V_{\text{T}})} \tag{4.66}$$

Parameters L, W, μ, C'_{ox}, and V_{T} are, respectively, the transistor channel length, the transistor channel width, the mobility, the oxide capacitance per surface unit, and the threshold voltage.

The PFAL inverter architecture is shown Fig. 4.17.

This functions in a way very similar to how the previous circuit functions.

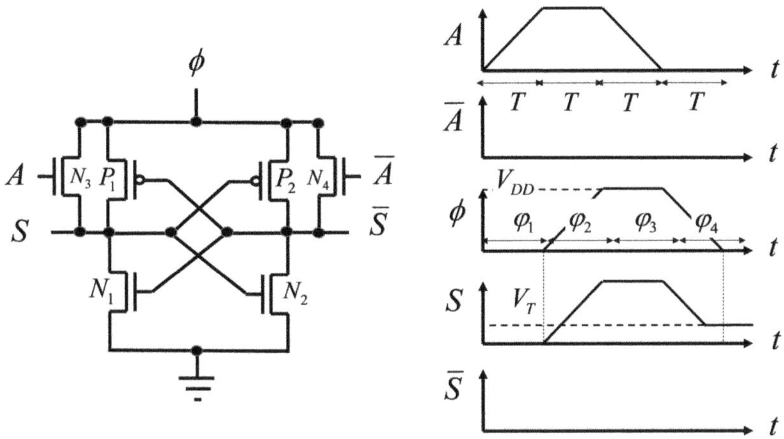

Figure 4.17 The PFAL inverter.

Phase 1: Input establishment phase

The input A is established. The complementary input is then at zero. The signal A is placed in the on state, which enables the N_3 transistor to become conductive.

Phase 2: Evaluation phase

Contrary to the previous case, the N_3 transistor is conductive from the beginning of the ramp rise as the gate-source voltage is initially V_{DD}, which is greater than the threshold. The output follows the supply without the threshold effect. The P_2, N_4, and N_2 transistors are initially blocked, and the voltage corresponding to $\bar{S}$ is at absolute zero. The P_1 transistor also becomes conductive when the supply voltage is greater than the threshold voltage.

Phase 3: Hold phase

The supply voltage is at the value of V_{DD} throughout the duration of this phase. The same goes for the signal of the output S. The input voltage A diminishes, but the P_1 transistor remains conductive throughout the duration of this phase. The N_2 transistor is conductive when the supply is greater than the threshold.

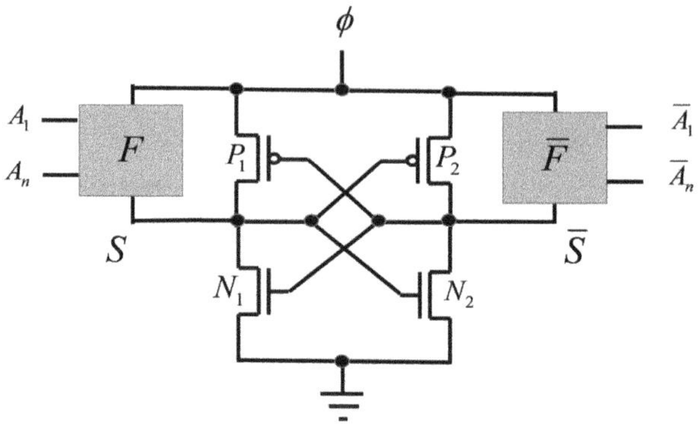

Figure 4.18 A general PFAL gate.

Phase 4: Resetting to zero

The supply voltage decreases toward zero, and the output voltage follows. The P_1 transistor remains so conductive that the supply is greater than the threshold. Then, the output node is electrically isolated like in the previous circuit, even though the output retains its value V_{TP}. Depending on the input logic state for the following event, this value will be maintained or connected at electric zero, which in this case engenders the nonadiabatic dissipation of $1/2C_{\mathrm{L}}V_{\mathrm{TP}}^2$.

So, both of these logic gate families bring about nonadiabatic dissipation linked to the threshold voltages, in addition to the adiabatic consumption of $RC_{\mathrm{L}}/T \cdot C_{\mathrm{L}}V_{\mathrm{DD}}^2$. This logic type will be called quasi adiabatic because it cannot reach optimal performance. PFAL gates are shown in a general way in the diagram of Fig. 4.18.

4.5.3 Calculation of Dissipation

Dissipation is calculated using PFAL (Fig. 4.19), but it can easily be transposed to other logic families.

Phase 1: Input establishment phase

All currents, including consumption, are at zero. Dissipation due to the input capacitance charge is counted in the command logic gate.

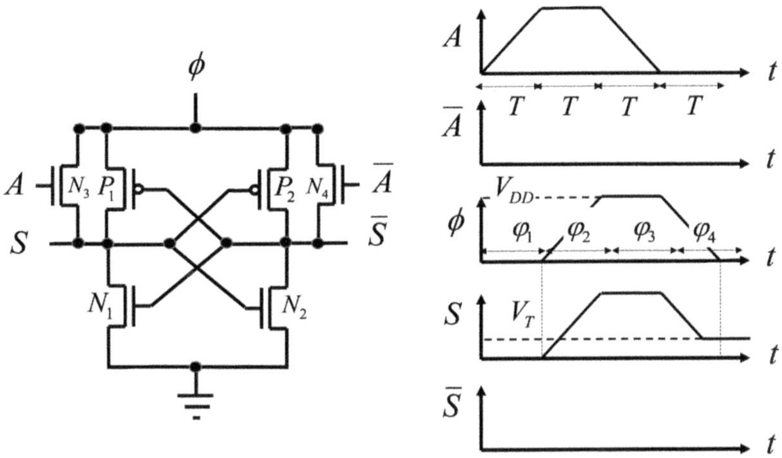

Figure 4.19 PFAL inverter-buffer.

Phase 2: Evaluation phase

The transistors that are performing the F are active. The transistor P_1 is also active when the "power clock" signal rises because the gate is at zero potential. The N_1 transistor is blocked, but there is a permanent current I_{N1} circulating below the threshold. The adiabatic dissipated energy in both the transistor channel and the interconnect resistance, as well as the series contact, is written:

$$E_{P1} = \frac{LC_{\mathrm{L}}^2 V_{\mathrm{DD}}^2}{W\mu C'_{\mathrm{ox}}\left(V_{\mathrm{DD}} - 2V_{\mathrm{T}}\right)T} \tag{4.67}$$

Into this formula is integrated the NMOS channel resistance.

The parameters L, W, C_{L}, C'_{ox}, T, and μ are, respectively, the transistor channel length, the transistor width, the gate load capacitance, the capacitance per surface unit of the gate oxide, the ramp rise time, and the carrier mobility. This is an estimate as according to the function F, many transistors can be either in series or in parallel. The subthreshold current through N_1 is written as:

$$I_{N1} = \mu C'_{\mathrm{ox}}\frac{W}{L}\phi_{\mathrm{t}}^2 e^{\frac{-V_{\mathrm{T}}+\eta V_{\mathrm{S}}}{n_0\phi_{\mathrm{t}}}} e^{1.8}\left(1 - e^{-\frac{V_{\mathrm{S}}}{\phi_{\mathrm{t}}}}\right) e^{\frac{V_{\mathrm{GSN1}}}{n_0\phi_{\mathrm{t}}}} \tag{4.68}$$

This expression uses the subthreshold current from the BSIM3 model of the transistor. The output voltage is V_{S}, which is equal to $V_{\mathrm{DD}} \cdot t/T$ and the voltage V_{GSN1} is at absolute zero in this phase. The transistor parameters V_T, η, n_0, and ϕ_t are the threshold voltage, the DIBL coefficient, the subthreshold slope, and the thermal voltage, respectively.

The energy dissipated in the N_1 transistor is:

$$E_{N1} = \int_0^T \mu C'_{ox} \frac{W}{L} \phi_t^2 e^{1.8} e^{\frac{-V_T + \eta V_{DD} \cdot t/T}{n_0 \phi_t}} \left(1 - e^{-\frac{V_{DD} \cdot t/T}{\phi_t}}\right) V_{DD} \cdot t/T \, dt \tag{4.69}$$

There is a simple calculation that ignores the DIBL coefficient and assumes that V_{DD} is a lot greater than ϕ_t and leads to the following conclusion:

$$E_{N1} = \mu C'_{ox} \frac{W}{L} \phi_t^2 e^{\frac{-V_T}{n_0 \phi_t}} e^{1.8} \frac{V_{DD} T}{2} \tag{4.70}$$

The dissipation due to the current passing through the P_2 transistor also needs to be counted. As the source-gate voltage of P_2 is at zero, the energy dissipated in P_2 is:

$$E_{P2} = \int_0^T \mu C'_{ox} \frac{W}{L} \phi_t^2 e^{1.8} e^{\frac{-V_T}{n_0 \phi_t}} \left(1 - e^{-\frac{V_{DD} \cdot t/T}{\phi_t}}\right) V_{DD} \cdot t/T \, dt \tag{4.71}$$

This integral is not very different from:

$$\mu C'_{ox} \frac{W}{L} \phi_t^2 e^{\frac{-V_T}{n_0 \phi_t}} e^{1.8} \frac{V_{DD} T}{2} \tag{4.72}$$

The total energy dissipated during phase 2 is:

$$E_2 = \mu C'_{ox} \frac{W}{L} \phi_t^2 e^{\frac{-V_T}{n_0 \phi_t}} e^{1.8} \cdot V_{DD} T + \frac{L C_L^2 V_{DD}^2}{W \mu C'_{ox} \left(V_{DD} - 2V_T\right) T} \tag{4.73}$$

Phase 3: Hold phase

As the output voltage is constant, the calculation of the energy dissipated by the subthreshold currents circulating in the N_1 and P_2 transistors, is:

$$E_3 = \int_T^{2T} \mu C'_{ox} \frac{W}{L} \phi_t^2 e^{1.8} e^{\frac{-V_T}{n_0 \phi_t}} \left(1 - e^{-\frac{V_{DD}}{\phi_t}}\right) V_{DD} dt + \int_T^{2T} \mu C'_{ox} \frac{W}{L} \phi_t^2 e^{1.8} e^{\frac{-V_T}{n_0 \phi_t}} \left(1 - e^{-\frac{V_{DD}}{\phi_t}}\right) V_{DD} dt \tag{4.74}$$

$$E_3 = 2C'_{ox} \frac{W}{L} \phi_t^2 e^{\frac{-V_T}{n_0 \phi_t}} e^{1.8} V_{DD} T \tag{4.75}$$

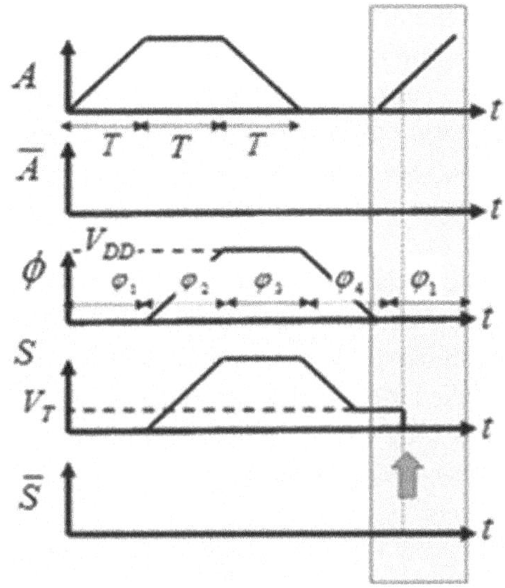

Figure 4.20 The following event in PFAL.

Phase 4: Resetting to zero

The power clock voltage returns toward zero. The P_1 gate linked to the complementary output at zero potential ensures that this transistor is conductive so that the value of the voltage ϕ is greater than the threshold voltage value V_T. Then, the transistor is no longer conductive and the output is electrically isolated. In reality, although the subthreshold currents circulating in the N_1 transistor are slowly discharging the output capacitance, the voltage variation is negligible during phase 3. The dissipation calculation is very similar to that in phase 1. The only difference is that the voltage excursion is $V_{DD} - V_T$ instead of V_{DD}. The dissipated energy, therefore, is:

$$E_4 = \mu C'_{ox} \frac{W}{L} \phi_t^2 e^{\frac{-V_T}{n_0 \phi_t}} e^{1.8} \cdot V_{DD} T + \frac{L C_L^2 (V_{DD} - V_T)^2}{W \mu C'_{ox} (V_{DD} - 2V_T) T} \quad (4.76)$$

The following event (Fig. 4.20)

The supply voltage has returned to zero, even though all of the transistors are blocked. The transistors performing the functions F

and $\bar{F}$ either become active or do not, depending on the following event logic state. If the function F is performed, a conduction path exists between the supply, which in this phase is at zero, and the output from the moment that the implicated transistor threshold voltages have been exceeded. The output voltage of the value V_T is then connected to the zero potential, which generates a dissipation of $1/2C_L V_{TP}^2$.

In this case, a nonadiabatic dissipation term needs to be included. Contrastingly, the energy corresponding to the function F is counted in the energy consumed by the command gates, which are not shown in the diagram. If the function F is not performed, the output voltage is maintained and only the static consumption is to be added. In summary, the gate average consumption is:

$$E = 4\mu C'_{ox}\frac{W}{L}\phi_t^2 e^{\frac{-V_T}{n_0\phi_t}} e^{1.8} \cdot V_{DD}T + \frac{2LC_L^2\left(V_{DD}-V_T\right)^2}{W\mu C'_{ox}\left(V_{DD}-2V_T\right)T} + \frac{1}{4}C_L V_T^2 \tag{4.77}$$

In the last term, the factor 1/4 accounts for the fact that the probability that the function F will be performed in the following event is assumed to be 0.5. Note that, by definition of the adiabatic protocol, the gate activity factor is always 1. The power clock signal rises and falls throughout the cycle duration $4T$ for all of the input values. Some authors present this property to be like that of a "micropipelined" structure, with each gate also being a register.

Calculating the energetic optimum is particularly simple in adiabatic logic because the dissipated energy function is made up of two terms: one that varies proportionally with the ramp rise time T and the other one varying proportionally to the inverse of the rise time. The minimum is reached when these two terms are equal, which corresponds to the optimal value of T:

$$T_{opt}^2 = \frac{L^2C_L^2\left(V_{DD}-V_T\right)^2}{2W^2\mu^2 C'^{\,2}_{ox}\left(V_{DD}-2V_T\right)\phi_t^2 e^{\frac{-V_T}{n_0\phi_t}} e^{1.8} \cdot V_{DD}} \tag{4.78}$$

$$T_{\mathrm{opt}} = \frac{LC_{\mathrm{L}}}{\sqrt{2}W\mu C'_{\mathrm{ox}}\phi_{\mathrm{t}}e^{\frac{-V_{\mathrm{T}}}{2n_0\phi_{\mathrm{t}}}}e^{0.9}}\sqrt{\frac{(V_{\mathrm{DD}}-V_{\mathrm{T}})^2}{(V_{\mathrm{DD}}-2V_{\mathrm{T}})\,V_{\mathrm{DD}}}} \tag{4.79}$$

From this, one can deduce the corresponding energy value:

$$E_{\mathrm{opt}} = \frac{2LC_{\mathrm{L}}}{\sqrt{2}W\mu C'_{\mathrm{ox}}\phi_{\mathrm{t}}e^{\frac{-V_{\mathrm{T}}}{2n\phi_{\mathrm{t}}}}e^{0.9}} \times\sqrt{\frac{(V_{\mathrm{DD}}-V_{\mathrm{T}})^2}{(V_{\mathrm{DD}}-2V_{\mathrm{T}})\,V_{\mathrm{DD}}}}4\mu C'_{\mathrm{ox}}\frac{W}{L}\phi_{\mathrm{t}}^2e^{\frac{-V_{\mathrm{T}}}{n\phi_{\mathrm{t}}}}e^{1.8}\cdot V_{\mathrm{DD}}+\frac{1}{4}C_{\mathrm{L}}V_{\mathrm{T}}^2 \tag{4.80}$$

$$E_{\mathrm{opt}} = \frac{8C_{\mathrm{L}}}{\sqrt{2}}\sqrt{\frac{(V_{\mathrm{DD}}-V_{\mathrm{T}})^2}{(V_{\mathrm{DD}}-2V_{\mathrm{T}})\,V_{\mathrm{DD}}}}\phi_{\mathrm{t}}e^{\frac{-V_{\mathrm{T}}}{2n_0\phi_{\mathrm{t}}}}e^{0.9}\cdot V_{\mathrm{DD}}+\frac{1}{4}C_{\mathrm{L}}V_{\mathrm{T}}^2 \tag{4.81}$$

It is now possible to optimize it with respect to V_{DD}, which leads to:

$$V_{\mathrm{DDopt}} = V_{\mathrm{T}}\frac{3+\sqrt{5}}{2} \tag{4.82}$$

This value is slightly different to the value $3V_{\mathrm{T}}$ in the classic case. The DIBL also can modify the result, but in this case there is no simple analytical solution. In the end, the minimal dissipated energy is written as:

$$\begin{aligned} E_{\mathrm{opt}} &= \frac{8C_{\mathrm{L}}1.6}{\sqrt{2}}\sqrt{\frac{2.6}{0.62}}\cdot\phi_{\mathrm{t}}e^{\frac{-V_{\mathrm{T}}}{2n_0\phi_{\mathrm{t}}}}e^{0.9}+\frac{1}{4}C_{\mathrm{L}}V_{\mathrm{T}}^2 \\ &= 44C_{\mathrm{L}}\phi_{\mathrm{t}}V_{\mathrm{T}}e^{\frac{-V_{\mathrm{T}}}{2n_0\phi_{\mathrm{t}}}}+\frac{1}{4}C_{\mathrm{L}}V_{\mathrm{T}}^2 \end{aligned} \tag{4.83}$$

Let us assume the following:

$$y = \frac{V_{\mathrm{T}}}{2n_0\phi_{\mathrm{t}}} \tag{4.84}$$

$$\begin{aligned} E_{\mathrm{opt}} &= 88n_0C_{\mathrm{L}}\phi_{\mathrm{t}}^2ye^{-y}+n_0^2\phi_{\mathrm{t}}^2C_{\mathrm{L}}y^2 \\ &= C_{\mathrm{L}}n_0^2\phi_{\mathrm{t}}^2\left(88\frac{1}{n_0}ye^{-y}+y^2\right) \end{aligned} \tag{4.85}$$

The $f(y) = mye^{-y}+y^2$ function is recognized here, as it has already been brought to light in CMOS logic. In this case, the energetic minimum is more obvious than in the nonadiabatic case because the parameter m value is defined (equal to 88) and is independent of

Table 4.3 Energy dissipation for different gate architectures

	Energy dissipation for a logical transition	Optimal threshold	Optimal period	Minimal dissipation
Conventional CMOS	$\frac{\alpha}{2} C_L \phi_t^2 k^2 n_0^2 \left(y^2 + mye^{-y}\right)$	$6n_0\phi_t \frac{1}{1-k\eta}$	Defined by k	$\frac{\alpha}{2} C_L \phi_t^2 k^2 n_0^2 (36 + 0.014.m)$
Adiabatic CMOS	$4\mu C'_{ox} \frac{W}{L} \phi_t^2 e^{-\frac{V_T}{n_0 \phi_t}} e^{1.8} V_{DD} T + \frac{2LC_L^2 (V_{DD}-V_T)^2}{W\mu C'_{ox}(V_{DD}-2V_T)T} + \frac{1}{4} C_L V_T^2$	$4n_0\phi_t \frac{1}{1-k\eta}$	$\frac{4LC_L}{\sqrt{2}W\mu C'_{ox}\phi_t e^{-\frac{V_T}{2n_0\phi_t}}} \sqrt{\frac{(V_{DD}-V_T)^2}{(V_{DD}-2V_T)V_{DD}}}$	$C_L n_0^2 \phi_t^2 \left(88 \frac{1}{n_0} ye^{-y} + y^2\right)$
Adiabatic subthreshold CMOS	$C_L \phi_t^2 \left(\frac{2n_0^2}{n_0+1} + \frac{2V_{DD}}{\phi_t} \sqrt{2n_0 (k+2)}\right)$	Not critical	$\frac{4C_L\phi_t}{I_0} \sqrt{\frac{2n_0}{k+2}}$	$C_L \phi_t^2 \left(\frac{2n_0^2}{n_0+1} + 12\sqrt{2n_0 (k+2)}\right)$

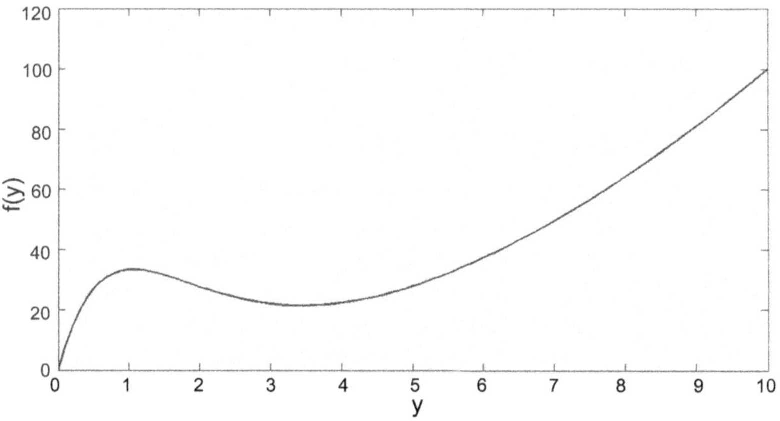

Figure 4.21 Energetic optimum in adiabatic logic.

the technology. The only effect the technology has comes from the coefficient n_0, that is modified between the values 1 and 1.5. The activity rate does not interfere because in adiabatic logic, it is always equal to 1. The optimal value of y is then 4, as shown in Fig. 4.21. The optimal values obtained for $n_0 = 1.3$ are then $V_{\text{Topt}} = 260$ mV, $V_{\text{DDopt}} = 670$ mV, and $E_{\text{opt}} = 0.046 \cdot C_{\text{L}}$ J.

Classical CMOS logic would dissipate energy of $E_{\text{DDopt}} = 0.44 \cdot C_{\text{L}}$ J at the voltage of 0.67 V. Therefore, there is a maximum gain of 10 possible, which is very far from the theoretical gain of the adiabatic method.

To summarize all these considerations about energy dissipation the different expressions are given in Table 4.3. Calculations for the subthreshold regime are detailed in Refs. [2] and [11].

Parameters L, W, μ, C'_{ox}, and V_{T} are, respectively, the transistor channel length, the transistor channel width, the mobility, the oxide capacitance per surface unit, and the threshold voltage.

I_0 is equal to $\mu C'_{\text{ox}} \phi_{\text{t}}^2 e^{1.8} W/L$ and ϕ_{t} is the thermal voltage (kT/q).

C_{L}, α, and T are, respectively, the mean load capacitance, the activity factor of the digital gate, and the ramping time in the adiabatic regime.

The parameter m is defined by $m = 2I_0 T/\alpha C_{\text{L}} k n_0 \phi_{\text{t}}$.

4.6 Interest of NEMS for Adiabatic Logic Implementation

4.6.1 History of Relay Technology

The previous section demonstrated the limitations of CMOS technology in adiabatic-type architectures. Will there be other technologies to bridge the gap? First of all, we want to exclude field-effect transistors (FinFET, FDSOI transistor, nanowire- or nanotube-based transistors, III-V-based transistors, graphene-based transistors, etc.). All of these devices have a subthreshold conduction and certain restrictions in terms of dissipation. So, in this publication, we have chosen to focus on nanorelay technologies that generally have extremely weak leakage currents. These devices have not yet been developed at the same scale as advanced transistors. The devices that have been developed are larger in size, and this technology can be considered a possible alternative. The nanorelay technology still encounters various problems due to contact reliability. Technological aspects have already been dealt with in various publications [9].

Using relays to perform logic functions is a very old idea, and in fact, the first ever computers were made in this way in the 1930s. The relay is the perfect device for Boolean applications because it has two perfectly differentiable states: an active state with a potentially high current if the generator allows it and an inactive state in which the current is almost at zero because it is limited by the surface leakage currents. The first computers were based on electromechanical technologies. In 1938, Konrad Zuse began constructing the first Z series, electromechanical computers containing a memory and affording limited programming. In 1941, Zuse constructed the Z3. It was based on 2 600 telephone relays, read programs on magnetic tapes, and worked perfectly, which made it the first functioning programmable computer. It had a clock frequency of 5.33 Hz and could perform 4 additions per second, or 15 multiplications in one minute. This was the first ever realization of complex relay-based logic.

Then, vacuum-tube-based technology brought in the electronic numerical integrator and computer (ENIAC), which is often incorrectly considered as the first computer. Afterward, semiconductor-based technology was developed. Relay-based logic has not completely disappeared however, as many power systems still use it when passing very high currents or very significant voltages. In 2011, a team at Berkeley published an article proving the feasibility of an integrated circuit based on electromechanical microrelay technology [9]. The dimensions were still large (a surface of 400 μ m^2), and the functioning voltage was very high ($\sim$10 V), but the foundation has been laid. Microrelays work electrostatically, and the distance between the mobile and fixed structures is typically 200 nm. For around a decade, research continued in laboratories in the Americas, Asia, and Europe. When the size of the devices was reduced, numerous difficulties appeared due to the relay size. It seems difficult to guarantee its reliability. The technology therefore has not progressed today sufficiently to provide a short- or long-term solution.

4.6.2 Different Actuation Technologies

Electromechanical relays are all based on the movement of a mobile structure that can electrically come into contact with two conductors after being displaced. There are multiple optional ways to create this movement: by applying a magnetic force, a thermal or piezoelectric effect, and an electrostatic force. It is quite difficult to make a global comparison of all these effects, but it needs to be admitted that the electrostatic effect is the most commonly used. It is actually quite simple to implement and does not require particular materials to be developed.

- **Magnetic relay:** They were able to obtain an actuation voltage of 5 V and a switching energy of 100 μJ on a surface of 20 mm^2 [10]. A few magnetic relays have been created in a smaller size, but slow movement and high switching energy are the main obstacles against use in electronic logic.
- **Thermal relay:** A thermal relay generally has four ports: two for providing thermal power and two for enabling the switching function [11]. Materials expand under the effect of temperature in this structure. The fact that this technique is

dedicated to particular applications can be explained by its high dissipated energy and the relatively slow physical effect.

- **Piezoelectric relay:** This technique is based on the deformation of certain materials when an electric field is applied [12]. It is mainly studied for electronic applications and often compared to electrostatic actuation. Some researchers consider this technique to be the most effective. PZT ($Pb(Zr_x,Ti_{1-x})O_3$) is one of the most commonly used. AlN and ZnO are also popular. Prepolarization techniques reduce the actuation voltage.
- **Electrostatic relay:** This is the most studied technique. We will detail it from the perspective of electronic logic. The principle is simple: in a capacitor made of two conducting plates, the electric field is sufficient to move an electrode if it is mobile and deformable enough. Note that the force is always attractive because the charges are opposite signs. When a voltage is applied to the central part of the device, the attractive force moves the membrane toward the fixed electrode. The two stops at each end restrict the displacement of the membrane, which otherwise could become stuck on the fixed electrode. The electrical contact is then established between the two external electrodes called the source and the drain, in reference to CMOS technology. When the voltage returns to zero, the springs play their role and the membrane returns to its initial position.

There are other possible technologies. The membrane can be replaced by a mobile beamer whose recall function is ensured by the beamer's own elasticity. If the recall force is not high enough to overcome the electric force with the relay in the closed position, the relay behavior may be bistable. The relay remains in the closed position when the voltage returns to zero. In general, the relays show an intermediate behavior. The relay opens again when the voltage has returned to a weaker value than that allowed by the closure. The diagram in Fig. 4.22 depicts this behavior and gives definitions of two relay-type characteristic voltages: the pull-in voltage (Vpi) and the pull-out voltage (Vpo).

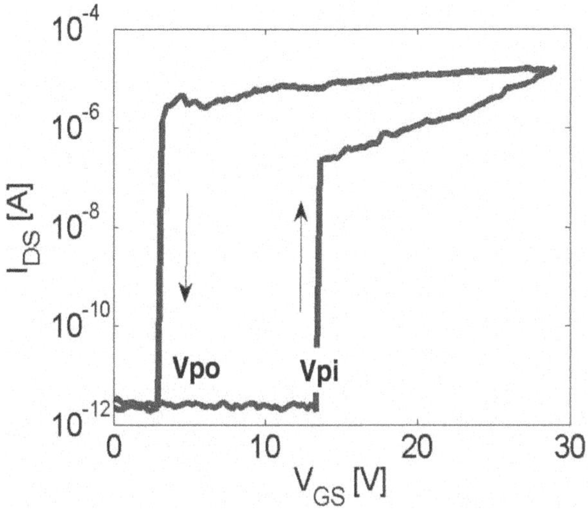

Figure 4.22 Characteristic curve of a microrelay.

So, it does seem possible to use microrelays in logic instead of transistors. The extremely weak level of the current when it is in the off-state (3 orders of magnitude below a transistor subthreshold current) guarantees a very weak static consumption. It remains to be verified, however, if the functioning voltage can be further reduced in the future. It will also need to be ascertained that the contact time (time that the mobile structure takes to come into contact with the fixed electrode) can also be reduced, in order to become compatible with reasonable operating frequencies.

4.6.3 Relay-Based Adiabatic Architectures

Microrelays do not have the major disadvantage of CMOS because the off-state leakage current is a few orders of magnitude lower than a transistor subthreshold current. On first analysis, the pull-in voltages could be considered equivalent to threshold voltages in adiabatic architectures. A more refined analysis of the microrelay-based logic gate architectures reveals that the pull-out voltage, not the pull-in voltage, plays the same role as the threshold voltage. If we can admit that the pull-out voltage can be reduced to a weaker value without causing a major inconvenience, we can prove the usefulness of microrelays.

In adiabatic gates, the resistance value of the switch in the active state directly contributes to the energetic dissipation. balance. Values lower than kΩ are needed when significant gains are envisaged. It actually seems very difficult to obtain a strong functioning endurance (measured by the number of logical changes before defects appear) and a fairly weak contact resistance at the same time. Contact resistance values lower than 1 kΩ can be obtained with soft materials (gold, silver), but the endurance is very poor. Hard materials (titanium, tungsten, silicon carbide, etc.) can perform high reliability, but the obtained contact resistance values are mostly greater than 1 kΩ. The major difficulty lies in this trade-off.

Pipeline-type solutions have been investigated. A reversible pipeline is possible, but the overhead in terms of the number of switches remains a major drawback. In this chapter, we have chosen to focus on the gate architectures inspired by quasi-adiabatic pipeline-type design. The results that have been identified so far come from simulation studies carried out by the European project NEMIAC and CEA-LETI in Grenoble [13]. The general functioning principle of an adiabatic gate in a pipeline is depicted in Fig. 4.23.

The function is ensured by the relays A and B connected in series (an AND function in the example). The output is maintained and is returned to zero by the supplementary switch T. There are many ways this relay can be commanded: either by an external logic element (the complementary gate in a dual-rail logic, for example) or by a signal coming from the following gates in the pipeline (in the case of a reversible pipeline) or even by the output signal of the gate itself. This last mode is the simplest and will be detailed in this section. The schema shown in Fig. 4.23 is then obtained. When the voltages at the inputs exceed the pull-in voltage, the A and B relays close. The supply voltage can then be adiabatically transmitted to the output S. When the output voltage exceeds the pull-in voltage (V_{pi}), the relay T closes and stays closed as long as the command voltage is greater than the pull-out voltage (V_{po}). The ramp descending transmits the supply voltage to the output as long as it is greater than the voltage V_{po}. Then, the output node is insulated and the output

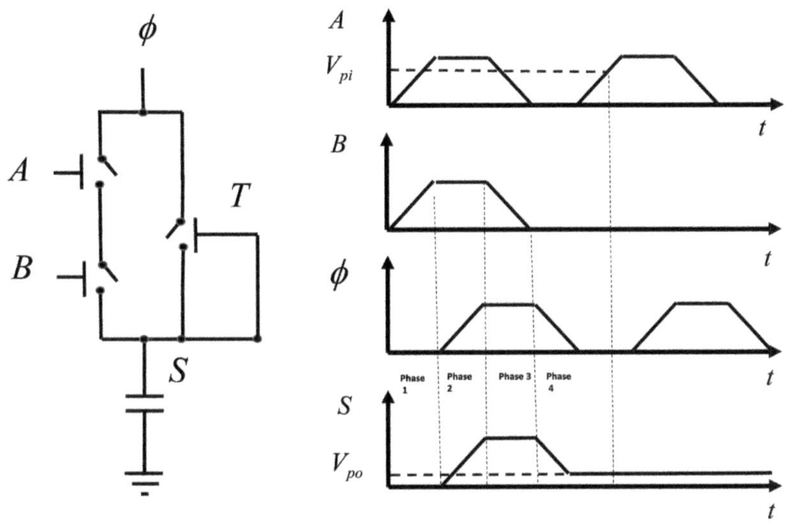

Figure 4.23 Microrelay-based adiabatic gate.

voltage stays at the pull-out voltage, awaiting the next event. In this case, there is no supplementary nonadiabatic dissipation. This is not the case in the second example, where the OR function is performed and which is shown Fig. 4.24.

The output capacitor is loaded when one of the inputs is in the on-state. The relay becomes conductive and remains that way for as long as the supply voltage has not dropped to values lower than the pull-out voltage. Then, it is maintained as the output node is insulated. The behavior is different in the following event. For example, when the input moves to the on-state, a conduction path at electric zero between the output and the supply opens. The output capacitor then abruptly discharges in the supply, which creates a nonadiabatic dissipation of $1/2 \cdot C_L V_{po}^2$. One must keep in mind that in a transistor-based gate, nonadiabatic dissipation is $1/2 \cdot C_L V_T^2$.

Now let us go back to microrelay technology. A general nanorelay design is depicted in Fig. 4.25. The pull-out voltage is easily obtained by writing that in the closed position, the restoring force is just sufficient to overcome the electric force and the adhesive force (F_A).

$$kg_s \geq F_A + \frac{\varepsilon_0 A}{2\,(g_e - g_s)^2} V_{po}^2 \tag{4.86}$$

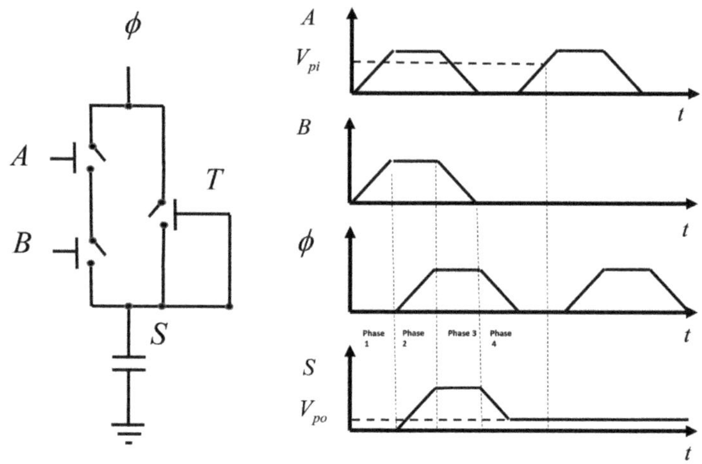

Figure 4.24 Circuit with nonadiabatic dissipation.

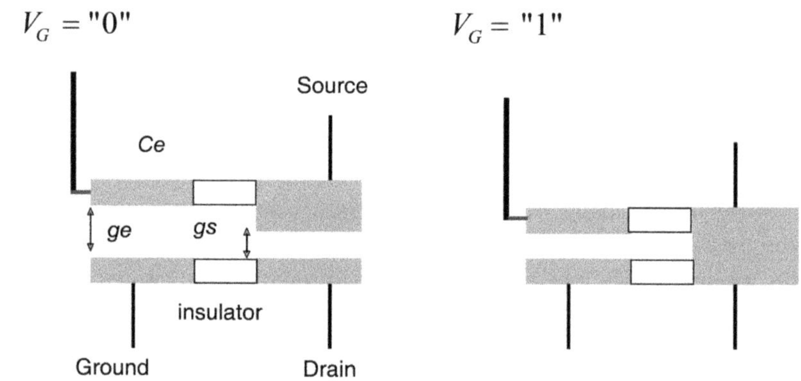

Figure 4.25 Architecture of a nanorelay.

From Eq. 4.86, the following can be deduced:

$$V_{\text{po}} = \sqrt{\frac{2\,(kg_{\text{s}} - F_{\text{A}})\,(g_{\text{e}} - g_{\text{s}})^2}{\varepsilon_0 A}} (4.87) \quad (4.87)$$

Therefore, it is possible to fix this voltage to any value by playing on the geometric parameters of the microrelay and the adhesive forces. It is even possible to render it null, which introduces a new gate architecture based on a bistable type relay.

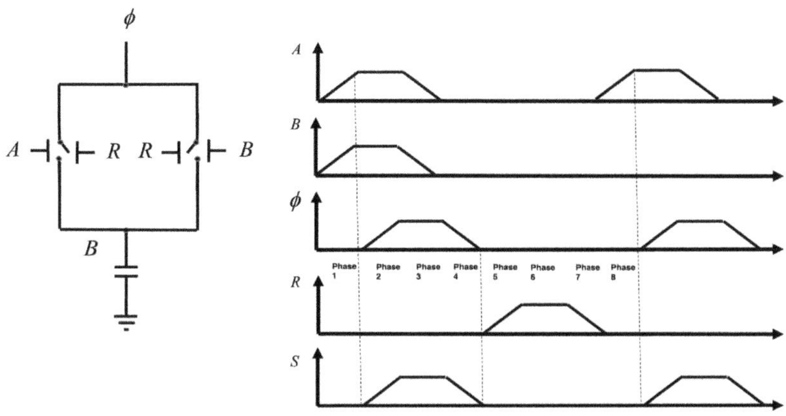

Figure 4.26 OR gate with bistable microrelays.

Nonadiabatic dissipation in a bistable-based gate is then zero, but a fourth electrode needs to be introduced to move the state from on to off because the restoring force is insufficient. One needs to account for a supplementary energy for the command of this electrode. However, as this operation can be done in each cycle, it is not necessary to implement logic in order to perform this operation. If we call r the resistance between the supply and the command electrode that is supposedly equivalent to a capacitance of C_r, the supplementary energy is $rC_r/T \cdot C_r V_{DD}^2$ when the command is adiabatic. As the resistance r can be weak because it only results from the interconnect distributing this signal to the different circuit gates, the supplementary dissipation can be considered negligible. The physical restriction is the value of the adhesive energy, which cannot be recovered. The energy consumption due to friction also needs to be counted if the relay is not working in vacuum.

The implementation of a bistable microrelay-based gate therefore seems to be an excellent one for years to come. One disadvantage to be aware of is that the signals are no longer symmetrical, as was the case in the first structure. The reason behind this is the need to insert into the cycle a signal that has returned to zero. All of these considerations are illustrated Fig. 4.26.

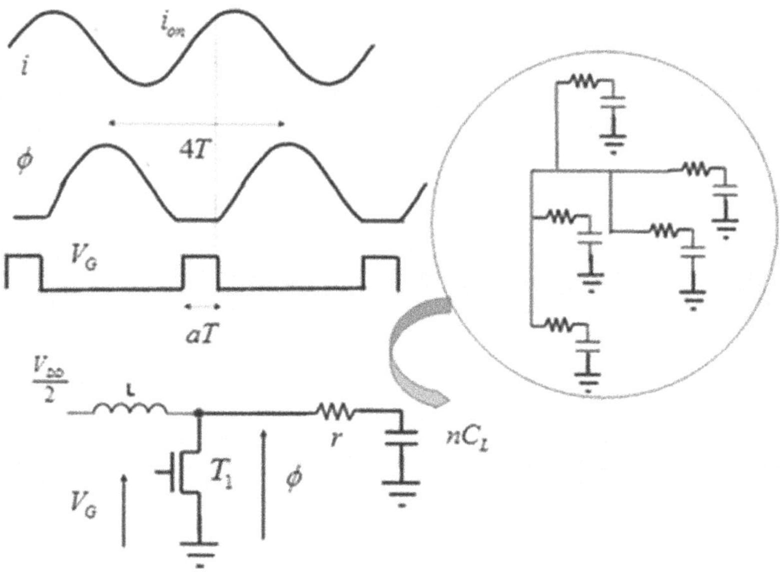

Figure 4.27 The inductor-based energy recovery supply.

4.6.4 Comparison between CMOS Logic and Microrelays

This last section compares the simulation performance results of the systems based either on technology involving FETs or on technology involving microrelays [13]. Microrelay-based technology does not yet exist on the scale of advanced CMOS transistors. So, a comparison based on experimental results is not possible and only simulations are given. For a realistic comparison it is necessary to include the effect of the power clock supply. In this chapter, we consider the most popular design of this supply: the inductor-based energy recovery supply [8].

The principle of inductive supplies is to move the energy from the gate capacitances to the inductance by creating a resonant circuit, as shown in Fig. 4.27. Another way to understand this diagram is to consider that the current flowing through an inductance is quasi constant, which enables the capacitor charge to be optimized [8]. The trapeze is the optimal shape for some hypothesis, but a staircase approximation is a satisfying low-frequency solution. The trapeze approximation by a sinusoid is also a solution. In this case the quarter of the period of the sinusoid ($4T$) is equivalent to the ramping time T of a trapezoidal signal. The T_1 transistor provides each cycle with energy so as to keep the circuit oscillating. The resistance r represents the sum of the inductance series resistance and the interconnect resistance. C_L is the equivalent average capacitance value of a load gate, and $n$gates are connected to the inductor.

In the first phase, the transistor is closed during a fraction aT of the circuit oscillation period. A current circulates in the inductance and varies linearly during this time because the voltage at the inductance terminals is constant. This energy is enough to excite the resonant circuit and to initiate the first oscillation cycle. The oscillation frequency depends on the inductance value and the equivalent capacitance of the load gates. By defining the oscillation period as $4T$, we can write:

$$\frac{4\pi^2}{16T^2}nLC_L = 1 \tag{4.88}$$

In the following analysis we assume that the ratio between the series resistance of the coil and the inductance of the coil has a constant value λ. This is a useful approximation.

Synthesizing the results obtained for CMOS technology in the usual regime

The energy dissipated on average per logic gate in a binary transition is the chosen metrics for the comparison. The logic gate is assumed to be charged by a capacitance of C_L. Then:

$$E_{CMOS} = \frac{\alpha}{2}C_L\phi_t^2k^2n_0^2\left(y^2 + mye^{-y}\right) \tag{4.89}$$

The parameter y is defined on the basis of the threshold voltage by:

$$y = \frac{V_T}{n_0\phi_t} \tag{4.90}$$

The factor k is the ratio between the supply voltage and the threshold voltage. It is chosen according to the operating frequency. One shall keep in mind that this function has a minimal threshold voltage value between 300 and 400 mV. The parameter α is the gate activity rate.

Synthesizing the results obtained for CMOS adiabatic logic [11]

This result integrates the supply performances into the energy recovery using inductances. The optimal value is retained. It corresponds to a frequency of $1/4T$ with:

$$T_{\text{opt}} = \frac{\pi}{2}\sqrt{\frac{RC_{\text{L}}}{\lambda}} \tag{4.91}$$

$$E_{\text{CMOS min}} = \frac{\pi}{2}C_{\text{L}}V_{\text{DD}}^2\sqrt{\lambda RC_{\text{L}}} + \frac{1}{2}C_{\text{L}}V_{\text{T}}^2 \tag{4.92}$$

The coefficient λ expresses the ratio between the resistance of the inductance and its value. R is the transistor resistance in the on state. The energy dissipated per gate during the period $4T$ is written as:

$$E_{\text{CMOSadia}} = \frac{\pi}{4}C_{\text{L}}V_{\text{DD}}^2\sqrt{\lambda RC_{\text{L}}}\left(\frac{T}{T_{\text{opt}}} + \frac{T_{\text{opt}}}{T}\right) + \frac{1}{2}C_{\text{L}}V_{\text{T}}^2 \tag{4.93}$$

Synthesizing the results obtained for adiabatic microrelay technology [11]

This result integrates the supply performances to the energy recovery using inductances. By ignoring the nonadiabatic dissipation linked to the pull-out voltage or by using a bistable type relay, one obtains:

$$E_{\text{relayadia}} \approx \frac{\pi^2}{8}\frac{C_{\text{L}}R}{T}C_{\text{L}}V_{\text{DD}}^2 + \frac{\lambda T}{2}C_{\text{L}}V_{\text{DD}}^2 \tag{4.94}$$

In this case, the value R is the microrelay resistance in the active state. The optimal value of T is retained. It corresponds to a frequency of $1/4T$.

$$T_{\text{opt}} = \frac{\pi}{2}\sqrt{\frac{RC_{\text{L}}}{\lambda}} \tag{4.95}$$

$$E_{\text{relayadia min}} \approx \frac{\pi}{2}C_{\text{L}}V_{\text{DD}}^2\sqrt{\lambda RC_{\text{L}}} \tag{4.96}$$

The average energy dissipated per gate during the period is written as:

$$E_{\text{relayadia min}} \approx \frac{\pi}{4}C_{\text{L}}V_{\text{DD}}^2\sqrt{\lambda RC_{\text{L}}}\left(\frac{T}{T_{\text{opt}}} + \frac{T_{\text{opt}}}{T}\right) \tag{4.97}$$

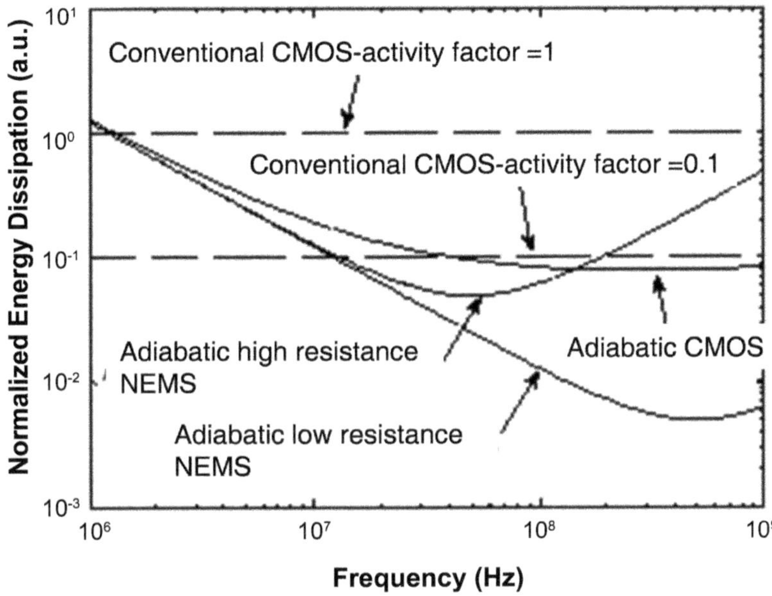

Figure 4.28 Comparison of field-effect-transistor-based adiabatic solutions and microrelay-based adiabatic solutions.

The set of these results is summarized in Fig. 4.28, showing the average energy normalized, meaning divided by the term $C_{\mathrm{L}}V_{\mathrm{DD}}^2$ for the different solutions depending on the operating frequency defined by $1/4T$.

This curve shows the superiority of microrelay technology if one knows how to perform relays that have a weak contact resistance (below 1 kΩ). Generally high-reliability relay technologies provide high contact resistance (more than 1 MΩ). The problem of compromising between compatibility and reliability remains. If technology cannot resolve this problem, the logic gate principle would need to be revisited in depth. Furthermore, microrelays not as switches but as variable capacitances would be needed. This very prospective research is being undertaken by CEA-LETI in Grenoble.

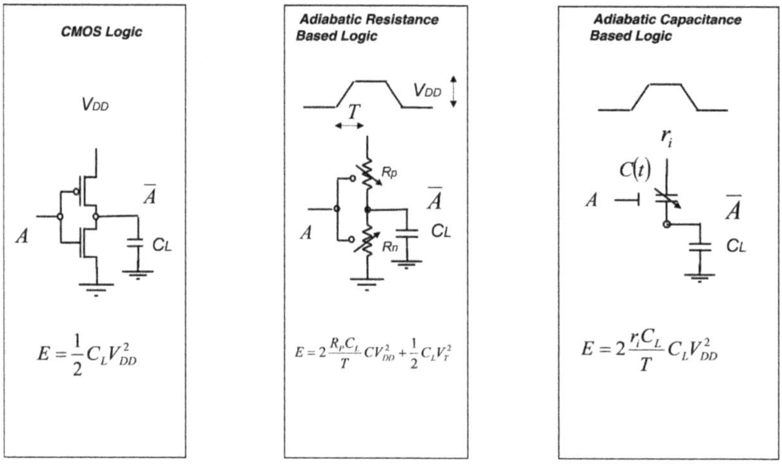

Figure 4.29 Capacitance-based adiabatic logic.

4.7 Capacitance-Based Adiabatic Logic

Capacitance-based adiabatic logic (CAL) was introduced in 2016 [14] as a new architectural paradigm in order to eliminate the contact problems inherent to the nanorelay technology. Figure 4.29 shows the basic comparison of a resistive-based and a capacitance-based inverter gate in a logic operation. The output load C_L represents the capacitance of the next gates and interconnect. The timing sequence is similar to adiabatic logic where the power supply is a trapezoidal voltage, called the power clock, divided into four phases. In this approach, the elementary device is a variable capacitor $C(t)$, which is controlled by an independent gate terminal V_G. For a positive-type device the capacitance increases when the gate voltage V_G increases, and for a negative-type device the capacitance value decreases when the gate voltage V_G increases.

If we write the circuit equations, it is easy to understand the dependence between the output value and the input voltage A. Neglecting the series resistance voltage in the circuit, it is possible to write:

$$C\frac{d(V_{DD}-V_S)}{dt}+(V_{DD}-V_S)\frac{dC}{dt}=C_L\frac{dV_s}{dt} \tag{4.98}$$

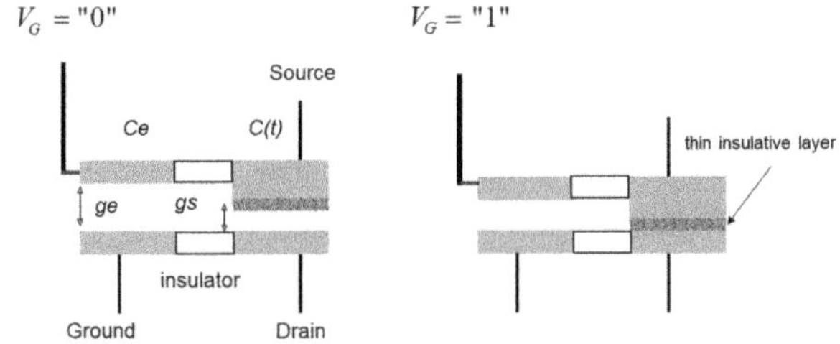

Figure 4.30 Example of a variable capacitance.

$$\frac{d}{dt}(C V_{\rm DD}) = \frac{d}{dt}(C_{\rm L} V_{\rm S}) + \frac{d}{dt}(C V_{\rm S}) \quad (4.99)$$

Finally, we obtain the basic equation of the capacitance-based logic.

$$V_{\rm s}(t) = \frac{C(t)}{C(t) + C_{\rm L}} V_{\rm DD}(t) \quad (4.100)$$

Introduction of series resistance in the calculation leads to a second-order differential equation, and it is impossible to have a simple analytical expression of the output voltage. It is easy to understand that changing the value of the variable capacitance $C(t)$ modifies the output. Two different voltages are possible corresponding to the two logic states. This architecture combines two advantages, a zero leakage current and a low active dissipation, because the adiabatic factor depends on the low interconnect resistance value. Figure 4.30 describes a possible electromechanical device one can use as a variable capacitance in an adiabatic gate.

Figure 4.31 gives the main simulation results for a series of four buffers for a (1,1,0,0,1,1,0) input. Energies and time are normalized as follows:

$$E_{\rm norm} = E / k g_{\rm e}^2 \quad (4.101)$$

$$t_{\rm norm} = t / \sqrt{\frac{m}{k}} \quad (4.102)$$

Here, k, $g_{\rm e}$, and m are, respectively, the stiffness of the beam, the input gap, and the mass of the beam. $Es_{\rm norm}$ is the normalized energy supplied by the generator, $Ed_{\rm norm}$ is the normalized damping energy, and $Ej_{\rm norm}$ is the normalized Joule dissipation.

We can see in Fig. 4.31 that the most important part of the energy involved in the logic transition is recovered. The energy involved in

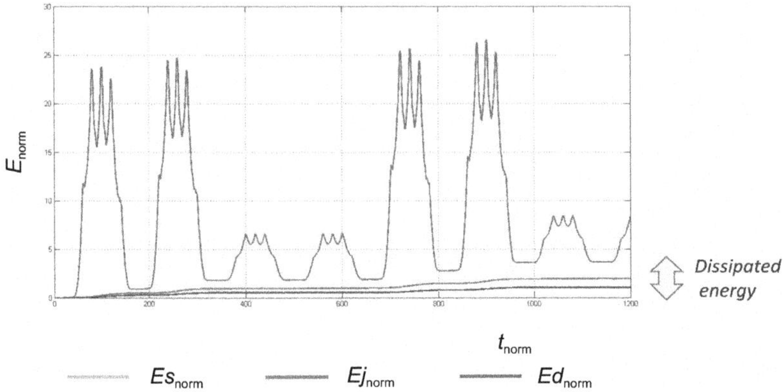

Figure 4.31 Energies in a series of four capacitance-based adiabatic buffers.

the logic transition is the peak of the *Es* signal (around 25), but the dissipated energy ($Ej + Ed$) is around 1 for the four gates. The gain of the adiabatic solution compared to a conventional CMOS logic is 100 in this example.

4.8 Conclusion

This chapter has proved the importance of the adiabatic principle, which is the solution to reduce drastically the power consumption of integrated logic. Unfortunately, CMOS technology is not compatible with ultra-low-power adiabatic architectures because a nonadiabatic dissipation proportional to the square of the threshold voltage appears in irreversible logical gates. Using nanorelays instead of transistors is a possible solution to reduce nonadiabatic dissipation, but the reliability of contact is considered a technological challenge. This explains the on-going research work on a new gate architecture paradigm: the CAL.

References

1. Jones, T. B., and Nenadic, N. G. (2013). *Electromechanics and MEMS*, Cambridge University Press, Cambridge.
2. Calhoun, B. H., Wang, A., and Chandrakasan, A. (2005). Modeling and sizing for minimum energy operation in subthreshold circuits, *IEEE J. Solid-State Circuits*, **40**(9), pp. 1778–1786.
3. Paul, S., Schlaffer, A. M., and Nossek, J. A. (2000). Optimal charging of capacitors, *IEEE Trans. Circuits Syst.*, **47**, pp. 1009–1016.
4. Gershenfeld, N. (1996). Signal entropy and thermodynamics of computation, *IBM J.*, **35**, pp. 577–586.
5. De Vos, A. (2010). *Reversible Computing: Fundamentals, Quantum Computing, and Applications*, Wiley-VCH Verlag GmbH & Co. KGaA.
6. Svensson, L. J., and Koller, J. G. (1994). Driving a capacitive load without dissipating fCV/sup 2/low power electronics, in *Proceedings of 1994 IEEE Symposium on Low Power Electronics*, pp. 100–101.
7. Landauer, R. (1961). Irreversibility and heat generation in the computing process, *IBM J. Res. Dev.*, **5**(3), pp. 183–191.
8. Teichmann, P. (2012). *Adiabatic Logic: Future Trend and System Level Perspective*, Springer, Netherlands.
9. Kam, H., King, T. J., and Liu, S. V., et al. (2011). Design, optimization, and scaling of MEM relays for ultra-low power digital logic, *IEEE Trans. Electron Devices*, **58**, pp. 236–250.
10. Ruan, M., Chen, J., and Wheeler, C. B. (2001). Latching microelectromagnetic relays, *Sens. Actuators A*, **91**, pp. 346–350.
11. Loh, O. Y., and Espinosa, H. D. (2012). Nanoelectromechanical contact switches, *Nat. Nanotechnol.*, **7**, pp. 283–295.
12. Sinha, N., Jones, T. S., Guo, Z., and Piazza, G. (2009). Body-biased complementary logic implemented using AlN piezoelectric MEMS switches, in *IEEE International Electron Devices Meeting (IEDM)*, pp. 813–816.
13. Houri, S., Billiot, G., Belleville, M., Valentian, A., and Fanet, H. (2015). Limits of CMOS technology and interest of NEMS relays for adiabatic logic applications, *IEEE Trans. Circuits Syst. I: Regul. Pap.*, **62**(6), pp. 1546–1554.
14. Fanet, H. (2016). *Ultra Low Power Electronics and Adiabatic Solutions*, Wiley-ISTE.

Part II

Revised Actuation, Sensing, and Data Storage Modes in Emerging Devices for Alternative Computing

Chapter 5

Control of Single Spin in CMOS Devices and Its Application in Quantum Bits

R. Maurand,[a] D. Kotekar-Patil,[a] A. Corna,[a] H. Bohuslavskyi,[a,b] A. Crippa,[a] R. Laviéville,[a,b] L. Hutin,[a] S. Barraud,[a] M. Vinet,[a] S. De Franceschi,[a] X. Jehl,[a] and M. Sanquer[a]

[a]*Université Grenoble Alpes, CEA-INAC-PHELIQS, F-38000 Grenoble, France*
[b]*CEA-LETI MINATEC Campus, F-38000 Grenoble, France*
marc.sanquer@cea.fr

We show how to measure and manipulate a single spin in a CMOS device fabricated in a preindustrial 300 mm CMOS foundry. The device can be used as a spin quantum bit working at a very low temperature. The spin manipulation is done by a microwave electric field applied directly on a gate.

The presented results are a proof-of-principle demonstration of the possibility to define qubits by means of conventional industrial fabrication processes.

5.1 Introduction

Since the pioneering work of Scott-Thomas et al. [1] it has been observed that carriers can be manipulated one by one in the channel

Emerging Devices for Low-Power and High-Performance Nanosystems: Physics, Novel Functions, and Data Processing
Edited by Simon Deleonibus

ISBN 978-981-4800-11-2 (Hardcover), 978-0-429-45873-6 (eBook)
www.panstanford.com

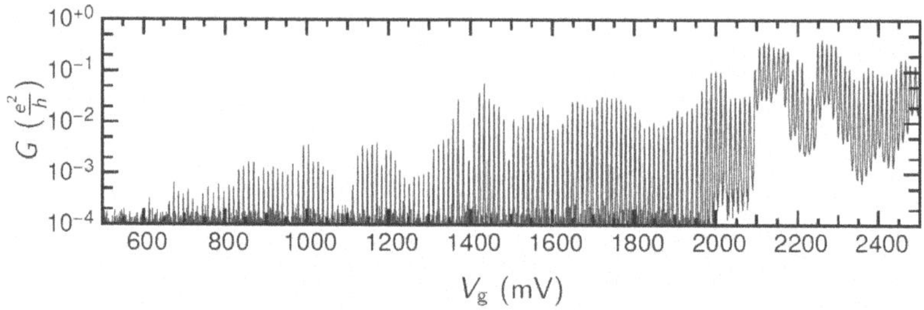

Figure 5.1 Periodic Coulomb oscillations of the drain-source conductance (in quantum units $e^2/h \approx 1/25{,}800\ \Omega$) versus gate voltage observed in a MOS-FET with large nitride spacers at $T = 60$ mK. The drain voltage V_d is small enough to be in the linear I_d-V_d regime. The length of the channel is 30 nm, and the width of the channel is 40 nm. The MOS-FET is done at the CEA-LETI using a 24-nm-thick silicon-on-insulator nanowire trigate technology.

of silicon field-effect devices. However, that takes a long time to be achieved in a controlled way in complementary metal-oxide semiconductor (CMOS) devices. The first time it was realized on purpose in silicon-on-insulator channel was by Ali and Ahmed [2] and Takahashi et al. [3]. The first time it was obtained in a CMOS transistor was by Boeuf et al. [4]. Since that time—at Grenoble—we have optimized a way to change a standard CMOS field-effect transistor (FET) into a metal-oxide semiconductor single-electron transistor (MOS-SET). The MOS-SET inherits from the main figures of merit of its companion device, that is, excellent electrostatic control by the gate voltage, compactness, standardization, excellent variability and yield, and fast and energy efficient operation [5]. An example is shown in Fig. 5.1. Periodic Coulomb oscillations of the drain-source conductance are observed in a metal-oxide semiconductor field-effect transistor (MOS-FET) with large 50-nm-thick nitride spacers (nonoverlapped geometry [4], sample similar to the one presented in Ref. [5]). The length of the channel is 30 nm, and the width of the channel is 40 nm. The MOS-FET is done at the CEA-LETI using a 24-nm-thick silicon-on-insulator nanowire trigate technology. More than 200 oscillations can be recorded corresponding to the addition of electrons one-by-one in the channel from zero to a density of few 10^{13} carriers/cm^2 (1.6

10^{13} e/cm^2 $\approx$ 200e/[30 nm $\times$ 40 nm]). The period in V_g is given by $\Delta V_g = e/C_g$, where C_g is approximately given by the planar gate-channel capacitance.

This technique made the MOS-SET an ideal platform for analog applications, as the silicon electron pump [6] or hybrid single-electron transistor–field-effect transistor (SET-FET) amplifiers [7] among many others [8, 9]. We have already shown that classical CMOS analog electronics can be fully cointegrated with a MOS-SET [10].

Nevertheless, the most radical application of the MOS-SET is probably its use as the elementary brick to build an all-silicon quantum computer.

One of the earliest—and today the most studied—proposals for quantum computation in semiconductors envisioned arrays of electrostatically defined dots, each containing a single electron whose two spin states provide a qubit [11]. Quantum logic is accomplished by changing voltages on the electrostatic gates to move electrons closer to and further from each other, activating and deactivating the exchange interaction. Most of the spin qubits studied are focused on qubits in both GaAs/AlGaAs [12] and Si/SiGe heterostructures [13] embedding a buried 2D electron gas whose properties are tailored by band diagram engineering. These structures can be fabricated using lab-scale lithography tools. From s? from large-scale integration perspective, however, III-V materials are not yet an option and the use of SiGe demands an adaptation with respect to conventional CMOS processes. Fortunately for this large-scale integration perspective a very large coherence time for the spin of electrons trapped on quantum dots in silicon, which can be isotopically purified [14], has been recently demonstrated [15]. As envisaged in the Loss–DiVincenzo proposal [11], two qubit gates for electron spins in isotopically purified silicon quantum dots have been further realized [16].

Despite these fantastic advances, the silicon qubit is still a challenger in the race toward quantum computing [17] compared to superconducting solid-state qubits [18] or trapped-ion qubits [19]. Nevertheless, silicon qubits present important advantages in terms of size (a few 10 nm instead of microns), scalability, and cointegration. This is particularly true if the silicon qubit can be realized in a full CMOS line.

The goal of this chapter is to describe first how we control a single spin in a CMOS device and second how we use it as a quantum bit. This is not the first qubit nor the most efficient, but as already stated, when it comes to a crucial issue such as large-scale integration, however, the range of possible choices for the qubit becomes much narrower and the CMOS spin qubit becomes a serious option. The quantum computer is certainly a long-term goal, but the mere fact that a single spin can be controlled, manipulated, and read out in a MOS-FET device fabricated in the same foundries used for standard microelectronics is remarkable and was not anticipated at all only a few years ago.

5.2 Control of the Single Spin in CMOS Devices

The charge of a single electron was measured a long time ago [20]. Unfortunately, the tiny spin of a single carrier corresponds to an extremely small magnetic moment whose magnetization cannot be measured directly. It is necessary to first perform a spin-to-charge conversion. In the context of electronic devices this is done at very low temperature by using either a spin-selective or an energy-selective tunneling event. In our measurement we used the spin-selective scheme known as the Pauli blockade [21]. The energy-selective scheme, known as the Elzerman protocol [22], relies on the Zeeman energy difference between an electron with the spin up or spin down. It is used, for instance, in the experiments by Veldhorst et al. [15, 16].

The Pauli blockade prevents the tunneling of one electron between two dots corresponding to the transition T(1,1)→T(0,2) when the triplet T(0,2) is too high in energy; (n,m) denotes the number of carriers in (dot1,dot2) [21]. The T(0,2) state is above the S(0,2) state because of the Pauli exclusion principle. Two electrons with parallel spins on dot 2 should occupy two different orbital states, the ground state and an excited state if the ground state is not degenerate. This costs an additional kinetic energy term, the energy difference between the ground and excited states (minored by the possible exchange energy). In our dots, this energy difference is of the order of 1 meV (about $k_B \times 12$ K), because our dots are extremely small, and therefore the T(0,2) state is typically 1 meV above the S(0,2) state. At a low energy difference the transition from the (1,1) state to the (0,2) state is then prevented when the (1,1) state is a triplet state. This Pauli blockade scenario is well established when there is no or weak spin-orbit interaction [21], which is true for electrons in silicon. In that case during the tunneling event from one dot to the next the spin of the electron is conserved.

If there is a significant spin-orbit interaction—as is the case for holes in silicon—the spin is possibly not conserved during the tunneling, but a reminiscent Pauli blockade persists [23, 24]. This spin blockade is partly dependent on the magnetic field. The spin blockade is stronger at $\mathbf{B} = 0$ compared to a finite magnetic field due to time-reversal symmetry. As a result, a current dip at $\mathbf{B} = 0$ is expected [23, 24]. This is in contrast with the case of small spin-orbit coupling where the spin blockade can be partially removed at $\mathbf{B} = 0$ due to hyperfine [25] or cotunneling mechanisms [26].

The Pauli blockade is used to initialize and read out the spin located in one of the two quantum dots. This is illustrated in Fig. 5.2. Initialization: both gate 1 and gate 2 are polarized such that a carrier (a hole in the present experiment [27]) is sitting under gate 2 and the two (0,1) states, the four (1,1) states, and the S(0,2) state are energy degenerate. Without spin consideration there will be a finite drain-source current at these gate voltages through the (0,1)→(1,1)→(0,2)→(0,1) sequence (at a finite drain-source voltage). S(1,1) = T(1,1) are almost degenerate because the exchange coupling between the two dots is small. In the presence of a static magnetic field the spin on dot 2 is in the ground state—say spin down. As long as a spin up enters in dot 1 it is transmitted into the drain. As soon as a spin down enters, the transfer is blocked. After a certain time of initialization, we can therefore be certain that the initial state is T(1,1), that is, spin down in both cases. This sequence corresponds to the first two steps of Fig. 5.2 (upper part).

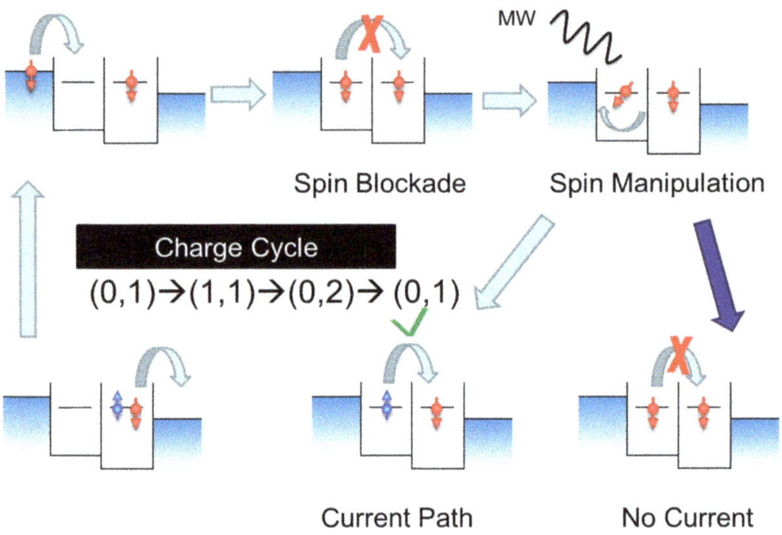

Figure 5.2 Principle of the Pauli spin blockade in two quantum dots in series and the principle for ESR or EDSR detection by the source-drain current. The diagram is made for electrons. (Top left): (0,1)(1,1) transition; the current is blocked as soon as a carrier with a spin down enters in dot 1 (top center). (Top right): Thanks to an electric (resp. magnetic) radiofrequency voltage (at the right frequency $h \cdot \nu = g \cdot \mu_B \cdot \mathbf{H}_0$) applied on gate 1, the spin rotates. For an RF signal applied *continuously* the spin is up after some time and a DC drain-source current due to EDSR (resp. ESR) is detected (bottom center and left, the indicated charge cycle with light blue arrows is continuously working). (Bottom right, dark-blue arrow): The radiofrequency voltage is applied during a *fixed* time τ (RF burst). If τ corresponds to a $2N\pi$ rotation the spin is still down at the end of the burst and there is no current.

The Pauli blockade is also used for read-out:

First the (magnetic) electron spin resonance (ESR) or the electric dipole spin resonance (EDSR) is detected when a continuous microwave signal at the right frequency $h \cdot \nu = g \cdot \mu_B \cdot \mathbf{H}_0$ is applied on gate 1. This corresponds to the cycle shown by the light-blue arrows in Fig. 5.2. Second the Rabi oscillations can be recorded if a microwave burst during a *fixed* time τ is applied when the two dots are in the Coulomb blockade regime, as explained in the next section and featured by the dark-blue arrow. If τ corresponds to a

$2N\pi$ rotation the spin is still down at the end of the microwave burst and there is no current. The initialization and read-out are done in the Pauli blockade regime (as depicted in Fig. 5.2). On the contrary the ESR or EDSR burst is applied when the two carriers are in the *Coulomb blockade* regime, to avoid a transient leakage during the burst and/or to detect several rotations. This means that the spin manipulation step is done when two levels below the Fermi energy in the drain (not shown in Fig. 5.2). In the Pauli blockade regime a current will appear as soon as $\tau = 2\pi$—if there is no Coulomb blockade—and the regime where $N > 1$ will not be accessible.

The Pauli spin blockade then offers a way to turn (or to convert) the information on a single spin orientation on dot 1 into a charge transfer event, possibly a measurable current after the integration of many such events.

Note that the Pauli blockade is efficient as long as the temperature is less than the energy separation between S(0,2) and T(0,2) ($\sim$1 meV). On the contrary the Elzerman protocol necessitates that $k_B T < g\mu_B H$ and the Zeeman splitting is much smaller than the singlet-triplet energy separation in general. With the Pauli blockade scheme it is possible to detect the spin at a temperature of about 1 K. As for the Elzerman protocol [22] one should go below $T = 0.1$ K. This could make a significant difference in the power consumption budget to have qubits and their peripherals working at $T = 1$ K instead of 0.1 K.

With the Pauli blockade it is possible to initialize and read out a single hole or electron spin and with ESR or EDSR it is possible to manipulate the spin. To be useful for applications the spin orientation should be stable enough between successive manipulations. Electron spins in silicon are remarkably insensitive to their electrostatic environment, which makes them very interesting as quantum bits. This has been observed for long in macroscopic silicon crystals doped with donors and studied by ESR standard techniques (where a huge number of spins are manipulated coherently). The spin relaxation time T_2 can reach 0.4 s in very diluted Si:P at 1.2 10^{14} cm^{-3} if the silicon crystal is purified from the ^{29}Si isotopes [28]. The nuclear spin relaxation time can be much longer: $T_{2N} = 192$ s at $T =$ 1.8 K in ^{28}Si:P (5 10^{11} cm^{-3}) [29]. Therefore, an isotopically purified

^{28}Si crystal can be considered as a "silicon vacuum" for electron spins and an excellent platform for spin qubits.

In macroscopic silicon crystals the electrons are localized on donors. In nanoscopic transistors the carriers can be localized on either quantum dots or donors. Remarkably for electrons in silicon quantum dots, T_2 reaches 28 ms [15]. Even if this time is smaller (by a factor of 10) compared to a macroscopic lightly n-doped ^{28}Si crystal, it is remarkably long if one considers that electron lives in an artificial nanostructure with nearby gates, electrodes, interfaces, defects, etc. The T_2 and the inhomogeneous T_2^* relaxation times are typically 10^3 times shorter in natural silicon, pointing to the importance of the hyperfine coupling with nuclear spins to explain the electron spin relaxation time limitation at low temperatures in natural silicon nanostructures. These remarkably long spin coherence times observed for electrons in silicon nanostructures pushed us to look after spin quantum bits in CMOS nanostructures fabricated in a preindustrial platform devoted to classical nanoelectronics.

In the following, we will concentrate on carriers confined in quantum dots rather than on dopants because it is difficult to control a single dopant with enough precision. This is possible by scanning tunneling microscopy (STM)-assisted nanopatterning [30], but this elegant technique is not scalable and hardly compatible with standard CMOS techniques, which are our main objective. We notice nevertheless that standard CMOS techniques associated with controlled channel doping permitted to build single-atom and coupled-atom transistors [31–34]. Therefore, it would be possible to use dopants to develop new functionalities for the silicon qubit, for instance, to store quantum information on nuclear spins that have a much longer coherence time than electron spins [35, 36].

Furthermore, we choose the fully-depleted silicon-on-insulator (FDSOI) technology and its variant—the trigate technology—to build our qubits. This is illustrated in Fig. 5.3.

The silicon nanowire field-effect transistors (NW-FETs) are fabricated on a 300 mm silicon-on-insulator (SOI) processing line [37]. First, a silicon nanowire is etched from a SOI wafer with a 10-nm-thick, undoped silicon device layer. The nanowire channel is

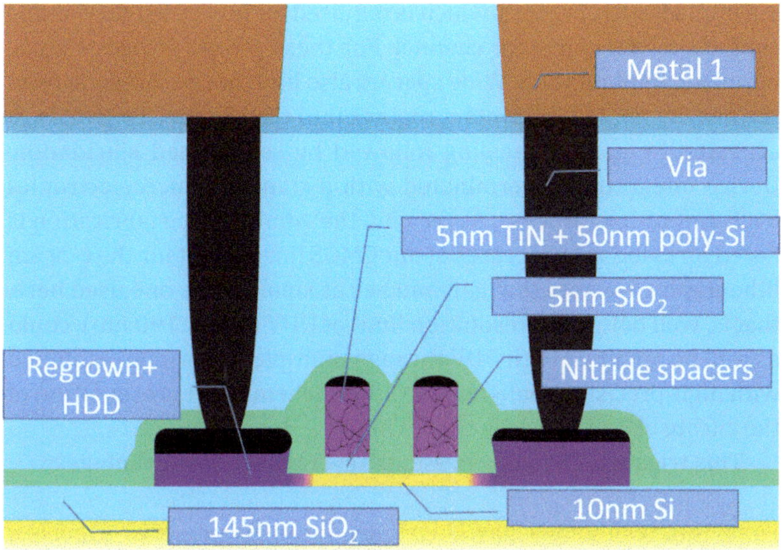

Figure 5.3 Schematic view of the coupled MOS-SET used for the CMOS hole spin quantum bit experiment. The thin silicon channel (yellow) is covered by two top gates. The distance between the two gates is 35 nm. The regrown source drains are heavily doped (violet), silicided (black), and in contact with metallic VIAs to the metal 1 layer. The oversized nitride spacers are featured in green. The 145-nm-thick (not to scale) buried SiO_2 oxide and the encapsulation oxide are featured in blue. The silicon substrate, on which the substrate bias is applied, is represented in yellow.

oriented along the [110] direction. Initially defined by deep-ultraviolet (DUV) lithography, its width W is trimmed down to about 15 nm by a controlled oxidation and etching process. Two parallel top gates, ~35 nm wide and with ~30 nm spacing between them, are successively patterned by means of a combined DUV and e-beam lithography. The latter enables us to achieve the necessary small spacing between the gates. The gate stack consists of a thin (~5 nm) TiN layer followed by a much thicker (~50 nm) polysilicon layer. Gate electrical isolation is ensured by a dielectric stack consisting of a SiO_2 layer of 7 nm and a Hf-based high-κ dielectric layer of 2 nm. Insulating SiN spacers are deposited all around the gates. Their width is deliberately large in order to fully cover the nanowire channel between the two gates and protect it from the successive ion implantation process, which is required for low-resistance ohmic contacts to the nanowire channel. For these p-type devices we use boron ion implantation. Wide spacers also limit boron diffusion from the heavily implanted contact regions into the channel. Dopants are activated by spike annealing followed by self-aligned silicidation. Device integration is terminated with a standard microelectronics back-end-of-line process. At the end, the whole device fabrication is based on standard processes of our CMOS line, except for the e-beam lithography. We note that gate pitches as small as the one used here, that is, well below the resolution limit of DUV (about 190 nm), could as well be obtained with DUV through multiple patterning combined with high-precision realignment [38]. A schematic representation of the encapsulated device is shown in Fig. 5.3.

This trigate technology presents several decisive advantages:

First, this technology has been introduced to preserve a perfect control of the electrostatic potential in the silicon channel below the gate, even at a small gate length. This control is decisive also in our spin qubits because we manipulate the spin with voltages applied on the gates (explained later). The better the control of the quantum dot potential with the gate, the lower the signal strength to control the spin will be. Moreover, the excellent "electrostatic integrity" delivered by these devices means that the potential of the dot is less sensitive to the potential applied on nearby gates, which is crucial for upscaling the qubit. It is important that the signal applied on gate 1 does not rotate simultaneously the spin located below gate 2. Moreover, we plan to measure the qubit by dispersive gate reflectometry in the future [39] (see Section 5.4) and this technique relies on optimal coupling between the gate and the qubit.

Second, this technology allows us to use the substrate bias to control the electrostatic potential in the nanowire in three dimensions. By applying a positive substrate bias (across the buried oxide [BOX]) we can, for instance, push the holes near the top of the nanowires where the first holes appear in the corners of the nanowire below the front gate (see Section 5.4) [40]. This can be used to create two quantum dots in parallel along the channel, which can be used either to encode a qubit (single-triplet qubit) or to detect the spin orientation using Pauli blockade and gate reflectometry [39].

Third, in the case of two gates in series along the nanowire channel, the substrate bias is used to control the tunneling rate between the two dots (and between the dots and the source/drain) by modulating the potential in the parts of the nanowire that are not covered by the front gates. The tunneling rates should be adjusted such that the carriers can be transmitted between the source, the two dots, and the drain at a sufficiently high rate to detect the source-drain current. The tunneling rates cannot be too large as well because the wave functions should be localized below each gate and not form a molecular-like delocalized state shared between the two dots (as far as we consider the single spin qubit and not a singlet-triplet coupled dot qubit with an adjustable exchange term, of course).

The trigate nanowire technology enables the manipulation of either electrons or holes, depending on the nature of the source-drain doping (n- or p-type) and on the polarity of the gate voltage. For our qubit we decided to use holes instead of electrons. Up to now there is no report of a hole qubit whatever the semiconductor material. Continuous EDSR of holes has been observed by Pribiag et al. [41]. The choice for holes is then rather extreme but motivated by a decisive advantage that a hole spin can be manipulated by an electric field as an electron spin cannot in principle. Because in CMOS devices the use of magnetic fields and magnetic coupling is very unusual (excepted for MRAM) as the electric field manipulation is the common rule, this makes the hole spin qubit much more compatible with the standard CMOS than the electron spin qubit.

The penalty to use holes rather than electrons in silicon comes from the larger sensitivity of holes to static disorder (the lower hole mobility compared to electron mobility is the signature of its effect). We noticed that holes are more prone to be localized by residual disorder at the top of the valence band compared to electrons at the bottom of the conduction band. This is probably due to the larger hole effective mass but possibly also due to the nature and charge states of the defects in the gate stack or in the spacers. As a consequence it is relatively easy to control the first electron in the MOS-SET, as it is much more difficult to certify that the first hole in a metal-oxide semiconductor single-hole transistor (MOS-SHT) is detected through transport measurement: the first hole could be localized in a shallow asperity of the electrostatic confinement potential and too weakly coupled to the electrodes to allow the detection of any current.

Compared to electrons, holes present both a much stronger spin-orbit coupling and an absence of valley-orbit coupling. Both features are important for the spin qubit. For a spin quantum bit it is important that the orbitals be nondegenerate (except for the twofold spin degeneracy, of course) because intraorbital transitions can spoil the qubit integrity. For electrons at the Si/SiO_2 interface in the presence of a vertical electric field, the ground state orbitals are doubly degenerate (without counting the spin degeneracy, four times degenerate in total). Fortunately for the electron spin qubit, this degeneracy is lifted by 3D electric field components in quantum dots or around a dopant. The typical energy spacing between the orbitals is 0.1–1 meV but very sensitive to the exact, and sometimes uncontrolled, microscopic configuration.

Nevertheless, for holes in bulk silicon the valence band is doubly degenerate (heavy and light holes) [42] and this degeneracy is lifted by strain or electric field. In 2D planar structures the ground state is spin-3/2 heavy-hole-like for instance [43]. The minimal hypothesis in our structures is that the ground state is a Kramer's doublet with a mixing of the characters of spin-3/2 heavy and spin-1/2 light holes.

The p-type character of holes in the valence band is responsible for the large spin-orbit coupling but also implies reduced hyperfine coupling [44]. This makes the hole spin less sensitive to the nuclear spin in the silicon channel than the electron spin is. Therefore, the isotopic purification of the silicon crystal can have less influence on hole spin coherence than it has on electron spin qubits, where it increases the spin decoherence time by a factor larger than 10^3. Nevertheless, the silicon isotopic effect for hole spins localized on boron acceptors in silicon has been studied by Steger et al. [29]: local fluctuations of the valence-band edge due to different isotopic configurations in the vicinity of the boron acceptors account for inhomogeneous broadening effects of the ESR line in natural silicon crystals. Therefore, hole spin resonance measurements in isotopically purified silicon quantum dot are very desirable in the future.

5.3 Hole Spin Qubit in CMOS Devices

The main advantage of the hole spin is that it can be manipulated by an electric field at a fast rate. The hole spin is coupled with the electric field through spin-orbit coupling (even in the bulk case):

$$\mathrm{H}_{\mathrm{so}} = \frac{\hbar}{4m^2c^2} \cdot E_{\mathrm{z}}\,\boldsymbol{\sigma}\,.(\mathbf{z} \times \mathbf{p}) \tag{5.1}$$

Here $\mathbf{E}_z$ is an electric field applied along the z direction (perpendicular to the surface), $\mathbf{p}$ is the hole wave vector, and $\boldsymbol{\sigma}$ is the Pauli vector associated to the Pauli matrices. This makes it possible to change the spin orientation with an electric field and to perform EDSR. A more specific EDSR effect can exist in our nanowire p-type metal-oxide semiconductor (PMOS), called the g-tensor modulation resonance (g-TMR). It corresponds to the case where the Landé g-factor is anisotropic and varies with the electric field [45]. Both the anisotropy and the voltage gate dependence of the g-factor have been observed in PMOS nanowire [46] and a Rabi frequency of up to 500 MHz has been predicted in that case. This Rabi frequency for EDSR is 10–100 times larger than the Rabi frequency obtained for ESR, about 100 kHz [15]. This is a decisive advantage for the spin qubit where one should have a large T_2/T_{Rabi} ratio that controls the number of operations that can be performed on the spin qubit without losing quantum coherence by spin relaxation (T_2). Also $1/T_{\mathrm{Rabi}}$ should not be too small compared to classical calculators (typically working at 1 GHz), to make the quantum computers attractive enough.

The g-TMR can be qualitatively explained as follows with a waving hand argument: In standard ESR one applies a radio frequency (RF) magnetic field $\mathbf{H}_{\mathrm{RF}}$ perpendicular to a static magnetic field $\mathbf{H}_0$. The ESR happens when $\upsilon\mu h = g_{\mathrm{B}}\mathbf{H}_0$, where υ is the RF and μ_{B} the Bohr magneton. For EDSR, in the presence of a RF *electric* field $\delta\mathbf{E}_{\mathrm{RF}}$, the spin "sees" an equivalent *magnetic* field $\mathbf{H}_{\mathrm{RF}}$ given by

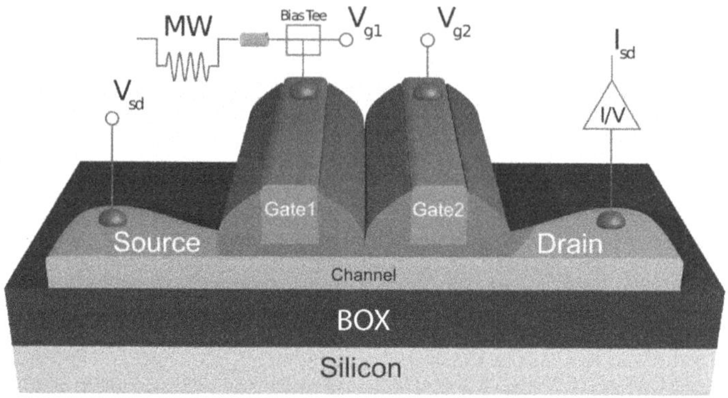

Figure 5.4 Schematic view of the electrical connections toward the coupled MOS-SET to control and read out the hole spin qubit under gate 1. Gate 2 is used to induce a Pauli blockade (spin-selective tunneling from dot 1 to dot 2). The read-out is done by the measurement of the DC drain-source current when a repetitive series of initialization/manipulation/read-out sequence is performed on gates 1 and 2 (repetition rate about 2 MHz). From Ref. [27].

$g\ \mathbf{H}_{RF} = (\delta g/\delta \mathbf{E}) \times \delta \mathbf{E}_{RF}\ \mathbf{H}_0$ (and because of the g-factor anisotropy, $\mathbf{H}_{RF}$ has a component perpendicular to $\mathbf{H}_0$).

The measurement scheme for the spin qubit is described in Fig. 5.2 and Fig. 5.4. It consists in measuring the source-drain current I_{SD} for a given V_{SD} in a few-millivolt range at a very low temperature ($T < 0.1$ K), with DC and microwave gate voltages applied in several sequences.

First, DC voltage is applied to gate 1 and to gate 2 in such a way that the two last empty levels in the dots are between the Fermi energy in the source and the drain for the occupation numbers (0,1), (1,1), and (0,2) (see Fig. 5.2). With a *hole* quantum dot we cannot say that the occupation numbers are absolute numbers and, in fact, "0 holes" means $2N$ holes with $N = 10 \approx 20$. This distinction is not important as long as the orbital-level spacing in the dot is much larger than the temperature and the Zeeman energy. This is realized in our hole quantum dots thanks to their very small size and their small density of states (compared to metals). These conditions—states (0,1), (1,1), and (0,2) lying in the Fermi energy window—are realized in a "triangular region" in (V_{gate1} and V_{gate2}) plot at finite V_{ds} (see Fig. 5.5): the empty hole states are indicated by the horizontal

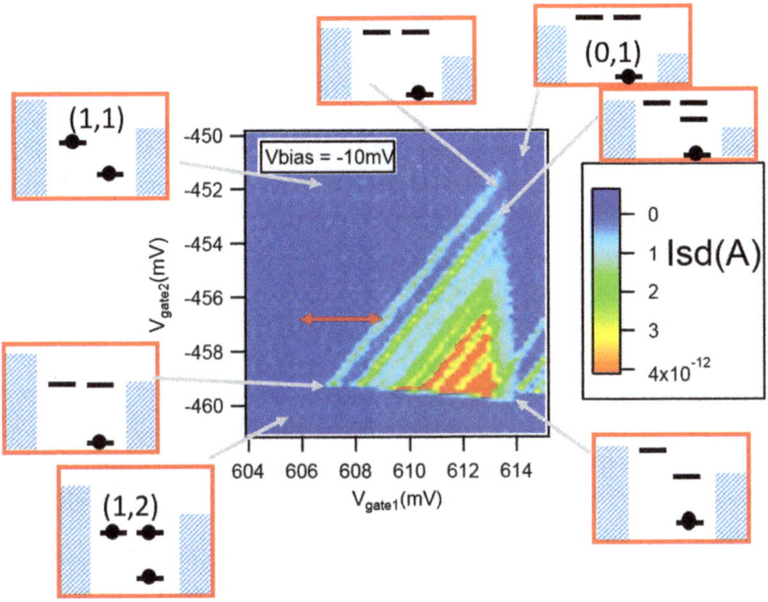

Figure 5.5 Drain-source current versus V_{gate1} and V_{gate2} at $T = 10$ mK and in absence of microwave on gate 1 ($V_{\text{ds}} = 10$ mV). The "triangular region" defines the region where the two last empty levels in the dots are between the Fermi energy in the source and the drain for the occupation numbers (0,1), (1,1), and (0,2) (see Fig. 5.2). The base of the triangle (along the diagonal) corresponds to the alignment of the two last empty levels in the dot, as the tip of the triangle (lower-right corner) corresponds to the maximum allowed detuning between these states: the last empty state of dot 1 aligned with the Fermi energy in the source and the last empty state of dot 2 aligned with the Fermi energy in the drain. The lines parallel to the base indicate when the last empty state of dot 1 is aligned with an excited state of dot 2. The tips of the red arrow indicate the points where by changing V_{gate1} we shift from the Coulomb blockade regime, that is, the (1,1) state below the Fermi energies both in source and drain (left tip), to the Pauli blockade regime—where the triplet T(1,1) is between the two Fermi energies but T(0,2) is not (right tip).

bar. The occupied hole states are marked with a black circle. The base of the triangle (along the diagonal) correspond to the alignment of the two last empty levels in the dot, as the tip of the triangle (lower-right corner) corresponds to the maximum allowed detuning between these states: the last empty state of dot 1 aligned with the

Fermi energy in the source and the last empty state of dot 2 aligned with the Fermi energy in the drain. The lines parallel to the base indicate when the last empty state of dot 1 is aligned with an excited state of dot 2. Between these lines there is current only if inelastic processes between the ground state of dot 1 and one state of dot 2 are permitted (by the concomitant emission of photons/phonons). The small extra figure in the lower-right corner, where there is a finite drain-source current, is a replica of the main triangle due to the presence of an offset charge.

Then we fix V_{gate2} and vary V_{gate1}, as indicated by the red arrow in Fig. 5.5. For $V_{\text{gate1}} \approx 606$ mV we are in the Coulomb blockade regime where the (1,1) state is the stable state: no empty state exists between the Fermi levels in the source and the drain. For $V_{\text{gate1}} \approx 609$ mV we are in the Pauli blockade regime, where the triplet T(1,1) lies between the two Fermi energies but T(0,2) does not. We apply a static magnetic field of about 0.144 T.

The sequence on V_{gate1} is as follows: In the initialization phase we put $V_{\text{gate1}} \approx 609$ mV and we wait for the time required ($\sim$150 ns) to ascertain that the initial state is $T^{-}(1,1)$, as explained in Section 5.2 (Fig. 5.2, top-center panel). Then we put $V_{\text{gate1}} \approx 606$ mV to be in the Coulomb blockade regime. We apply a microwave signal on V_{gate1} during τ_{burst} (Fig. 5.2, top-right panel: spin manipulation done in the Coulomb blockade regime). The burst can be applied at the beginning or at the end of the Coulomb blockade sequence (duration 175 ns) without observed difference, indicating that the inelastic spin scattering time T_1 (for flipping the spin orientation along the static magnetic field) is much longer than 175 ns. The mixing with the DC signal is done using a bias tee represented in Fig. 5.4. The microwave frequency is varied across the resonant frequency for the EDSR $\upsilon\mu h = g_{\text{B}}\mathbf{H}_0$ that is, 8.938 GHz. During the RF burst the spin of the hole located under gate 1 is rotating on the Bloch sphere. Then we return after $\sim$175 ns (much larger than τ_{burst}) on the Pauli blockade regime, $V_{\text{gate1}} \approx 609$ mV (Fig. 5.2, bottom panels). Depending on the respective spin orientation for the hole on dot 1 and on dot 2, a hole is either transferred or not: if the state after the burst is T(1,1) there is no hole transfer, but if the state is S(1,1), one carrier is transmitted into the drain. For $\tau_{\text{burst}} \approx 0$, there is no transfer because the initial state is T(1,1). We choose a read-

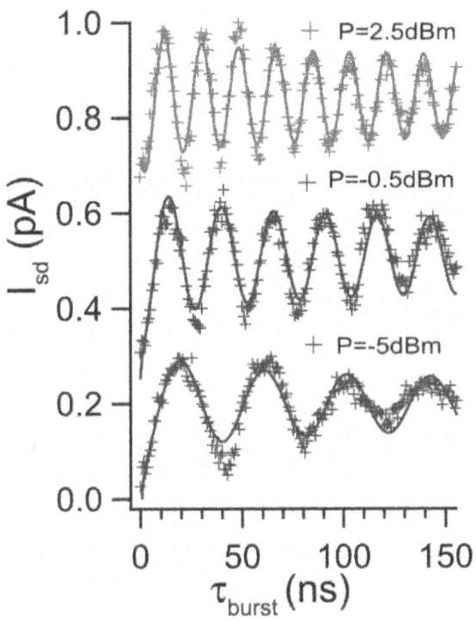

Figure 5.6 Rabi oscillation of the hole spin qubit under gate 1 as a function of the burst time during which an RF signal (8.938 GHz, $\mathbf{H} = 0.144$ T) is applied on gate 1, the coupled MOS-SET being in the Coulomb blockade regime. The solid lines are numerical fits as in Ref. [47]. Three powers for the RF signal are shown. As expected the Rabi frequency increases with the power, reaching 55 MHz for the largest applied power (2.5 dBm). Rabi frequencies are 24, 39, and 55 MHz for $P_{MW} = -5, -0.5$, and 2.5 dBm, respectively. From Ref. [27].

out time of about 150 ns such that the full sequence (initialization, manipulation, and read-out) lasts ~435 ns. This fixes a limit to our detection because one hole transferred from source to drain each 450 ns corresponds to a DC current of about 3.6 pA. Too long a time for the full sequence will result in a poor signal-to-noise ratio, as a shorter sequence will induce incomplete initialization or read-out.

The integrated DC current (integration time = 1 s) as a function of the RF burst duration is plotted in Fig. 5.6.

As expected, the detected current oscillates as a function of τ_{burst}. The inverse pseudo-period of the oscillations is the Rabi frequency. It increases linearly with the microwave voltage amplitude (the square root of the voltage power). For our largest power the Rabi frequency f_{Rabi} reaches 55 MHz (and even 85 MHz, not represented here [27]), which is much larger than for magnetic manipulation (ESR) of spin qubit in silicon ($f_{Rabi} \approx 100$ kHz, [15]).

Table 5.1 Characteristic times for the CMOS hole qubit and other electron spin qubits

$1/f_{\mathrm{Rabi}}$	$1/f_{\mathrm{Rabi}}$ (time for a 2π rotation)	T_2^* (shortest inhomogeneous dephasing time)	$T_{\mathrm{echo}}^{\mathrm{Hahn}}$
CMOS hole qubit (EDSR)	11 ns	60 ns	245 ± 12 ns
Natural silicon electron qubit using a P donor (ESR) [54]	0.3 us	55 ± 5 ns	206 ± 12 us
^{28}Silicon electron qubit using QuDot (ESR) [15]	3.3 us	120 us	28 ms

Various spin relaxation times can be evaluated by making echo type of experiments. The inhomogeneous dephasing time T_2* can be obtained by a Ramsey fringes-like experiment, which consists in applying two short, phase-coherent, $\pi/2$ MW pulses separated by a variable delay time ($\pi/2$ refers to $\pi/2$ rotation from the pole to the equator of the Bloch sphere). That way, a short T_2*, of ~60 ns (corresponding to the free evolution of the spin on the equator of the Bloch sphere), has been measured [27]. Dephasing the two short MW pulses can also be used to show that the hole spin can be rotated around two perpendicular axes on the Bloch sphere. This can be also done by changing the RF between the two pulses. If the source of dephasing fluctuates slowly on the timescale of the hole spin dynamics, spin echo techniques can extend spin coherence: a Hahn echo experiment, where a π pulse around the north pole is introduced halfway between the two $\pi/2$ pulses, permits to refocus the spin on the Bloch sphere, if the defocusing source is kept constant during the sequence. The fit of the amplitude of the oscillation decay versus the time between the two $\pi/2$ pulses gives a coherence time $T_{\mathrm{echo}}^{\mathrm{Hahn}} \approx 245 \pm 12$ ns [27]. Table 5.1 summarizes the various characteristic times for our qubit and comparison with the state of the art for ESR in silicon qubits.

T_2* and T_{echo} are relatively short (but much larger than the inverse Rabi frequency), and at least two mechanisms could be invoked as an explanation: the hyperfine coupling with a nuclear spin [24] (Overhauser noise) coming from ^{29}Si atoms, paramagnetic impurities, boron dopants in the channel, etc., and the charge noise. The distinction between the various mechanisms deserves further studies, but it is likely that the charge noise is dominating above the magnetic noise. In particular the gate is very well coupled to the qubit (a figure of merit for the CMOS technology), hence any kind of gate voltage noise is potentially affecting the qubit decoherence time. The sensitivity of the spin to the electric noise source is a penalty for using fast electrical manipulation of the hole spin.

The mechanism for EDSR is also not fully clarified in our hole spin quantum bit. It could be g-tensor modulation as explained at the beginning of this section [45]. It could also involve more standard Rashba spin-orbit couplings [48]. The magnetic-field angle dependence of spin blockade, EDSR, and Rabi frequency are currently studied to measure the g-tensor anisotropy and clarify the mechanism responsible for the observed EDSR.

5.4 Dispersive RF Gate-Reflectometry and Scalable 1D Linear Array Architecture

The next step toward a functional CMOS qubit is to perform fast, high-fidelity, *single-shot* qubit read-out. In Section 5.3, we measured only the spin orientation in dot 1 by integrating over $\sim$1 s the source-drain current at a repetition rate of 2 MHz. Moreover, each time we measure the spin orientation we destroy the phase and the qubit itself. There are two options for single-shot read-out of the charge transfer—therefore the single spin read-out—between dot 1 and dot 2. In the first option, we use an external high-bandwidth charge detector, for instance, a nearby single-electron transistor [15] or a quantum point contact [49]. In the second option, we measure the change of the quantum capacitance for dot 1 when a hole can be shared between the two dots—at the S(1,1)/S(0,2) degeneracy. The quantum capacitance is directly proportional to the density of states $C_{\mathrm{q}} \propto e^2 g(E_{\mathrm{F}})$, which becomes larger at the degeneracy point [39]. This quantum capacitance contributes to the geometrical gate capacitance of dot 1 and dot 2. By detecting

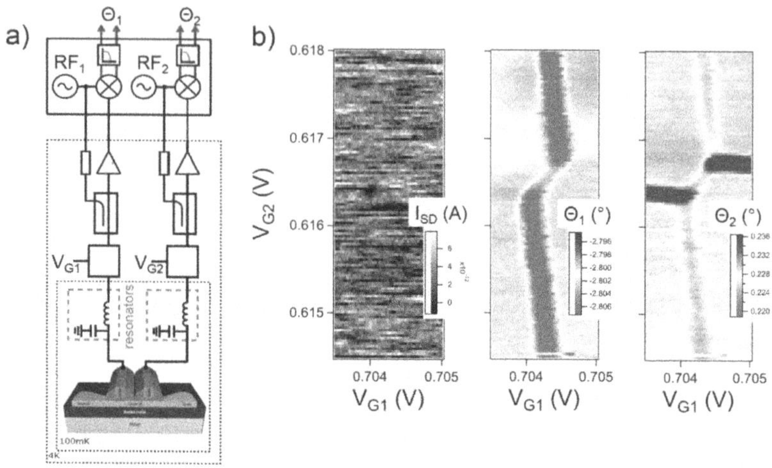

Figure 5.7 (a) Schematic of a dual-port reflectometry setup. Each read-out channel is connected to one of the two top gates of the device, so that two independent charge transfer detectors are realized. (b) Either of the sensors monitors the single charge transitions involving the dot underneath, though no net source-drain current flows through the transistor; Θ_1 and Θ_2 represent the phase of the signal back reflected by gate 1 and gate 2, respectively.

the small change in the gate capacitance of dot 2 it is possible to detect whether the double-dot system is in the S(1,1) or T(1,1) state [39]. We already noticed that the FDSOI trigate technology results in perfect control of the electrostatic potential in the silicon channel by the gate at least for our geometry: gate length = 30 nm, channel width = 15 nm, gate oxide thickness = 5 nm, and thick spacers. This means the gate electrode is very well connected to the qubit. Therefore, our devices are well adapted to this type of gate capacitance measurements [39, 50]. The gate capacitance can be measured with a large bandwidth, thanks to the so-called dispersive RF reflectometry technique [51, 52].

A typical RF reflectometry setup and measurement are shown in Fig. 5.7. Figure 5.7a reports a schematic of the dual-port reflectometry circuit, where the two gates of the devices represent two independent read-out channels; either circuit comprises a

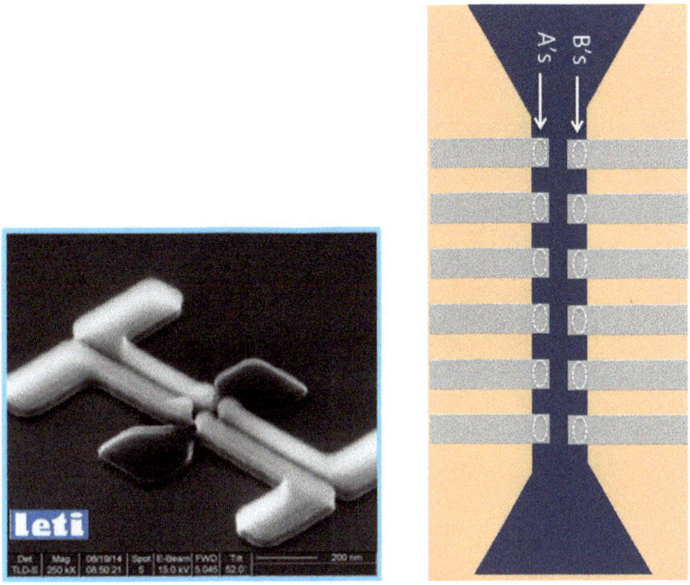

Figure 5.8 A linear array of CMOS spin qubits. (Right panel): the FDSOI nanowire (in blue) is covered with couples of split gates (in gray), forming coupled corner quantum dots indicated by dotted circles [40]. The quantum dots in row A can form the qubits as the quantum dots in row B are used to detect the Pauli blockade by dispersive RF gate reflectometry [53]. The A qubits can exchange their spin by nearest-neighbor exchange interaction, modulated by the back-gate voltage applied across the buried oxide to the substrate. (Top panel): A SEM view of a FDSOI nanowire covered by two split gates, fabricated in Grenoble at CEA-LETI.

resonator, to maximize the RF signal delivered to the sample, and electronics for the amplification of the signal back reflected. The latter is the demodulated-to-baseband signal, so the variations in its phase and amplitude point to the modifications in the device admittance. The potential of such a technique is shown by Fig. 5.7b: the phase signal recorded appears also when no source-drain carrier transport takes place, for instance, if the tunneling rates are too small. This property of gate-dispersive read-out allows one to get rid of multiple reservoirs in a few-qubit architecture, thereby leading to a tighter qubit pitch on the order of 70 nm, as in Fig. 5.8. Figure 5.7b also demonstrates that a combination of traces simultaneously

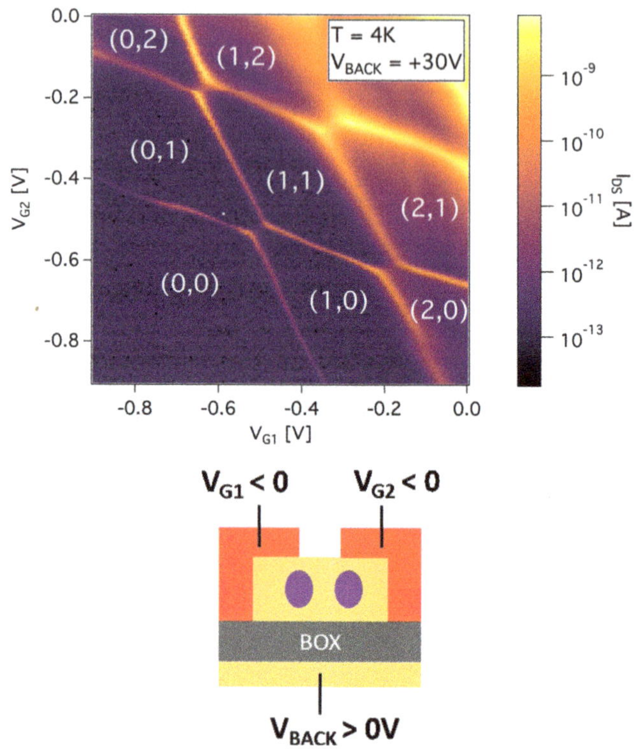

Figure 5.9 (Top): Stability diagram for two face-to-face electron corner dots: the drain-source current is recorded as a function of the two gate voltages applied on the split gate ($T = 4.2$ K, $V_{DS} = 3$ mV, and V_{back} (substrate bias) $= +30$ V). The current is zero when the electron number on both dots is fixed. The current lines delimitate regions where the number (n,m) of electrons on both dots is constant. (Right): Sketch of the cross section of the silicon nanowire (yellow) covered by the split gate (red). The two coupled quantum dots are featured as violet circles. From Ref. [53].

acquired from different gate sensors permits the reconstruction of the full honeycomb structure of a double-quantum-dot system where to perform qubit operations.

Nevertheless, single-shot read-out of the spin qubit has not been yet obtained using the dispersive RF reflectometry technique. This is currently under study in Grenoble. It will permit the measurement of the spin qubit on dot 1 using the gate reflectometry on dot 2, for instance, without the need for the drain reservoir (the source reservoir provides holes during the initialization step). This method permits one to envision a scalable qubit 1D linear array using the face-to-face arrangement of corner dots along a nanowire [40, 53], as shown in Fig. 5.8.

Figure 5.9 shows the stability diagram for two face-to-face electron corner dots as in Fig. 5.8 recorded at $T = 4.2$ K. The stability diagram is recorded at $V_b = +30$ V, for which value the two corner dots are relatively strongly coupled. The honeycomb lattice typical for two interacting quantum dots is clearly visible. The two dots are in parallel such that a drain-source current is measured along the line where the number of electrons is degenerate on one of the two dots (in contrast, for two dots in series with a negligible cotunneling effect the current will appear only at the intersections of the lines, the so-called triple points, evolving to triangles at finite V_{ds}, as in Fig. 5.5). For electrons it is possible to know the exact occupation number of the corner dots, which constitute therefore a clean platform for implementing an electron spin CMOS qubit and a 1D array as featured in Fig. 5.8.

5.5 Summary

By adapting the most advanced CMOS technology—available only on preindustrial and industrial platforms—we have been able to make a CMOS device work as a spin qubit at a low temperature. The spin qubit is based on a PMOS type of NW-FET that allows for the first time the realization of a hole spin qubit. The decisive advantage of using holes is that the spin can be manipulated by an electric voltage applied on standard gates, thanks to spin-orbit coupling. This breakthrough provides an entirely new way to envision a CMOS quantum computer core that is fully cointegrable with classical CMOS cryoelectronic peripherals. The realization of a scalable multiqubit layout remains to be demonstrated. One important, possibly indispensable, ingredient is a means to extend the spatial distance over which qubits can be coupled. Another topic that requires continued attention for all semiconductor qubits is the realization of high-fidelity qubit manipulation, particularly for two-qubit gates, and scalable read-out.

Acknowledgment

The research leading to these results has been supported by the European Community's H2020 Framework under the project MOS-Based Quantum Information Technology (MOSQUITO) Grant Number 688539.

References

1. Scott-Thomas, J. H. F., Field, S. B., Kastner, M. A., Smith, H. I., and Antonadis, D. A. (1989). Conductance oscillations periodic in the density of a one-dimensional electron gas, *Phys. Rev. Lett.*, **62**, pp. 583–586.
2. Ali, D., and Ahmed, H. (1994). Coulomb blockade in a silicon tunnel junction device, *Appl. Phys. Lett.*, **64**, pp. 2119–2120.
3. Takahashi, Y., Nagase, M., Namatsu, H., Kurihara, K., Iwdate, K., Nakajima, Y., Horiguchi, S., Murase, K., and Tabe, M. (1995). Fabrication technique for Si single-electron transistor operating at room temperature, *Electron. Lett.*, **31**, p. 136.
4. Boeuf, F., Jehl, X., Sanquer, M., and Skotnicki, T. (2003). Controlled single-electron effects in non overlapped ultra-short silicon field effect transistors, *IEEE Trans. Nanotechnol.*, **2**, pp. 144–146.
5. Hofheinz, M., Jehl, X., Sanquer, M., Molas, G., Vinet, M., and Deleonibus, S. (2006). A simple and controlled single electron transistor based on doping modulation in silicon nanowires, *Appl. Phys. Lett.*, **89**, p. 143504.
6. Jehl, X., Voisin, B., Charron, T., Clapera, P., Ray, S., Roche, B., Sanquer, M., Djordjevic, S., Devoille, L., Wacquez, R., and Vinet, M. (2013). A hybrid metal/semiconductor electron pump for practical realization of a quantum ampere, *Phys. Rev. X*, **3**, p. 021012.
7. Lavieville, R., Barraud, S., Arvet, C., Vizioz, C., Corna, A., Jehl, X., Sanquer, M., and Vinet, M. (2016). Demonstration of single hole transistor and hybrid circuits for multi-valued logic and memory applications up to 350K using CMOS silicon nanowires, *Adv. Electron. Mater.*, **2**, pp. 1500244–1500251.
8. Takahashi, Y., Ono, Y., Fujiwara, A., and Inokawa, H. (2002). Silicon single-electron devices, *J. Phys.: Condens. Matter*, **14**, pp. R995–R1033.
9. Gautier, J., Jehl, X., and Sanquer, M. (2009). Single electron devices and applications, in *Nano-CMOS Era* (Deleonibus, S., ed.), Pan Standford Publishing, Singapore, pp. 279–297.

10. Clapera, P., Ray, S., Jehl, X., Sanquer, M., Valentian, A., and Barraud, S. (2015). Co-integrated voltage controlled oscillators and a MOS-single electron transistor, *Phys. Rev. Appl.*, **4**, p. 044009.
11. Loss, D., and DiVincenzo, D. P. (1998). Quantum computation with quantum dots, *Phys. Rev. A*, **57**, pp. 120–126.
12. Shulman, M. D., Dial, O. E., Harvey, S. P., Bluhm, H., Umansky, V., and Yacoby, A. (2012). Demonstration of entanglement of electrostatically coupled singlet-triplet qubits, *Science*, **336**, pp. 202–205.
13. Kawakami, E., Scarlino, P., Ward, D. R., Braakman, F. R., Savage, D. E., Lagally, M. G., Friesen, M., Coppersmith, S. N., Eriksson, M. A., and Vandersypen, L. M. K. (2014). Electrical control of a long-lived spin qubit in a Si/SiGe quantum dot, *Nat. Nanotechnol.*, **9**, pp. 666–670.
14. Itoh, K. M., and Watanabe, H. (2014). Isotope engineering of silicon and diamond for quantum computing and sensing applications, *MRS Commun.*, **4**, pp. 143–157.
15. Veldhorst, M., Hwang, J., Yang, C., Leenstra, A., de Ronde, B., Dehollain, J., Muhonen, J., Hudson, F., Itoh, K., Morello, A., and Dzurak, A. S. (2014). An addressable quantum dot qubit with fault-tolerant control-fidelity, *Nat. Nanotechnol.*, **9**, pp. 981–985.
16. Veldhorst, M., Yang, C. H., Hwang, J. C. C., Huang, W., Dehollain, J. P., Muhonen, J. T., Simmons, S., Laucht, A., Hudson, F. E., Itoh, K. M., Morello, A., and Dzurak, A. S. (2015). A two-qubit logic gate in silicon, *Nature*, **526**, pp. 410–414.
17. Ladd, T. D., Jelezko, F., Laflamme, R., Nakamura, Y., Monroe, C., and O'Brien, J. L. (2010). Quantum computers, *Nature*, **464**, pp. 45–53.
18. Lucero, E., Barends, R., Chen, Y., Kelly, J., Mariantoni, M., Megrant, A., O'Malley, P., Sank, D., Vainsencher, A., Wenner, J., White, T., Yin, Y., Cleland, A. N., and Martinis, J. M. (2012). Computing prime factors with a Josephson phase qubit quantum processor, *Nat. Phys.*, **8**, pp. 719–723.
19. Lanyon, B. P., Hempel, C., Nigg, D., Müller, M., Gerritsma, R., Zähringer, F., Schindler, P., Barreiro, J. T., Rambach, M., Kirchmair, G., Hennrich, M., Zoller, P., Blatt, R., and Roos, C. F. (2011). Universal digital quantum simulation with trapped ions, *Science*, **334**, pp. 57–61.
20. Millikan, R. A. (1911). On the elementary electrical charge and the Avogadro constant, *Phys. Rev.*, **32**, pp. 349–397.
21. Ono, K., Austing, D. G., Tokura, Y., and Tarucha, S. (2002). Current rectification by Pauli exclusion in a weakly coupled double quantum dot system, *Science*, **297**(5585), pp. 1313–1317.

22. Elzerman, J. M., Hanson, R., Van Beveren, L. W., Witkamp, B., Vandersypen, L. M. K., and Kouwenhoven, L. P. (2004). Single-shot read-out of an individual electron spin in a quantum dot, *Nature*, **430**(6998), pp. 431–435.
23. Li, R., Hudson, F. E., Dzurak, A. S., and Hamilton, A. R. (2015). Pauli spin blockade of heavy holes in a silicon double quantum dot, *Nano Lett.*, **15**, pp. 7314–7321.
24. Bohuslavskyi, H., Kotekar-Patil, D., Maurand, R., Corna, A., Barraud, S., Bourdet, L., Hutin, L., Niquet, Y.-M., Jehl, X., De Franceschi, S., Vinet, M., and Sanquer, M. (2016). Pauli blockade in a few-hole PMOS double quantum dot dominated by spin-orbit interaction, *Appl. Phys. Lett.*, **109**, pp. 193101–193104.
25. Koppens, F. H. L., Folk, J. A., Elzerman, J. M., and Hanson, R. (2005). Control and detection of singlet- triplet mixing in a random nuclear field, *Science*, **309**, pp. 1346–1350.
26. Qassemi, F., Coish, W. A., and Wilhelm, F. K. (2009). Stationary and transient leakage current in the Pauli spin blockade, *Phys. Rev. Lett.*, **102**, p. 176806.
27. Maurand, R., Jehl, X., Kotekar Patil, D., Corna, A., Bohuslavskyi, H., Laviéville, R., Hutin, L., Barraud, S., Vinet, M., Sanquer, M., and De Franceschi, S. (2016). A CMOS silicon spin qubit, *Nat. Commun.*, **7**, p. 13575.
28. Witzel, W. M., Carroll, M. S., Morello, A., Cywinski, L., and Das Sarma, S. (2010). Electron spin decoherence in isotope-enriched silicon, *Phys. Rev. Lett.*, **105**, p. 187602.
29. Steger, M., Saeedi, K., Thewalt, M. L. W., Morton, J. J. L., Riemann, H., Abrosimov, N. V., Becker, P., and Pohl, H. J. (2012). Quantum information storage for over 180 s using donor spins in a 28Si "semiconductor vacuum," *Science*, **336**, p. 1280.
30. O'Brien, J. L., Schofield, S. R., Simmons, M. Y., Clark, R. G., Dzurak, A. S., Curson, N. J., Kane, B. E., McAlpine, N. S., Hawley, M. E., and Brown, G. W. (2001). Towards the fabrication of phosphorus qubits for a silicon quantum computer, *Phys. Rev. B*, **64**, p. 161401.
31. Zwanenburg, F. A., Dzurak, A. S., Morello, A., Simmons, M. Y., Hollenberg, L. C., Klimeck, G., Rogge, S., Coppersmith, S. N., and Eriksson, M. A. (2013). Silicon quantum electronics, *Rev. Mod. Phys.*, **85**, p. 961.
32. Sellier, H., Lansbergen, G. P., Caro, J., Rogge, S., Collaert, N., Ferain, I., Jurczak, M., and Biesemans, S. (2006). Transport spectroscopy of a single dopant in a gated silicon nanowire, *Phys. Rev. Lett.*, **97**, p. 206805.

33. Pierre, M., Wacquez, R., Jehl, X., Sanquer, M., Vinet, M., and Cueto, O. (2010). Single donor ionization energies in a nanoscale CMOS channel, *Nat. Nanotechnol.*, **5**, pp. 133–137.

34. Roche, B., Dupont-Ferrier, E., Voisin, B., Cobian, M., Jehl, X., Wacquez, R., Vinet, M., Niquet, Y. M., and Sanquer, M. (2012). Detection of a large valley-orbit splitting in silicon with two-donor spectroscopy, *Phys. Rev. Lett.*, **108**, p. 206812.

35. Muhonen, J. T., Dehollain, J. P., Laucht, A., Hudson, F. E., Kalra, R., Sekiguchi, T., Itoh, K. M., Jamieson, D. N., McCallum, J. C., Dzurak, A. S., and Morello, A. (2014). Storing quantum information for 30 seconds in a nanoelectronic device, *Nat. Nano*, **9**, pp. 986–991.

36. Morton, J. J. L., Tyryshkin, A. M., Brown, R. M., Shankar, S., Lovett, B. W., Ardavan, A., Schenkel, T., Haller, E. E., Ager, J. W., and Lyon, S. A. (2008). Solid-state quantum memory using the 31P nuclear spin, *Nature*, **455**, pp. 1085–1088.

37. Barraud, S., Coquand, R., Cassé, M., Koyama, M., Hartmann, J. M., Maffini-Alvaro, V., Comboroure, C., Vizioz, C., Auissenac, F., Faynot, O., and Poiroux, T. (2012). Performance of omega-shaped-gate silicon nanowire MOSFET with diameter down to 8 nm, *IEEE Electron Device Lett.*, **33**, pp. 1526–1528.

38. Natarajan, S., et al. (2014). A 14nm logic technology featuring 2nd-generation FinFET interconnects, self-aligned double patterning and a 0.0588 μm^2 SRAM cell size, in *IEEE International Electron Devices Meeting*, pp. 71–73.

39. Betz, A. C., Wacquez, R., Vinet, M., Jehl, X., Saraiva, A. L., Sanquer, M., Ferguson, A. J., and Gonzalez-Zalba, M. F. (2015). Dispersively detected Pauli spin-blockade in a silicon nanowire field-effect transistor, *Nano Lett.*, **15**(7), pp. 4622–4627.

40. Voisin, B., Nguyen, V., Renard, J., Jehl, X., Barraud, S., Triozon, F., Vinet, M., Duchemin, Y., Niquet, Y. M., De Franceschi, S., and Sanquer, M. (2015). Few-electron edge-state quantum dots in a silicon nanowire field-effect transistor, *Nano Lett.*, **14**, p. 2094.

41. Pribiag, V. S., Nadj-Perge, S., Frolov, S. M., van den Berg, J. W. G., van Weperen, I., Plissard, S. R., Bakkers, E. P. A. M., and Kouwenhoven, L. P. (2013). Electrical control of single hole spins in nanowire quantum dots, *Nat. Nanotechnol.*, **8**, pp. 170–174.

42. Hensel, J. C., and Feher, G. (1963). Cyclotron resonance experiments in uniaxially stressed silicon: valence band inverse mass parameters and deformation potentials, *Phys. Rev.*, **129**, pp. 1041–1062.

43. Winkler, R. (2003). *Spin–Orbit Coupling Effects in Two-Dimensional Electron and Hole Systems*, vol. 191, Springer, New York.
44. Testelin, C., Bernardot, F., Eble, B., and Chamarro, M. (2009). Hole-spin dephasing time associated with hyperfine interaction in quantum dots, *Phys. Rev. B*, **79**, p. 195440.
45. Kato, Y., Myers, R. C., Driscoll, D. C., Gossard, A. C., Levy, J., and Awshalom, D. D. (2003). Gigahertz electron spin manipulation using voltage-controlled g-tensor modulation, *Science*, **299**, pp. 1201–1204.
46. Voisin, B., Maurand, R., Barraud, S., Vinet, M., Jehl, X., Sanquer, M., Renard, J., and De Franceschi, S. (2016). Electrical control of the first hole spin in a silicon nanowire MOSFET, *Nano Lett.*, **16**, p. 88.
47. Koppens, F. H. L., Klauser, D., Coish, W. A., Nowack, K. C., Kouwenhoven, L. P., Loss, D., and Vandersypen, L. M. K. (2007). Universal phase shift and non-exponential decay of driven single-spin oscillations, *Phys. Rev. Lett.*, **99**, p. 106803.
48. Golovach, V. N., Borhani, M., and Loss, D. (2006). Electric-dipole-induced spin resonance in quantum dots, *Phys. Rev. B*, **74**, p. 165319.
49. Wu, X., Ward, D. R., Prance, J. R., Kim, D., Gamble, J. K., Mohr, R. T., Shi, Z., Savage, D. E., Lagally, M. G., Friesen, M., Coppersmith, S. N., and Eriksson, M. A. (2014). Two-axis control of a singlet–triplet qubit with an integrated micromagnet, *Proc. Natl. Acad. Sci. U. S. A.*, **111**, pp. 11938–11942.
50. Crippa, A., Maurand, R., Kotekar-Patil, D., Corna, A., Bohuslavskyi, H., Orlov, A. O., Fay, P., Laviéville, R., Barraud, S., Vinet, M., Sanquer, M., De Franceschi, S., and Jehl, X. (2017). Level spectrum and charge relaxation in a silicon double quantum dot probed by dual-gate reflectometry, *Nano Lett.*, **17**(2), pp. 1001–1006.
51. Ciccarelli, C., and Ferguson, A. (2011). Impedance of the single-electron transistor at radio-frequencies, *New J. Phys.*, **13**, pp. 093015–093023.
52. Colless, J., Mahoney, A., Hornibrook, J., Doherty, A., Lu, H., Gossard, A., and Reilly, D. (2013). Dispersive readout of a few-electron double quantum dot with fast rf gate sensors, *Phys. Rev. Lett.*, **110**, pp. 046805–046809.
53. De Franceschi, S., Hutin, L., Maurand, R., Bourdet, L., Bohuslavskyi, H., Corna, A., Kotekar-Patil, D., Barraud, S., Jehl, X., Niquet, Y. M., Sanquer, M., and Vinet, M. (2016). SOI technology for quantum information processing, in *IEEE International Electron Devices Meeting (IEDM)*, pp. 339–342.

54. Pla, J. J., Tan, K. Y., Dehollain, J. P., Lim, W. H., Morton, J. J. L., Jamieson, D. N., Dzurak, A. S., and Morello, A. (2012). A single-atom electron spin qubit in silicon, *Nature*, **489**, p. 541.
55. Jehl, X., and Sanquer, M. (2013). Single donor transport spectroscopy in ultimate silicon transistors, in *Single Atom Nanoelectronics* (Prati, E., and Shinada, T., eds.), Pan Stanford Publishing, Singapore, pp. 231–253.
56. Stegner, A. R., Tezuka, H., Andlauer, T., Stutzmann, M., Thewalt, M. L. W., Brandt, M. S., and Itoh, K. M. (2010). Isotope effect on electron paramagnetic resonance of boron acceptors in silicon, *Phys. Rev. B*, **82**, p. 115213.

Chapter 6

Physically Defined Coupled Silicon Quantum Dots Containing a Few Electrons for Electron Spin Qubits

Tetsuo Kodera, Kosuke Horibe, and Shunri Oda

Tokyo Institute of Technology, 2-12-1 Ookayama, Meguro, Tokyo, 152-8552, Japan

kodera.t.ac@m.titech.ac.jp

Scalable architectures for quantum computing such as solid-state qubits are necessary to perform useful computations. Research on Si qubits has accelerated the demonstration of 2-qubit logic gates with enhanced coherence time using isotopically purified substrates. The main unresolved issue is the large-scale integration of qubits. The advantage of Si-based systems is that they can employ CMOS process technologies for large-scale integration. We describe the fabrication and electrical properties of physically defined coupled quantum dots prepared on silicon-on-insulator substrates, which are promising for high-density integration. Then, we present the measurement results of charge sensing and describe few-electron regimes, which are essential for multiple qubit integration.

Emerging Devices for Low-Power and High-Performance Nanosystems: Physics, Novel Functions, and Data Processing

Edited by Simon Deleonibus

ISBN 978-981-4800-11-2 (Hardcover), 978-0-429-45873-6 (eBook)

www.panstanford.com

6.1 Introduction

Over the past several decades, the semiconductor industry has achieved miniaturization of device sizes and great improvement in operating speed, power consumption, and integration of devices on chips. The size of transistors has decreased to nanometers, entering the world of quantum mechanics. There are a number of approaches aimed at further development, such as continuation of miniaturization, introduction of new materials and structures, and use of new architectures. One promising approach is to utilize quantum mechanics. Indeed, research and development of quantum computers has recently accelerated. A decade ago the research community predicted that quantum computing would be realized in 30 years or more. Yet quantum computing has already been realized: D-Wave's latest computers are based on quantum annealing [1] and superconducting devices. Practical applications that would be difficult for current supercomputers may be demonstrated by quantum computers within several years. The first quantum computation of simple factorization of the number 15 was demonstrated using 7-qubit nuclear magnetic resonance (NMR)-based devices in 2001 [2]. However, it is difficult to scale up NMR beyond a few qubits. Scalable architectures, such as solid-state qubits, are necessary to perform useful computations. Superconducting qubits [3] and semiconductor quantum dots (QDs) are being investigated. GaAs-based QDs are being studied extensively; however, nuclear spins in Ga and As atoms fatally limit the coherence time of electron spins in GaAs systems. On the other hand, natural silicon contains only 4% isotope with nuclear spins, and thus, long electron-spin coherence times are expected. Moreover, using isotopically engineered Si [4], qubits with long electron-spin coherence times were demonstrated [5, 6]. Research on Si qubits has accelerated the demonstration of 2-qubit logic gates [7]. The main outstanding issue is the large-scale integration of qubits. The advantage of Si-based systems is that they can employ complementary metal-oxide semiconductor (CMOS) process technologies for large-scale integration.

In this chapter, we first give a brief introduction of quantum computing and quantum bits in QDs. Then, we describe the fabrication and electrical properties of physically defined coupled QDs prepared on silicon-on-insulator (SOI) substrates [8–15], which are promising for high-density integration. Finally, we present the measurement results of charge sensing and discuss few-electron regimes, which are essential for multiple qubit integration.

6.2 Quantum Computing and QDs

6.2.1 Quantum Computers

Research in the field of quantum computing started in the 1960s, but it was unclear whether the quantum computer would be superior in computation speed compared to any classical computer. Therefore, only theoretical approaches were being followed at that time. However, in 1994, the quantum algorithm, which can solve prime factorizations with a very small error probability and at particularly high speed, was proposed by Shor [16]. The difficulty of solving prime factorizations of very large integers ensures the viability of encrypted communications, such as Rivest–Shamir–Adleman (RSA) cryptography. Indeed, it is impossible to solve such prime factorizations within realistic computation times even by using today's best-performance supercomputers. However, if a quantum computer with Shor's quantum algorithm comes out, prime factor decomposition can be carried out significantly more rapidly, which would have an extremely large social impact. Thus, experiments on quantum computers in real physical systems have been accelerated since the discovery of the quantum algorithm.

6.2.2 Qubits in QDs

A bit in quantum computing, termed a "quantum bit" (qubit), is realized by the superposition of two-level quantum states. Quantum computing requires two well-defined and controllable states to qualify a qubit, which possesses the stability and ability to construct a quantum logic gate. So far, various two-level systems have been investigated theoretically and experimentally to build qubits, such as NMR on molecules in solution [2], trapped ions, and linear optical quantum effects [17]. However, for the purpose of large-scale integration, qubits based on solid-state quantum systems are more promising. In particular, charges or spins of electrons, confined in semiconductor QDs, have attracted attention as candidate state variables for defining qubits.

One example of a charge qubit is a single electron confined in a tunnel-coupled double quantum dot (DQD), in which we use the superposition of the probability of presence of the single electron in each QD. Charge qubit operation has been achieved in a study using a 2D electron gas in a compound semiconductor heterostructure (GaAs/AlGaAs) [18]. However, charge qubits typically have short coherence times (several ns), which is not enough to perform several thousand quantum gate operations before decoherence occurs.

On the other hand, longer coherence times of several microseconds have been expected in spin qubits using spins of electrons confined in semiconductor QDs [19–21]. It has been shown that the basic quantum gates can be obtained by single spin manipulation combined with 2-qubit operations. Loss and DiVincenzo have shown that an exclusive-or (XOR) gate can be constructed by combining single spin rotations with SWAP gate operations swapping two qubits [19]. Since then, experiments on spin qubits using QDs have intensified. Electron spins confined in QDs have been manipulated by the exchange interaction between tunnel-coupled QDs [22] or by electron spin resonance (ESR) using the effective oscillating external magnetic field [23, 24]. Measurements of spin states have been performed by spin-to-charge conversion using the Pauli spin blockade [25] or the energy-selective spin readout [26].

Early studies on spin qubits using spins of electrons in QDs have been conducted with the GaAs/AlGaAs heterostructure because it has a relatively small electron effective mass ($m* = 0.067$ m_0 for GaAs, where m_0 is the mass of a free electron). The small electron effective mass contributes to the apparent quantum effect experimentally at cryogenic temperatures. Electron beam lithography (EBL) was used to fabricate devices with dimensions suitable for constructing clear quantum states and investigating quantum properties. However, progress in quantum operation has revealed that the hyperfine interactions between electron spins and nuclear spins, in compound semiconductors, destroy the coherence of the spin qubits [20, 27]. Therefore, in recent years, the study of group IV semiconductors, which naturally have fewer nuclear spins, has attracted attention. Si is a particularly promising material for elaborating spin qubits because it has extremely long coherence times, owing to weak spin-orbit coupling, free from piezoelectric electron-phonon coupling, and the capability of isotopic purification for the preparation of nuclear-spin-free Si substrates to remove hyperfine coupling [4]. The other important advantage of Si is that the fabrication process for Si qubits is similar to conventional Si device fabrication technologies, such as the CMOS process.

Recently, significant progress has been made in electron spin qubits in silicon. A singlet-triplet qubit in a DQD based on Si/SiGe heterostructures has been demonstrated [5]. In the singlet-triplet qubit, manipulations of spins by exchange interactions have been demonstrated with a spin dephasing timeT_2* of 360 ns. In another study, a longitudinal spin relaxation time T_1 of 10 ms at zero magnetic field is extended to $T_1 = 3$ s for triplet states at $\mathbf{B} = 1$ T in singlet-triplet qubits [28]. A singlet-triplet qubit has also been studied in a DQD based on the coupled phosphorus dimers in Si, and T_1 was measured to be 4 ms [29]. Single-electron spin qubits in QDs based on Si/SiGe heterostructures [30, 31], Si MOS structures [32, 33], and phosphorus donors in Si [34] have been reported. In Ref. [32], manipulation of single-electron spin qubits has been achieved by using ESR, with a long dephasing time T_2* of 63 ns. Furthermore, a controlled-NOT gate of 2-qubit operation has been realized in a DQD based on isotopically enriched Si MOS structures [7].

6.2.3 Coupled Si QDs Containing a Few Electrons

To implement quantum logic gates based on electron spins, it is necessary to adjust the number of electrons in an individual QD to a few electrons or a single electron to create spin states that are energetically well defined and separated from other states. However, it was difficult to form single-electron occupancy of QDs in Si devices because of the relatively large effective mass m^* of electrons (for the minor axis direction $m^* = 0.19\ m_0$, and for the major axis direction

$m^* = 0.916\ m_0$) and their small tunneling rates compared to those in the GaAs/AlGaAs heterostructure. The few-electron regime has been achieved in silicon QDs in Si/SiGe heterostructures and Si MOS structures by forming smaller QDs using sophisticated fabrication technologies.

6.2.4 Valley-Orbit States in Silicon QDs

The conduction band structure of Si is more complex than that of GaAs because electrons in the conduction band of silicon have valley degrees of freedom [35]. In recent years, valley physics in silicon has been studied both experimentally and theoretically. Bulk (3D) silicon has sixfold degenerate conduction band minima (valleys) in the Brillouin zone (Fig. 6.1a). This degeneracy is lifted by strain or the presence of an interface. Confinement of electrons in the z direction, or 2D silicon, lifts the sixfold valley degeneracy to four Δ-valleys and twofold Γ-valleys, as shown in the right image of Fig. 6.1a and the center part of Fig. 6.1b. Here, the four Δ-valleys and two Γ-valleys have a light effective mass ($m^* = 0.19\ m_0$) and a heavy effective mass ($m^* = 0.916\ m_0$) parallel to the interface, respectively. Therefore, the four Δ-valleys have an energy several 10 meV higher than the two Γ-valleys. The sharp and flat Si/SiO_2 interface produces a potential step in the z direction and lifts the degeneracy of the Γ-valleys to the valley splitting, E_v, which is theoretically predicted to be on the order of 0.1 to 0.3 meV [36]. The experimental E_v of a metal-oxide semiconductor (MOS) inversion layer varies from 0.3 to 1.2 meV [37]. A very large E_v of 23 meV has been observed in SiO_2/Si (100)/SiO_2 quantum well transistors using a SIMOX (separation by implantation of oxygen) substrate [38]. In a silicon single QD based on a MOS structure, an E_v of 0.3 to 0.8 meV has been measured by using magnetic spectroscopy or a steep increase in the relaxation of electrons (hot spot) [33]. In Si/SiGe QDs, an E_v of 0.12 to 0.27 meV has been observed [39]. Hybridization between valleys and orbits has been predicted [40], making it inappropriate to define distinct orbital and valley quantum numbers. Whether the valley-orbit level behaves like valleys or orbits depends on the degree of mixing. The "valley blockade" is the blockade between different valley states. The valley blockade

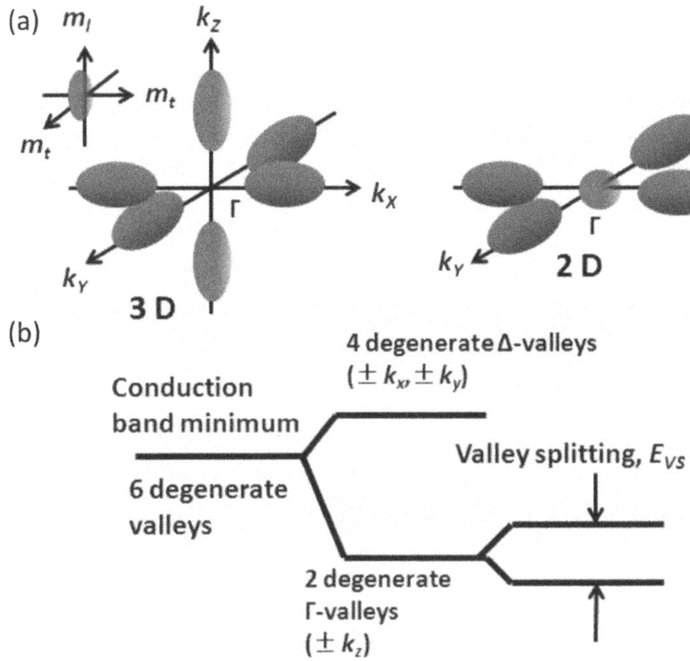

Figure 6.1 (a) Schematic diagram of conduction band minima (valleys) in bulk silicon. Six ellipsoids of equal energy in the Brillouin zone are shown. Each ellipsoid has two light traverse masses (m_t) and a heavy longitudinal mass (m_l). Confinement of electrons in the z direction, or 2D silicon, lifts the six-fold valley degeneracy to four Δ-valleys and twofold Γ-valleys. (b) Schematic of six degenerate valleys in bulk silicon (left), lifting of the valley degeneracy under z direction confinement (center), and valley splitting due to a potential step at the Si/SiO_2 interface (right).

has also been observed in a carbon-nanotube-based QD [41]. In silicon DQDs based on a MOS structure, the valley blockade has been observed in spin-valley blockade configurations [32]. Typically, larger valley splitting is desirable for electron spin qubits because larger splitting prevents different valleys from mixing and forming a valley blockade. The elements of valley physics described here, valley splitting, valley-orbit states, and valley blockade, require further investigation to reveal how they influence electron spin qubit operation.

6.2.5 Charge Sensing

Single-electron transistors (SETs) have been studied because their high charge sensitivity allows them to be used as charge sensors (CSs) for reading out qubit states in QDs [42]. In QD systems with GaAs/AlGaAs and Si/SiGe heterostructures, many experiments have been carried out with quantum point contact (QPC) CSs [43, 44]. Recent studies have focused on the use of a SET CS in order to improve the signal-to-noise ratio (SNR) during qubit readout [45]. In qubit measurements, the so-called spin-to-charge conversion allows the spin state of electrons in QDs to be detected by sensing the electron occupancy states of the QDs using a nearby QPC or SET as a CS.

6.3 Device Structure and Fabrication

6.3.1 Structure

Figures 6.2a and 6.2b show top-view scanning electron micrographs of a single QD device and a DQD device, respectively, while Fig. 6.2c shows a schematic cross section of the device structure. QDs and side gates are formed in the SOI layer by EBL and reactive ion etching (RIE). The QDs and side gates are passivated by thermal oxidation. Then, an SiO_2 layer is deposited by low-pressure chemical vapor deposition (LPCVD) as a gate insulator. A poly-Si top gate (TG), deposited by LPCVD, is fabricated on the gate insulator. The device has a MOS structure. A 2D electron gas (2DEG) layer is induced near the Si/SiO_2 interface by the electric field generated by the TG. The source and drain regions are heavily n-type doped by P ion implantation and an activation and drive-in process. During ion implantation, the TG acts as a mask to self-align the source and drain. Thus, QDs are located in the undoped region.

6.3.2 Fabrication

The device is fabricated by the following process flow: thinning of the SOI layer, forming of the QDs and side gates using EBL, passivation of the QDs, forming of the TG contact, and forming of the

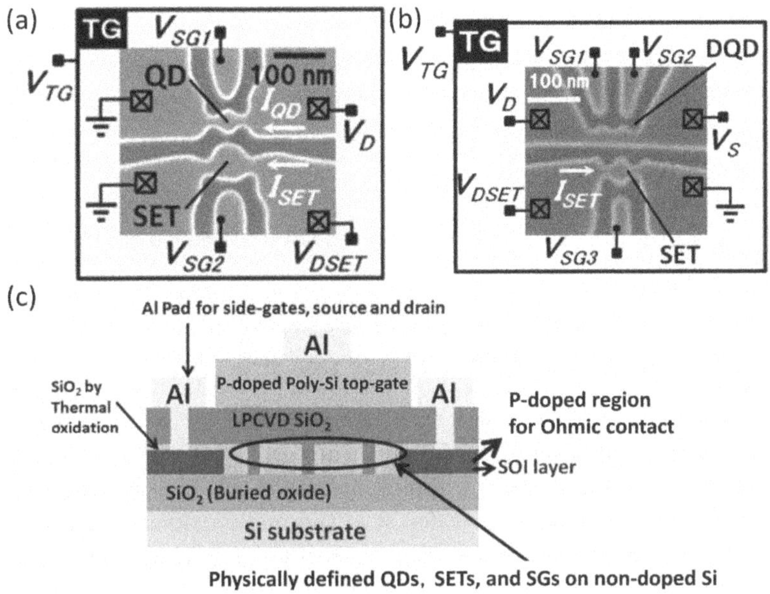

Figure 6.2 (a) (b) Scanning electron micrographs of (a) a single QD device and (b) a DQD device. (c) Schematic cross section of the device structure.

metal contact pads. Next, we introduce the details of the fabrication process.

6.3.2.1 Wafer cleaning

In the fabrication of silicon devices, process cleanliness is extremely important. Therefore, wafer cleaning is performed after cutting of the initial SOI wafer and between each fabrication step. Particles and impurities are removed from the wafers with, in sequence, pure water, SPM (H_2SO_4:H_2O_2 = 1:1) at 100°C (to remove organic and metallic impurities), SC1 (NH_3:H_2O_2:pure water = 1:5:25) at 150°C (to remove organic impurities and silicon particles), SC2 (HCl:H_2O_2:pure water = 1:1:5) at 100°C (to remove metal impurities), and 1.5% HF (to remove oxide and metal impurities). Here, SPM stands for sulfuric acid–peroxide mixture, SC1 stands for standard cleaning 1, and SC2 stands for standard cleaning 2.

6.3.2.2 Thinning of the SOI layer

To observe electric transport characteristics and quantum effects, substantial charging energy or quantum orbital-level spacing is required. Thus, a thin SOI layer is advantageous because it allows a small total capacitance and strong quantum confinement to be achieved along the perpendicular direction. The thickness of the SOI layer is adjusted to around 30 nm by repeated thermal oxidation and HF etching.

6.3.2.3 Formation of QDs and side gate structures

QDs, a proximal SET, and a side gate structure are formed in the SOI layer by EBL and RIE. We performed thermal oxidation prior to EBL to reduce electron-trapping defects at the Si/SiO_2 interface. After SOI thinning of the wafer, it is thermally oxidized at 900°C for 5 min. to create a 3-nm-thick protective SiO_2 layer on the SOI surface before EBL. This thin SiO_2 layer protects the SOI surface from the EBL process and suppresses the formation of electron-trapping defects. The physically defined constrictions in the silicon channel act as tunnel barriers for the QD and SET, owing to bandgap enlargement in the narrower regions. The QD pattern shape and the distributions of topical exposure were optimized by adopting proximity-effect correction for the forward scattering in EBL to form smaller QDs. After EBL, we performed RIE with Cl_2 or SF_6 gas.

6.3.2.4 Gate oxide formation

A 3-nm-thick SiO_2 layer was formed to passivate the surface of the SOI layer by thermal oxidation at 900°C for 5 min. Then, a 50-nm-thick SiO_2 layer was deposited for the gate insulator by LPCVD at 840°C.

6.3.2.5 Top gate formation

To form the TG, we deposit a 200-nm-thick amorphous Si layer by LPCVD at 500°C. After the deposition, the TG pattern is formed by using photolithography and plasma etching.

6.3.2.6 Impurity doping

For low-temperature measurements of devices, an ohmic contact is required between the SOI layer and the metal pads. To create this ohmic contact, impurity doping is performed on the SOI wafer by ion implantation. On implanting P^+ ions into the wafer using phosphine, the SOI layer becomes n-type. The projected range of the ions is set to be around the interface between Si and SiO_2, which is controlled by setting the accelerating voltage to 50 keV. The amorphous Si TG behaves as a mask, preventing the ions from being injected into the QD region. After ion implantation, an activation and drive-in anneal is performed at 1000°C for 30 min. to repair the crystal damage in the SOI layer and to steer the P atoms into substitutional lattice sites, which activates the carriers in the SOI layer. This drive-in process contributes to the crystallization of the amorphous Si TG.

6.3.2.7 Metal contact pad formation

To perform low-temperature measurements of device characteristics, metal contact pads are required to create ohmic contacts. Al contact pads are formed by a photolithography–electron beam evaporation lift-off process sequence. Between the development of the photoresist and the deposition of Al, we performed a wet etch with buffered hydrogen fluoride (BHF) to remove the SiO_2 layer on the contact pads. Finally, forming gas annealing (FGA) was conducted at 420°C for 30 min.

6.4 Measurements

6.4.1 Charge Sensing Measurements and Observation of Few-Electron Silicon Single and Double QDs

Charge sensing measurements are essential for the observation of spin states of electrons confined in QDs. To perform charge sensing measurements, we fabricate a silicon single-QD device with a SET as a CS. The electron occupancy of the single QD is for the first time clearly measured using a charge sensing SET in physically defined QD devices.

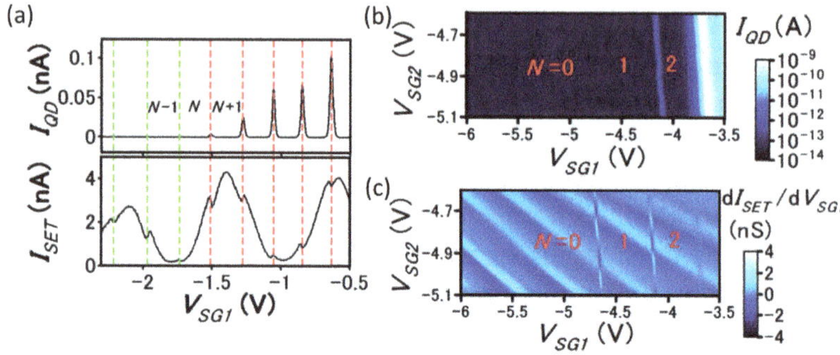

Figure 6.3 (a) Current through QD (I_{QD}) and SET (I_{SET}) as a function of side gate voltage V_{SG1}; $V_D = 0.3$ mV, $V_{DSET} = 6$ mV, $V_{TG} = 5.414$ V, and $V_{SG2} = -4$ V. Dashed lines indicate the positions of coulomb peaks in I_{QD} that represent the boundaries between different numbers of electron in the QD. (b, c) Plots of (b) I_{QD} and (c) SET transconductance dI_{SET}/dV_{SG1} as functions of V_{SG1} and V_{SG2}, measured simultaneously; $V_D = 1$ mV, $V_{DSET} = 5$ mV, and $V_{TG} = 3.8$ V. Modified from Ref. [9].

Measurements for single QDs (Fig. 6.2a) and DQDs (Fig. 6.2b) are performed at 4 K and 300 mK, respectively. The current flowing through the QD (I_{QD}) and SET (I_{SET}) are measured by connecting their drain contact to the source measurement unit of a semiconductor device parameter analyzer. The side gate voltages (V_{SG1} and V_{SG2}) are applied to control the electrochemical potentials in the QD and the SET. The SET is used as a CS to probe the change in the number of electrons confined in the QD.

Figure 6.3a shows simultaneous measurements of the currents flowing through the QD (I_{QD}) and SET (I_{SET}) as a function of the side gate voltage V_{SG1}. Conventional periodic coulomb peaks are observed in I_{QD} for the single QD system, in which the electrochemical potential of the QD is varied by sweeping V_{SG1}. The sharp coulomb peaks are caused by resonant tunneling between the electrochemical potentials of the QD and the Fermi energy in the source and drain electrodes. Sharp spikes are observed in the broad coulomb oscillation of I_{SET} at the V_{SG1} position (indicated by the five dashed lines in the high-voltage part in Fig. 6.3a) of the coulomb peaks of I_{QD}. This is because the stationary number of electrons (N)

confined in the QD changes at the V_{SG1} position of the I_{QD} coulomb peaks, which leads to a change in the electrochemical potential of the SET. Then shifts occur in the coulomb oscillation of I_{SET} on the V_{SG1} axis, which results in the inflection observed for I_{SET}. At the V_{SG1} position indicated by the three dashed lines in the low-voltage part in Fig. 6.3a, although the tunnel rates in the tunnel barriers are so small that coulomb peaks of I_{QD} are not observed for such negative values of V_{SG1}, the charge transport can be observed from the sharp spikes in I_{SET}. This result clearly shows that the number of electrons in the QD can be observed by a CS SET nearby.

In addition, observation of few-electron occupancy in the QD was also achieved with a large negative applied V_{SG1}. Figures 6.3b and 6.3c show simultaneously measured QD current I_{QD} and the SET transconductance dI_{SET}/dV_{SG1}, respectively, as functions of $V_{SG1}$and V_{SG2}, for another single QD sample. The clear lines in dI_{SET}/dV_{SG1}represent the change in the number of electrons confined in the QD and indicate that no further transitions occur for $V_{SG1}<-4.7$ V. Therefore, the QD contains no electrons below this gate voltage. This charge sensing measurement thus enables the number of electrons in the QD to be determined accurately as $N =$ 0, 1, and 2 charge occupancy. Although the transition line between $N = 0$ and $N = 1$ is clearly visible in the SET dI_{SET}/dV_{SG1} data of Fig. 6.3c, no coulomb peak corresponding to this transition is evident in I_{QD}in Fig. 6.3b, because the tunneling rates between the QD and the leads are too small to observe a current through the QD. The barrier thickness depends on the electrochemical potentials of the QD, which determine the tunneling probability.

Figure 6.4a plots dI_{SET}/dV_{SG1} as functions of V_{SG1} and V_{SG2} in a DQD device as shown in Fig. 6.2b. The thick (broad) oscillations in the background in Fig. 6.4a are coulomb oscillations of the SET itself. Figure 6.4b shows a schematic charge stability diagram for few-electron DQDs, that is, the equilibrium electron numbers (N_1, N_2) in each QD in the V_{SG1}-V_{SG2} plane [46]. In Fig. 6.4a, a structure similar to the charge stability diagram of DQD is obtained from the lines of charge sensing, or an abrupt change in I_{SET}. No transition lines exist in the lower-left region, where pinch-off of the DQD is realized.

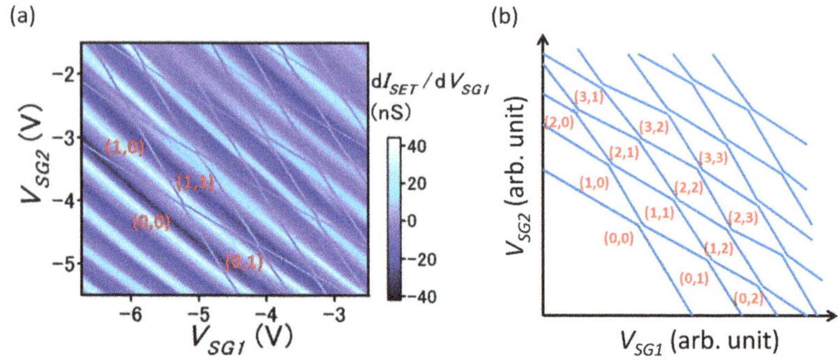

Figure 6.4 (a) Transconductance of the SET CS dI_{SET}/dV_{SG1} as functions of V_{SG1} and V_{SG2} in a DQD device as shown in Fig. 6.2b; $V_{TG} = 5.8$ V and $V_{SG3} = 0$ V, $V_D = V_S = 0.9$ V, and $V_{DSET} = 3$ mV. A well-defined charge stability diagram is obtained for the few-electron regime in the DQD. (b) Schematic charge stability diagram for few-electron DQDs. Modified from Ref. [9].

6.4.2 Capacitive Parameters for Designing SET CSs

The key capacitive parameters of SET CSs were investigated to determine optimum design guidelines for increasing the SNR in qubit measurements [9]. We used two single QDs coupled in parallel as shown in Fig. 6.2a to study the dependence of charge sensitivity and capacitive coupling strength on device configuration. The two QDs had different diameters. The diameters of the QDs and the distance between them were estimated by scanning electron microscopy (SEM). These estimates were presumably valid in the many-electron regime, where the absolute value of the side gate voltage is small. The estimated device dimensions were used to calculate the theoretical electrostatic coupling strength between the QDs. We estimated the difference in the charge sensitivities of the two QDs and identified the key capacitive parameter determining the SNR in qubit measurements. In addition, if the smaller QD was in the few-electron regime, we estimated the electrostatic coupling strength between the QDs and determined its increase with a decreasing number of electrons. Finally, we performed numerical calculations to estimate the dependence of the SNR in RF single-shot measurements in the MHz frequency domain on the distance between the qubit and the SET and the SET diameter.

Measured stability diagrams for QD1 and QD2 are shown in Figs. 6.5a and 6.5b, respectively. The regular coulomb diamonds obtained indicate the presence of single QDs. From the observed coulomb diamonds, charging energies of $E_{\mathrm{QD1}} \approx 14.0$ meV for QD1 and $E_{\mathrm{QD2}} \approx 6.2$ meV for QD2 were obtained. Figures 6.6a and 6.6b show contour plots of $I_{\mathrm{QD1}}$and I_{QD2}, respectively, as functions of V_{TG} and V_{G1}, measured simultaneously. The coulomb peak positions of I_{QD2} in Fig. 6.6b show zigzag patterns. The shifts appear at identical voltages to the coulomb peaks of I_{QD1}. Similarly, in Fig. 6.6a, the coulomb peaks of I_{QD1} exhibit small shifts at voltages corresponding to the coulomb peaks of I_{QD2}. Figure 6.7a illustrates the two single QDs coupled in parallel, with several electrical parameters indicated. QD1 (QD2) is coupled to the two leads through tunnel barriers represented by tunnel resistors R_{L1} and R_{R1} (R_{L2} and R_{R2}) and capacitors C_{L1} and C_{R1} (C_{L2} and C_{R2}) connected in parallel. The dots are coupled to each other by a capacitor C_C. The electrochemical potentials of an N-electron QD1 (μ_{QD1}) and an n-electron QD2 (μ_{QD2}) can be expressed as:

$$\mu_{\mathrm{QD1}}(N, n) = \frac{E_{\mathrm{QD1}}}{e}\left(Ne - \frac{e}{2} - C_{\mathrm{L1}}V_{\mathrm{L1}} - C_{\mathrm{R1}}V_{\mathrm{R1}} - \sum_i C_i V_i\right) + \frac{E_{\mathrm{QD1}}}{e}k_{\mathrm{QD1}}\left(ne - C_{\mathrm{L2}}V_{\mathrm{L2}} - C_{\mathrm{R1}}V_{\mathrm{R2}} - \sum_j C_j V_j\right) + E_N \tag{6.1}$$

and

$$\mu_{\mathrm{QD2}}(N, n) = \frac{E_{\mathrm{QD2}}}{e}\left(ne - \frac{e}{2} - C_{\mathrm{L2}}V_{\mathrm{L2}} - C_{\mathrm{R2}}V_{\mathrm{R2}} - \sum_j C_j V_j\right) + \frac{E_{\mathrm{QD2}}}{e}k_{\mathrm{QD2}}\left(Ne - C_{\mathrm{L1}}V_{\mathrm{L1}} - C_{\mathrm{R1}}V_{\mathrm{R1}} - \sum_i C_i V_i\right) + E_n. \tag{6.2}$$

$C_{\mathrm{QD1}} = C_C + C_{\mathrm{L1}} + C_{\mathrm{R1}} + \Sigma C_i$ and $C_{\mathrm{QD2}} = C_C + C_{\mathrm{L2}} + C_{\mathrm{R2}} + \Sigma C_j$ are the total capacitances of QD1 and QD2, respectively, where ΣC_i (ΣC_j) is the capacitance between QD1 (QD2) and other gates or environments for voltage V_i (V_j). E_N and E_n are the energies of the topmost filled single-particle states for QD1 and QD2, respectively. $E_{\mathrm{QD1}} = \alpha e^2/C_{\mathrm{QD1}}$ and $E_{\mathrm{QD2}} = \alpha e^2/C_{\mathrm{QD2}}$ are the charging energies of QD1 and QD2, respectively, where $\alpha = (1 - C_C^2/C_{\mathrm{QD1}}C_{\mathrm{QD2}})^{-1}$. Equations 6.1 and 6.2 are equivalent to the electrochemical potentials of two QDs coupled in series, as described in Ref. [46], except that there is no tunnel coupling between the parallel QDs. The term $ek_{\mathrm{QD2}} = eC_C/C_{\mathrm{QD1}}$ corresponds to the equivalent charge induced in QD2 when the number of electrons in QD1 decreases from N to $N - 1$. Similarly, $ek_{\mathrm{QD1}} = eC_C/C_{\mathrm{QD2}}$ is equivalent to the charge induced in QD1 when the number of electrons in QD2 decreases from n to $n-1$. We define S_{QD2} as the ratio of the peak shift magnitude to the peak-to-peak spacing in terms of V_{TG}. At the coulomb peaks of I_{QD2} in Fig. 6.6b, $S_{\mathrm{QD2}}$is given by $k_{\mathrm{QD2}}(1 + \Delta E_n/E_{\mathrm{QD2}})^{-1}$, where $\Delta E_n = E_n - E_{n-1}$. A $k_{\mathrm{QD2}} = S_{\mathrm{QD2}}$ value of 0.088 ± 0.005 is estimated from Fig. 6.6b, assuming that ΔE_n is negligibly small in the large-n regime. Similarly, a $k_{\mathrm{QD1}} = S_{\mathrm{QD1}}$ value of 0.038 ± 0.003 is obtained from Fig. 6.6a, where $S_{\mathrm{QD1}} = k_{\mathrm{QD1}}(1 + \Delta E_N / E_{\mathrm{QD1}})^{-1}$ is identical to k_{QD1} when ΔE_N is negligibly small in the large-N regime. Here, k_{QD2} is larger than k_{QD1} since the QD2 diameterL_{QD2} ($\sim$70 nm) is larger than the QD1 diameter L_{QD1} ($\sim$35 nm). Both k_{QD1} and k_{QD2} are calculated by solving a 3D Poisson equation and by assuming that QD1 and QD2 are both 2D conductive disks. Here, the distance, d, between the edges of QD1 and QD2 is constant (see Fig. 6.7b). Figure 6.7c shows the results as functions of $L_{\mathrm{QD2}}/L_{\mathrm{QD1}}$. An increase in L_{QD2} causes k_{QD2} to increase and k_{QD1} to decrease, and both k_{QD2} and k_{QD1} are larger when $d = 0.5\ L_{\mathrm{QD1}}$ than when $d = L_{\mathrm{QD1}}$. By using the device parameters $L_{\mathrm{QD1}} \approx 35$ nm, $L_{\mathrm{QD2}} \approx 70$ nm, and $d \approx 25$ nm for the calculations, we obtain $k_{\mathrm{QD2}} = 0.086$ and $k_{\mathrm{QD1}} = 0.036$. These are comparable to the measured values ($k_{\mathrm{QD2}} \approx 0.088$ and $k_{\mathrm{QD1}} \approx 0.038$), confirming the validity of the model and the calculation.

To study the charge sensitivity of QDs when they are used as SET CSs, the SET current, I_{SET}, and the low-frequency spectral density of the SET shot noise, S_{I}, were calculated using master equations [47–49] for charge states corresponding to $N - 1(n - 1)$ and N (n) excess electrons in QD1 (QD2). Assuming that $R_{\mathrm{L(R)}}$ is much larger than the quantized resistance $R_Q = \pi\hbar/2e^2$ and that the quasi-stationary condition where the carrier frequency ω for RF single-

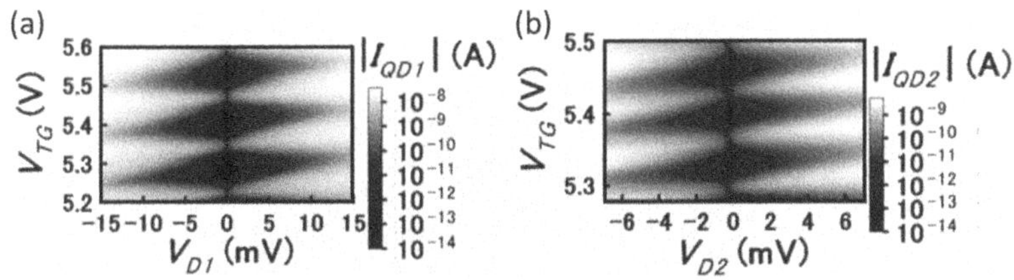

Figure 6.5 (a, b) Stability diagrams for QD1 and QD2. $V_{\mathrm{SG1}} = 0$ V and $V_{\mathrm{SG2}} =$ −4 V. Modified from Ref. [9].

shot measurements is much smaller than I_{SET}/e [48, 50],

$$I_{\mathrm{SET}} = e\frac{\Gamma_{L-}\Gamma_{R+} - \Gamma_{L+}\Gamma_{R-}}{\Gamma_{+} + \Gamma_{-}} \tag{6.3}$$

and

$$S_I = 2e^2 \frac{\sum\limits_{\pm} \Gamma_{\pm}^2 (2\Gamma_{L\mp}\Gamma_{R\mp} + \Gamma_{L\mp}\Gamma_{R\pm} + \Gamma_{L\pm}\Gamma_{R\pm})}{(\Gamma_{+} + \Gamma_{-})^2} \tag{6.4}$$

are obtained. Here, the electron tunneling rates are $\Gamma_{\mathrm{L(R)}\pm} = (1/e^2 R_{\mathrm{L(R)}})\gamma(\pm\,[-eV_{\mathrm{L(R)}} - \mu_{\mathrm{SET}}(N,\,n)])$, $\Gamma_{+} = \Gamma_{\mathrm{L+}} + \Gamma_{\mathrm{R+}}$, and $\Gamma_{-} = \Gamma_{\mathrm{L-}} + \Gamma_{\mathrm{R-}}$, where $\gamma(\varepsilon) = \varepsilon/[1 - \exp(-\varepsilon/k_{\mathrm{B}}T)]$ [48], and the electrochemical potential of the SET, $\mu_{\mathrm{SET}}(N,\,n)$, is identical to that given in Eq. 6.1 for QD1 and Eq. 6.2 for QD2. In the case of dc measurements, the charge sensitivity, δq, is given by $\delta q = (S_{\mathrm{I}})^{-1/2}/(\partial I_{\mathrm{SET}}/\partial Q)$ [51], where Q is the gate charge in the SET and can be expressed as the term $\Sigma C_i V_i$ in Eq. 6.1 or $\Sigma C_j V_j$ in Eq. 6.2. When QD1 and QD2 have the same resistance (i.e., the same parasitic and tunnel resistances, such that $R_{\mathrm{L1}} = R_{\mathrm{L2}}$ and $R_{\mathrm{R1}} = R_{\mathrm{R2}}$), the charge sensitivities, δq_1 and δq_2, are inversely proportional to the charging energies, E_{QD1} and E_{QD2}, respectively. Therefore, E_{QD1} and E_{QD2} are the key capacitive parameters to determine δq, and $\delta q_2/\delta q_1 = E_{\mathrm{QD1}}/E_{\mathrm{QD2}}$. The charging energy depends on the QD diameter, which means that a smaller QD can act as a more sensitive SET CS. It should be noted that increasing the SET diameter increases the total amount of induced charge but reduces the sensitivity. Thus, optimization of these two parameters involves a trade-off.

If k_{QD2} is much less than unity, which is usually the case for experiments with 2DEG systems, the SNR of the charge sensing signal due to a change in N during the dc measurement is

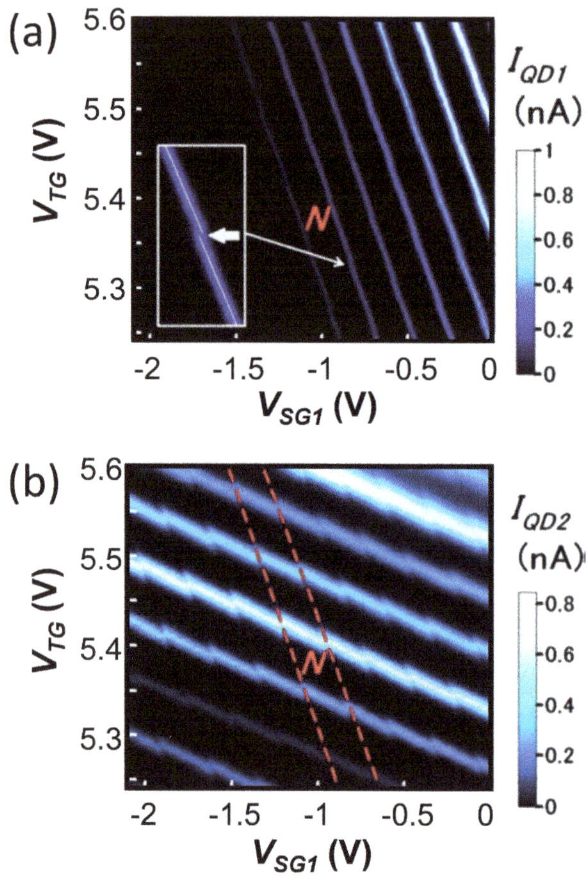

Figure 6.6 Contour plots of (a) $I_{QD1}$and (b) I_{QD2}as functions of V_{TG} and V_{SG1}. $V_{D1} = 1$ mV, $V_{D2} = 1$ mV, and $V_{SG2} = -4$V. The two dashed lines in (b) show the positions of two coulomb peaks in (a). (Inset) Magnified plot of a shift in the coulomb peak. Similar shifts occur at all crossing points between the coulomb peaks of QD1 and QD2. Modified from Ref. [9].

given by SNR $= 20\text{Log}_{10}[ek_{QD2}(\Delta f)^{-1/2}(\delta q_2)^{-1}]$, where Δf is the measurement bandwidth. Note that hereafter QD2 is assumed to function as the SET CS. If one considers the capacitive coupling, k_{QD2}, and the charge sensitivity, δq_2, the key capacitive parameter determining the SNR is the electrostatic coupling energy between the two QDs, $E_m = e^2\alpha k_{QD2}/C_{QD2}$, because $k_{QD2}/\delta q_2$ is proportional

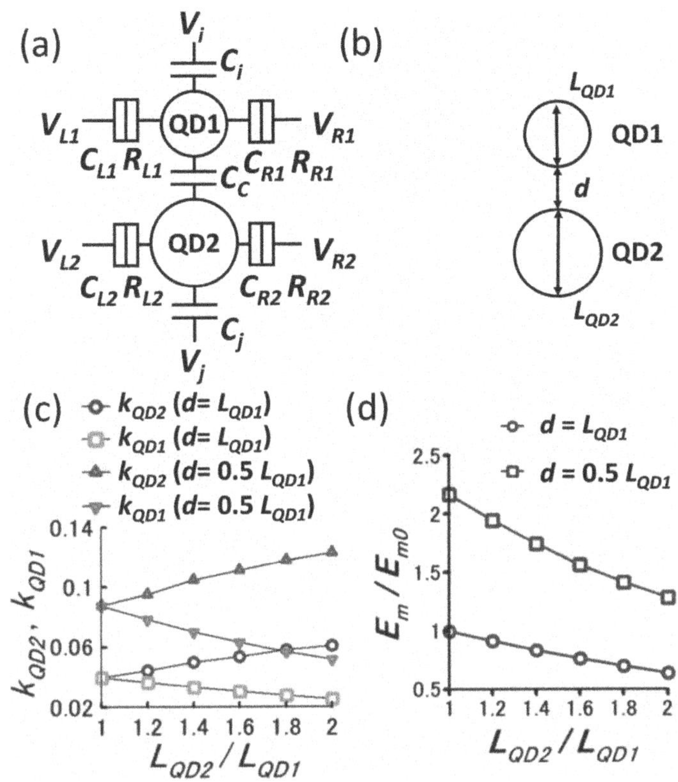

Figure 6.7 (a) Schematics of QD1 and QD2 with associated electrical parameters. (b) Dimensions of QD1 and QD2 used in numerical calculations. (c) Calculation results for k_{QD1} and k_{QD2} as functions of L_{QD2}/L_{QD1}. L_{QD1} is fixed. (d) Calculated E_m/E_{m0} values as a function of L_{QD2}/L_{QD1}. Here, E_{m0} is the value of E_m for $L_{QD2} = L_{QD1} = d$. The plots in (c) and (d) show calculation results for two cases, namely, $d = L_{QD1}$ and $d = 0.5\ L_{QD1}$, respectively. Modified from Ref. [9].

to E_m. Figure 6.7d plots calculated E_m/E_{m0} values as a function of L_{QD2}/L_{QD1}. Here, E_{m0} is the value of E_m when $L_{QD2} = L_{QD1} = d$. E_m decreases with increasing L_{QD2} or d, which indicates that a decrease in the SET diameter leads to an increase in the SNR. It is noteworthy that decreasing the SET diameter reduces the capacitive coupling value,k_{QD2}, but also induces a greater increase in the sensitivity, leading to an increase in the SNR.

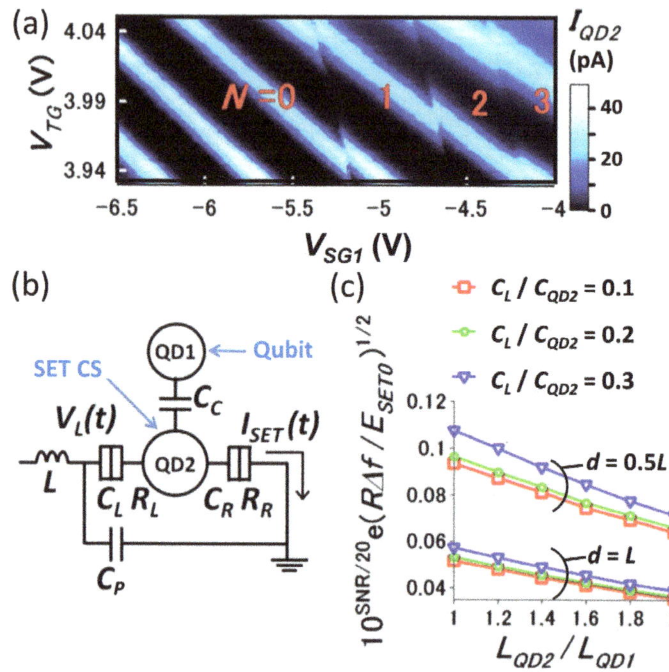

Figure 6.8 (a) Contour plots of I_{QD2} as a function of V_{TG} and V_{G1}. Here, $V_{D1} = 1$ mV, $V_{D2} = 5$ mV, and $V_{G2} = -5.8$ V. (b) Schematic of the RF measurement setup. (c) Calculation results for SNR as a function of L_{QD2} (in units of L_{QD1}) under optimized gate bias conditions. Using a y-axis value of $Y = 10^{SNR/20}\ e(R\Delta f/E_{SETO})^{1/2}$, the SNR is expressible as $SNR = 20\text{Log}_{10}[Y/\{e(R\Delta f/E_{SETO})^{1/2}\}]$. Here, $R_L = R_R = R$, $k_BT = 0.01\ E_{SETO}$ (where E_{SETO} is the QD2 charging energy for $L_{QD2} = L_{QD1}$ and $d = L_{QD1}$), $A = 0.2\ E_{SETO}/e$, $V_0 = 0$, and L_{QD1} is fixed. Calculations were performed for $C_L/C_{QD2} = 0.1, 0.2$, and 0.3. Calculation results for $d = L_{QD1}$ and $d = 0.5\ L_{QD1}$ are shown. Modified from Ref. [9].

Figure 6.8a shows the contour plots of I_{QD2} as a function of V_{TG} and V_{G1}, obtained from another device. In this device, QD1 is in the few-electron regime. In this low-N regime, k_{QD2} values of 1.10, 0.98, and 0.93 were obtained for the 0↔1, 1↔2, and 2↔3 transitions of N, respectively. It should be noted that n here was large. The value of k_{QD2} increased with decreasing N because the effective size of QD1 decreased to a point below its physically defined size owing to the large negative voltage to minimize the number of electrons.

This result indicates that operating in the few-electron regime will improve the SNR when reading qubit states in QDs with laterally coupled SET systems.

The dependence of L_{QD2} and d on the SNR was investigated for RF single-shot measurements under the conditions shown in Fig. 6.8b. L is the inductor in the setup and C_P is the parasitic capacitance of the QDs. Here, the exact resonance at the carrier frequency, $\omega = (LC_P)^{-1/2}$, and quasi-stationary conditions under which ω is much less than I_{SET}/e are assumed. By also assuming a low circuit impedance, Z, approximately equal to $L/C_P R_d$ and much less than R_0, where R_d is the differential resistance of the SET and R_0 the cable wave resistance, is $\sim$50 Ω, the SET bias voltage can be expressed by a simple sine wave: $V_L(t) = V_0 + A \sin \omega t$ [50]. Here, V_0 is the static component and A is the amplitude of the SET bias. By restricting the analysis to the first harmonic of the reflected wave, the relationship

$$\text{SNR} = 20\log_{10}\frac{\left|\langle I_{SET}(t)\sin\omega t\rangle - \langle I'_{SET}(t)\sin\omega t\rangle\right|}{\sqrt{\Delta\ f\langle S_I(t)\sin^2\omega t\rangle}} \tag{6.5}$$

is obtained. Here $\langle\ldots\rangle$ denotes the time average and $I?_{SET}$ is the current generated when a charge $\pm\ ek_{QD2}$ is induced in QD2. Calculation of the capacitive coupling between the QD and the SET demonstrates that increasing both L_{QD2} and d reduces the SNR, similar to the results obtained in the case of dc measurements. Figure 6.8c plots the calculated SNR values as functions of L_{QD2}/L_{QD1} using the optimized gate bias conditions (included in $\sum C_j V_j$ in Eq. 6.2). Here, L_{QD1} is fixed. These plots show that as L_{QD2} increases, the SNR decreases and that the SNR for $d = 0.5\ L_{QD1}$ is larger than that for $d = L_{QD1}$. It is also evident that an increase in C_L/C_{QD2} causes a slight increase in the SNR. Decreasing L_{QD2}/L_{QD1} from 2 down to 1 roughly doubles the signal power for both $d = 0.5\ L_{QD1}$ and $d = L_{QD1}$. Generally, experimental devices are designed so as to minimize the distance between the QD and the SET CS in order to enhance the capacitive coupling between the two. However, there is a lower limit on the QD-SET distance to prevent direct tunneling, so decreasing the SET diameter can also be adopted as an effective means of improving the SNR.

This study focuses on the capacitive parameters of the SET CS in order to establish guidelines for improving the SNR during qubit readout. However, reducing the parasitic resistance of the device also improves the SNR, as does asymmetric tuning of the tunnel resistance of the SET.

6.5 Summary

Physically defined Si QDs in MOS structures with undoped SOI substrates for quantum computing have been studied. These structures are promising for the large-scale integration necessary for quantum information processing. CSs, as tools for determining the absolute number of electrons in QDs as well as for observing the few-electron regimes, are described. The present study represents an important step toward the implementation of practical qubits.

Acknowledgment

This work was financially supported by JSPS Grants-in-Aid (Kakenhi Nos. 26249048, 26709023, 26630151, and 16F16806), the CREST program of the Japan Science and Technology Agency (JST), and the Project for Developing Innovation Systems of the Ministry of Education, Culture, Sports, Science and Technology (MEXT) of Japan.

The authors are grateful to H. Mizuta, K. Uchida, and T. Hiramoto for discussions and for allowing the authors the use of their facilities. EBL was performed in the Nanotechnology Platform at Quantum Nanoelectronics Research Center, Tokyo Institute of Technology.

References

1. Johnson, M. W., Amin, M. H. S., Gildert, S., Lanting, T., Hamze, F., Dickson, N., Harris, R., Berkley, A. J., Johansson, J., Bunyk, P., Chapple, E. M., Enderud, C., Hilton, J. P., Karimi, K., Ladizinsky, E., Ladizinsky, N., Oh, T., Perminov, I., Rich, C., Thom, M. C., Tolkacheva, E., Truncik, C. J. S., Uchaikin, S., Wang, J., Wilson, B., and Rose, G. (2011). Quantum annealing with manufactured spins, *Nature*, **473**, pp. 194–198.
2. Vandersypen, L. M. K., Steffen, M., Breyta, G., Yannoni, C. S., Sherwood, M. H., and Chuang, I. L. (2001). Experimental realization of Shor's quantum

factoring algorithm using nuclear magnetic resonance, *Nature*, **414**, pp. 883–887.

3. Nakamura, Y., Pashkin, Y. A., and Tsai, J. S. (1999). Coherent control of macroscopic quantum states in a single-Cooper-pair box, *Nature*, **398**, pp. 786–788.
4. Itoh, K. M., and Watanabe, H. (2014). Isotope engineering of silicon and diamond for quantum computing and sensing applications, *Mater. Res. Soc. Commun.*, **4**, pp. 143–157.
5. Maune, B. M., Borselli, M. G., Huang, B., Ladd, T. D., Deelman, P. W., Holabird, K. S., Kiselev, A. A., Alvarado-Rodriguez, I., Ross, R. S., Schmitz, A. E., Sokolich, M., Watson, C. A., Gyure, M., and Hunter, A. T. (2012). Coherent singlet-triplet oscillations in a silicon-based double quantum dot, *Nature*, **481**, pp. 344–347.
6. Veldhorst, M., Hwang, J. C. C., Yang, C. H., Leenstra, A. W., de Ronde, B., Dehollain, J. P., Muhonen, J. T., Hudson, F. E., Itoh, K. M., Morello, A., and Dzurak, A. S. (2014). An addressable quantum dot qubit with fault-tolerant control-fidelity, *Nat. Nanotechnol.*, **9**, pp. 981–985.
7. Veldhorst, M., Yang, C. H., Hwang, J. C. C., Huang, W., Dehollain, J. P., Muhonen, J. T., Simmons, S., Laucht, A., Hudson, F. E., Itoh, K. M., Morello, A., and Dzurak, A. S. (2015). A two-qubit logic gate in silicon, *Nature*, **526**, pp. 410–414.
8. Yamahata, G., Kodera, T., Mizuta, H., Uchida, K., and Oda, S. (2009). Control of inter-dot electrostatic coupling by a side gate in silicon double quantum dot operating at 4.5 K, *Appl. Phys. Express*, **2**(9), p. 095002.
9. Horibe, K., Kodera, T., Kambara, T., Uchida, K., and Oda, S. (2012). Key capacitive parameters for designing single-electron transistor charge sensors, *J. Appl. Phys.*, **111**, p. 093715.
10. Yamahata, G., Kodera, T., Churchill, H. O. H., Uchida, K., Marcus, C. M., and Oda, S. (2012). Magnetic field dependence of Pauli spin blockade: a window into the sources of spin relaxation in silicon quantum dots, *Phys. Rev. B*, **86**, p. 115322.
11. Kambara, T., Kodera, T., Arakawa, Y., and Oda, S. (2013). Dual function of single electron transistor coupled with double quantum dot: gating and charge sensing, *Jpn. J. Appl. Phys.*, **52**, pp. 04CJ01-1-4.
12. Yamada, K., Kodera, T., Kambara, T., and Oda, S. (2014). Fabrication and characterization of p-channel Si double quantum dots, *Appl. Phys. Lett.*, **105**, p. 113110.

13. Horibe, K., Kodera, T., and Oda, S. (2015). Lithographically-defined few-electron silicon quantum dots based on a silicon-on-insulator substrate, *Appl. Phys. Lett.*, **106**, p. 083111.
14. Horibe, K., Kodera, T., and Oda, S. (2015). Back-action-induced excitation of electrons in a silicon quantum dot with a single electron transistor charge sensor, *Appl. Phys. Lett.*, **106**, p. 053119.
15. Ihara, S., Andreev, A., Williams, D. A., Kodera, T., and Oda, S. (2015). Quantum dots in single electron transistors with ultrathin silicon-on-insulator structures, *Appl. Phys. Lett.*, **107**, p. 013102.
16. Shor, P. W. (1994). Algorithms for quantum computation: discrete logarithms and factoring, in *IEEE Symposium on Foundations of Computer Science*, p. 124.
17. Politi, A., Matthews, J. C. F., and O'Brien, J. L. (2009). Shor's quantum factoring algorithm on a photonic chip, *Science*, **325**, p. 1221.
18. Shinkai, G., Hayashi, T., Ota, T., and Fujisawa, T. (2009). Correlated coherent oscillations in coupled semiconductor charge qubits, *Phys. Rev. Lett.*, **103**, p. 056802.
19. Loss, D., and DiVincenzo, D. P. (1998). Quantum computation with quantum dots, *Phys. Rev. A*, **57**, p. 120.
20. Khaetskii, A. V., Loss, D., and Glazman, L. (2002). Electron spin decoherence in quantum dots due to interaction with nuclei, *Phys. Rev. Lett.*, **88**, p. 186802.
21. Golovach, V. N., Khaetskii, A., and Loss, D. (2004). Phonon-induced decay of the electron spin in quantum dots, *Phys. Rev. Lett.*, **93**, p. 016601.
22. Petta, J. R., Johnson, A. C., Taylor, J. M., Laird, E. A., Yacoby, A., Lukin, M. D., Marcus, C. M., Hanson, M. P., and Gossard, A. C. (2005). Coherent manipulation of coupled electron spins in semiconductor quantum dots, *Science*, **309**, p. 2180.
23. Koppens, F. H. L., Buizert, C., Tielrooij, K. J., Vink, I. T., Nowack, K. C., Meunier, T., Kouwenhoven, L. P., and Vandersypen, L. M. K. (2006). Driven coherent oscillations of a single electron spin in a quantum dot, *Nature*, **442**, p. 766.
24. Pioro-Ladrière, M., Obata, T., Tokura, Y., Shin, Y.-S., Kubo, T., Yoshida, K., Taniyama, T., and Tarucha, S. (2008). Electrically driven single-electron spin resonance in a slanting Zeeman field, *Nat. Phys.*, **4**, p. 776.
25. Ono, K., Austing, D. G., Tokura, Y., and Tarucha, S. (2002). Current rectification by Pauli exclusion in a weakly coupled double dot system, *Science*, **297**, pp. 1313–1317.

26. Hanson, R., Willems van Beveren, L. H., Vink, I. T., Elzerman, J. M., Naber, W. J. M., Koppens, F. H. L., Kouwenhoven, L. P., and Vandersypen, L. M. K. (2005). Single-shot readout of electron spin states in a quantum dot using spin-dependent tunnel rates, *Phys. Rev. Lett.*, **94**, p. 196802.
27. Kodera, T., Ono, K., Kitamura, Y., Tokura, Y., Arakawa, Y., and Tarucha, S. (2009). Quantitative estimation of exchange interaction energy using two-electron vertical double quantum dots, *Phys. Rev. Lett.*, **102**, pp. 146802-1-4.
28. Prance, J. R., Shi, Z., Simmons, C. B., Savage, D. E., Lagally, M. G., Schreiber, L. R., Vandersypen, L. M. K., Friesen, M., Joynt, R., Coppersmith, S. N., and Eriksson, M. A. (2012). Single-shot measurement of triplet-singlet relaxation in a Si/SiGe double quantum dot, *Phys. Rev. Lett.*, **108**, p. 046808.
29. Dehollain, J. P., Muhonen, J. T., Tan, K. Y., Saraiva, A., Jamieson, D. N., Dzurak, A. S., and Morello, A. (2014). Single-shot readout and relaxation of singlet and triplet states in exchange-coupled P31 electron spins in silicon, *Phys. Rev. Lett.*, **112**, p. 236801.
30. Kawakami, E., Scarlino, P., Ward, D. R., Braakman, F. R., Savage, D. E., Lagally, M. G., Friesen, M., Coppersmith, S. N., Eriksson, M. A., and Vandersypen, L. M. K. (2014). Electrical control of a long-lived spin qubitin a Si/SiGe quantum dot, *Nat. Nanotechnol.*, **9**, p. 666.
31. Takeda, K., Kamioka, J., Otsuka, T., Yoneda, J., Nakajima, T., Delbecq, M. R., Amaha, S., Allison, G., Kodera, T., Oda, S., and Tarucha, S. (2016). A fault-tolerant addressable spin qubit in a natural silicon quantum dot, *Sci. Adv.*, **2**(8), pp. e1600694-1-6.
32. Hao, X., Ruskov, R., Xiao, M., Tahan, C., and Jiang, H. (2014). Electron spin resonance and spin–valley physics in a silicon double quantum dot, *Nat. Commun.*, **5**, p. 3860.
33. Yang, C. H., Rossi, A., Ruskov, R., Lai, N. S., Mohiyaddin, F. A., Lee, S., Tahan, C., Klimeck, G., Morello, A., and Dzurak, A. S. (2013). Spin-valley lifetimes in a silicon quantum dot with tunable valley splitting, *Nat. Commun.*, **4**, p. 2069.
34. Tyryshkin, A. M., Tojo, S., Morton, J. J. L., Riemann, H., Abrosimov, N. V., Becker, P., Pohl, H.-J., Schenkel, T., Thewalt, M. L. W., Itoh, K. M., and Lyon, S. A. (2012). Electron spin coherence exceeding seconds in high-purity silicon, *Nat. Mater.*, **11**, p. 143.
35. Zwanenburg, F. A., Dzurak, A. S., Morello, A., Simmons, M. Y., Hollenberg, L. C. L., Klimeck, G., Rogge, S., Coppersmith, S. N., and Eriksson, M. A. (2013). Silicon quantum electronics, *Rev. Mod. Phys.*, **85**, p. 961.

36. Culcer, D., Cywiński, Ł., Li, Q., Hu, X., and Sarma, S. D. (2009). Realizing singlet-triplet qubits in multivalley Si quantum dots, *Phys. Rev. B*, **80**, p. 205302.
37. Köhler, H., and Roos, M. (2006). Quantitative determination of the valley splitting in n-type inverted silicon (100) MOSFET surfaces, *Phys. Status Solidi B*, **91**, p. 233.
38. Takashina, K., Ono, Y., Fujiwara, A., Takahashi, Y., and Hirayama, Y. (2006). Valley polarization in Si(100) at zero magnetic field, *Phys. Rev. Lett.*, **96**, p. 236801.
39. Borselli, M. G., Ross, R. S., Kiselev, A. A., Croke, E. T., Holabird, K. S., Deelman, P. W., Warren, L. D., Alvarado-Rodriguez, I., Milosavljevic, I., Ku, F. C., Wong, W. S., Schmitz, A. E., Sokolich, M., Gyure, M. F., and Hunter, A. T. (2011). Measurement of valley splitting in high-symmetry Si/SiGe quantum dots, *Appl. Phys. Lett.*, **98**, p. 123118.
40. Friesen, M., and Coppersmith, S. N. (2010). Theory of valley-orbit coupling in a Si/SiGe quantum dot, *Phys. Rev. B*, **81**, p. 115324.
41. Pei, F., Laird, E. A., Steele, G. A., and Kouwenhoven, L. P. (2012). Valley–spin blockade and spin resonance in carbon nanotubes, *Nat. Nanotechnol.*, **7**, p. 630.
42. Shnirman, A., and Schön, G. (1998). Quantum measurements performed with a single-electron transistor, *Phys. Rev. B*, **57**, p. 15400.
43. Lu, W., Ji, Z. Q., Pfeiffer, L., West, K. W., and Rimbeg, A. J. (2003). Real-time detection of electron tunnelling in a quantum dot, *Nature (London)*, **423**, p. 422.
44. Elzerman, J. M., Hanson, R., Willems van Beveren, L. H., Witkamp, B., Vandersypen, L. M. K., and Kouwenhoven, L. P. (2004). Single-shot read-out of an individual electron spin in a quantum dot, *Nature (London)*, **430**, p. 431.
45. Barthel, C., Medford, J., Marcus, C. M., Hanson, M. P., and Gossard, A. C. (2010). Interlaced dynamical decoupling and coherent operation of a singlet-triplet qubit, *Phys. Rev. Lett.*, **105**, p. 266808.
46. van der Wiel, W. G., De Franceschi, S., Elzerman, J. M., Fujisawa, T., Tarucha, S., and Kouwenhoven, L. P. (2002). Electron transport through double quantum dots, *Rev. Mod. Phys.*, **75**, p. 1.
47. Korotkov, A. N. (1994). Intrinsic noise of the single-electron transistor, *Phys. Rev. B*, **49**, p. 10381.
48. Hershfield, S., Davies, J. H., Hyldgaaed, P., Stanton, C. J., and Wilkins, J. W. (1993). Zero-frequency current noise for the double-tunnel-junction Coulomb blockade, *Phys. Rev. B*, **47**, p. 1967.

49. Hanke, U., Galperin, Y. M., Chao, K. A., and Zou, N. (1993). Finite-frequency shot noise in a correlated tunneling current, *Phys. Rev. B*, **48**, p. 17209.
50. Korotkov, A. N., and Paalanen, M. A. (1999). Charge sensitivity of radio frequency single-electron transistor, *Appl. Phys. Lett.*, **74**, p. 4052.
51. Zimmerli, G., Eiles, T. M., Kautz, R. L., and Martinins, J. M. (1992). Noise in the Coulomb blockade electrometer, *Appl. Phys. Lett.*, **61**, p. 237.

Chapter 7

Oxide Memristor and Applications

Mingyi Rao, Rivu Midya, and J. Joshua Yang
Department of Electrical and Computer Engineering, University of Massachusetts, Amherst, Massachusetts 01003, USA
jjyang@umass.edu

Memristors are effectively variable resistors with memory whose resistance can be modified electrically. The state of internal resistance is a function of the history of the applied voltage and current. Such devices have the ability to store and process data and provide for several attractive performance characteristics that make them more capable of solving computing problems encountered nowadays than conventional CMOS-based technology. Two-terminal resistance switches based on ionic motion are an important class of memristors. These devices are often built out of a simple conductor/insulator/conductor stack, and the insulator is often made of a metal oxide. Although the idea for such devices was conceived as early as the 1960s, only recent progress in the field has been able to exploit useful features such as fast operation speed, low energy consumption, and high endurance. Apart from that, these devices can be scaled down to less than 10 nm and stacked in three dimensions. In this chapter, we review recent progress in the development and understanding of memristive devices. We

Emerging Devices for Low-Power and High-Performance Nanosystems: Physics, Novel Functions, and Data Processing
Edited by Simon Deleonibus

ISBN 978-981-4800-11-2 (Hardcover), 978-0-429-45873-6 (eBook)
www.panstanford.com

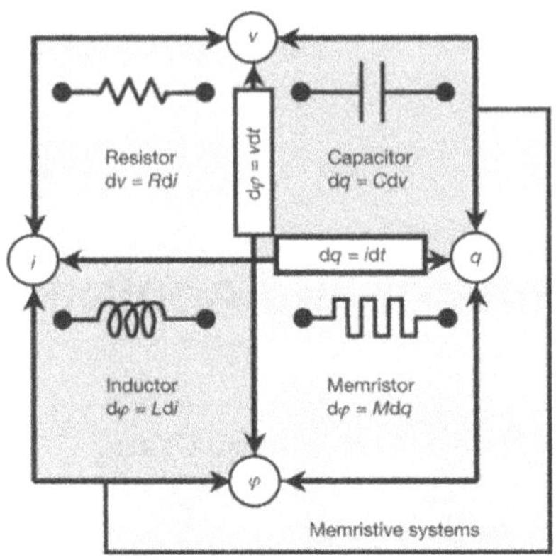

Figure 7.1 Four fundamental circuit elements and their relationship. Reprinted by permission from Macmillan Publishers Ltd: *Nature* (Ref. [2]), copyright (2008).

also discuss how oxide memristors and related devices are used in computing applications.

7.1 Concept

Resistors, capacitors, and inductors have long been considered the only three fundamental passive circuit elements. In more than a century of effort, physicists and electrical engineers have established a series of equations to describe relations between basic circuit variables. In 1971, L. Chua predicted that there existed a fourth fundamental passive circuit element, which bridged the magnetic flux φ and charge q [1]. Chua named the fourth element a "memristor," where memristance is defined as

$$d\varphi = Mdq. \tag{7.1}$$

A complete set of relations is shown in Fig. 7.1. It was not until 2008 that R. Stanley Williams and his colleagues from Hewlett-Packard Labs linked the memristor theory with real devices [2, 3]. They used a coupled variable-resistor model for a memristor in the original toy model, with mobile dopants under an electric field. Assuming the memristor has two regions—a doped, low-resistance region and an undoped, high-resistance region, an external bias $v(t)$ will move the boundary of the two regions by causing the charged dopant to drift. This phenomenon can be mathematically described as

$$v(t) = \left(R_{\mathrm{ON}} \frac{w(t)}{D} + R_{\mathrm{OFF}}(1 - \frac{w(t)}{D}) \right) i(t), \tag{7.2}$$

where $w(t)$ is the width of the doped area and D is the width of the entire region. R_{ON} and R_{OFF} represent the resistance of a fully doped device and an undoped device, respectively. If $R_{\mathrm{OFF}} \ll R_{\mathrm{ON}}$, the following equation can be deduced:

$$M(q) = R_{\mathrm{OFF}} \left(1 - \frac{\mu_V R_{\mathrm{ON}}}{D^2} q(t) \right) \tag{7.3}$$

Here $M(q)$ is the magnetic flux and μ_V is the average dopant mobility. Hence, there is a linear relationship between magnetic flux and charge. A more easily captured behavior of such a device is the nonlinear relationship between voltage and current denoted by Eq. 7.1. Furthermore, the memristor is static if no current or voltage is applied. If $I(t) = 0$, we find $q(t)$ and $M(t)$ are constants.

As mentioned in Ref. [2], "The fact that the magnetic field does not play an explicit role in the mechanism of memristance is one possible reason why the phenomenon has been hidden for so long." Such resistive switching phenomena have been widely studied ever since the early 1960s [4], but the great success of Si technology since then had suppressed interests in other technologies until the beginning of this century, when keeping up with Moore's law became more challenging than ever as the chip area was shrinking. In the past decade, memristive devices started to attract attention for memory and computing applications for their great potential in scalability, their remarkable electrical performance, and new functionalities.

7.2 Operation Modes

The operation modes of resistive switching ("memristive") devices can be classified into two categories, unipolar and bipolar, as shown

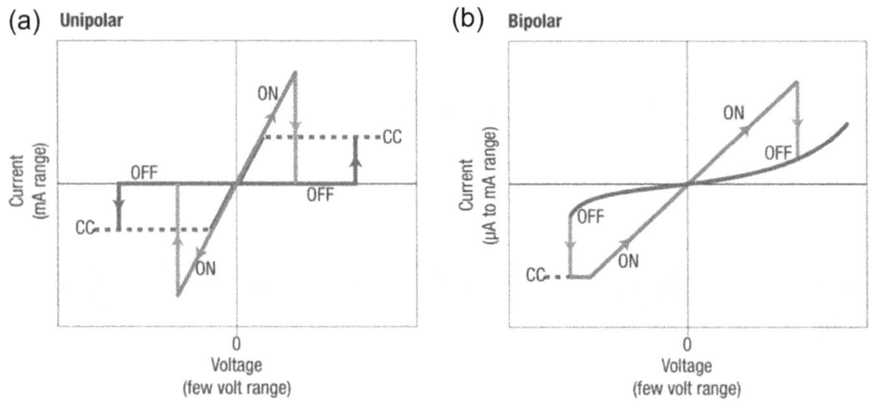

Figure 7.2 Two basic operation modes of a memristive device. (a) Unipolar (or nonpolar) and (b) bipolar. The CC denotes compliance current controlled by an external circuit. Reprinted by permission from Macmillan Publishers Ltd: *Nature Materials* (Ref. [5]), copyright (2007).

in Fig. 7.2 [5]. The process by which the device transitions from a high resistance state (HRS) to a low resistance state (LRS) is called set and the process by which the device transitions from LRS to HRS is called reset. For the majority of reported memristors, a larger voltage than that for normal subsequent switching is needed for the initial HRS to transition to LRS compared to the first set. This process is called forming. The reason some devices need this forming (or electroforming) process is because a larger voltage or more time is required to build a complete filament from the pristine state of a dielectric film, whereas reset and set will just rupture or restore part of the filament. In dielectric memristive materials (e.g., oxides or nitrides) operating on manipulating native dopants (e.g., oxygen or nitrogen vacancies), the electroforming process involves creating native dopants electrically while reset and set processes are dominated by moving the native dopants electrically. Unipolar switching means set and reset happen at the same polarity of applied voltages, while bipolar means set and reset happen at opposite polarities. If a device exhibits unipolar switching in both biasing polarities, it can also be referred to as a nonpolar memristor.

7.3 Memristive Switching Mechanism

The nonlinear relationship between voltage and current in a memristor translates to a change in the resistance of the memristor due to the application of an electrical bias. Reduction-oxidation (redox) reactions, phase change, and magnetoresistive effects are possible causes of resistive switching. This chapter will focus on redox-based resistive switching and its related applications.

Redox-based resistive switching can be grouped into two large categories: nonuniform and uniform. In nonuniform switching devices, the active region during switching is laterally localized rather than uniformly distributed across the entire device area. The localized region is often called the conductive filament (CF). The mechanisms of nonuniform switching include electrochemical mechanism, valence change mechanism, and thermochemical mechanism. The uniform switching occurs across the entire device area and usually at the interface between an electrode and the memristive materials, which is often referred to as interfacial switching.

7.3.1 Electrochemical Memristor

The most common structure of an electrochemical (also called conductive bridge) memristor is a dielectric film sandwiched between an electrochemically active metal electrode (Ag, Cu, Ni, etc.) and an electrochemically inert metal electrode (Pt, W, etc.). Such devices usually work in bipolar switching mode due to the formation and rupture of a CF composed of metal atoms. During the forming process, the applied voltage ionizes the active metal atoms. In conventional electrochemical memristors based on electrolyte materials with high ion conductivity, the metal ions tend to move across the switching layer and accumulate on the inert electrode. They can nucleate and grow toward the active electrode side [6]. If the ion conductivity is low the scenario is different. In this case, the filament usually originates from the active electrode and grows toward the inert electrode [7, 8]. Once two electrodes are connected by a conductive channel the device will be switched on. In the following reset process, a reversed biased voltage again ionizes the

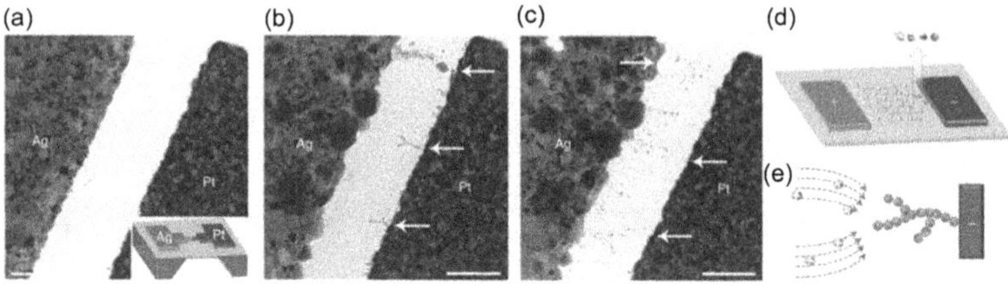

Figure 7.3 (a) TEM image of an as-fabricated resistive switching device on a SiNx membrane. (Inset): The device structure schematic. Scalar bar: 200 nm. (b) After the forming process, some filaments have been observed as indicated by arrows. Scalar bar: 200 nm. (c) After the reset process, the observation at the same site showed that the filament had dissolved. Scalar bar: 200 nm. (d) Schematic showing the metal ionization, migration, and crystallization processes. (e) Schematic of cation nucleation at the electrode to form a branch-shaped conductive filament. Reprinted by permission from Macmillan Publishers Ltd: *Nature Communications* (Ref. [9]), copyright (2012).

metal atoms from the inert electrode side and drives them in the opposite direction, resulting in a partially ruptured filament. Due to the existence of remnants of the ruptured filament, the subsequent set processes need a smaller voltage or a shorter time to rebuild the filament.

In 2012, Yang et al. observed that the conductive bridge of a Ag filament is not necessarily a continuous wire. Instead it can be composed of discrete nanoparticles (Fig. 7.3) [9]. The device is of a lateral structure type, with Ag and Pt as the active electrode and the inert electrode, respectively. Transmission electron microscopy (TEM) was employed to capture the dynamics of the filament. Furthermore, they formulated a relationship of the wait time (time needed for on switching) t_W, electric field, and temperature, showing that filament growth can be limited by cation mobility in the dielectric film as per the following equation:

$$t_W = \frac{1}{\upsilon} e^{E_a'/k_B T} = \frac{1}{\upsilon} e^{E_a/k_B T} \cdot e^{-aE/k_B T} \tag{7.4}$$

Here υ is the attempt frequency, k_B is the Boltzmann constant, and T is the absolute temperature. The real migration barrier E_a' is reduced from E_a to $E_a - aE$ due to the electric field **E**, where a is a constant. Similar TEM observation has been reported by [8, 10, 11].

A detailed simulation of conductive bridge–type resistive switching has been carried out by Onofrio et al. [12]. They used molecular dynamics simulation for this purpose. It has been confirmed that a CF is responsible for the switching if and only if metal ions aggregate into metallic clusters. The filament is stable and results in nonvolatile switching. This work also showed that ultrafast switching can be achieved in a conductive bridge memristive device.

7.3.2 Valence Change Memristor

Unlike electrochemical memristors, the mobile species in valence change memristors are oxygen (or nitrogen) vacancies rather than active metal ions. A large number of dielectric films, including $SrTiO_3$ [13], TiO_2 [14], HfO_2 [15], Ta_2O_5 [16], and AlN [17], have been reported to show such kind of resistive switching. Pt, Pd, TiN, etc., are typical inert electrode materials serving as the anode during reset (off) switching operation, while the counter electrode is normally an oxidizable metal, leading to an asymmetrical device structure that favors a bipolar switching behavior.

The switching mechanism has been studied by Yang et al. [3]. The switching process can be described as follows. In the virgin state, there are more oxygen vacancies near the oxidizable electrode. Since oxygen vacancies serve as n-type dopants in TiO_2 [18], when a negative voltage is applied to the inert electrode, positively charged vacancies will be attracted to it. Once the conductive channel composed of oxygen vacancies penetrates through the film, the device is switched on. On the other hand, when a positive voltage is applied to the inert electrode, oxygen vacancies are repelled away from that electrode and the original electronic barrier (e.g., a Schottky barrier between electrode and dielectric) is recovered. The device is then switched off.

TEM observation of the filament in a valence change memristor was conducted by Li et al. [20]; the corresponding phase map is shown in Fig. 7.4. The phase map featuring $\Delta\varphi^{\text{bias}}(x, y)$ reflected the electric potential map because their difference is a constant [20]. Experimental results supported the conclusion that the forming and set processes are a result of generated oxygen vacancies bridging both electrodes, while reset is the result of recombination of oxygen

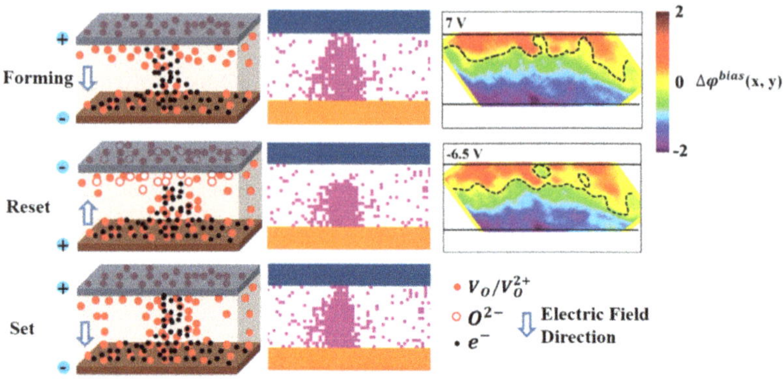

Figure 7.4 Switching mechanism of a HfO_2-based resistive switching device. From top to bottom are the forming, reset, and set process. The left column is the schematic of switching dynamics. The middle column shows the simulation result of a conductive filament using s random circuit breaker network model. The right column is the potential map of the resistive switching region, where dark red is for higher potential and dark blue is for lower potential. The scalar bar means phase values, which is proportional to electric potential. The dashed line is the boundary between positive and negative potential regions. 7 V formed the device, while −6.5 V reset the device. From Ref. [19]. Copyright © 2017. Reproduced with permission of John Wiley & Sons.

vacancies in the filament and oxygen ions. A random circuit breaker network model was used to simulate the process and further supported their theory.

7.3.3 Thermochemical Effect Memristor

Thermochemical effects can coexist with the valence change mechanism or electrochemical mechanism, though it alone can also cause resistive switching in dielectric films. This type of switching behavior is a result of heat-induced formation and rupture of CFs. It usually does not depend on external voltage polarity, so it generally demonstrates unipolar behavior. Transition metal oxides, including CuO [21], NiO [22], and HfO_2 [23], can show unipolar switching.

Applying an electric field to the transition metal oxide leads to local heat generation. The on switching is a result of oxygen diffusing out of the high-temperature region under a thermophoresis effect,

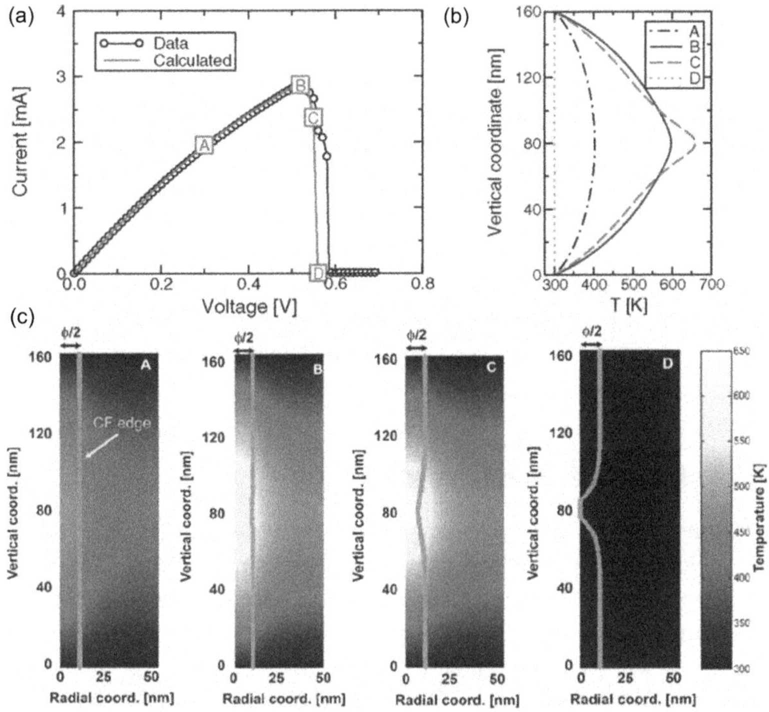

Figure 7.5 (a) Experimental and simulated I-V curve during the reset process of a NiO film with n-Si as the bottom electrode and the top electrode. (b, c) Simulation of the thermal dissolution of the filament corresponding to the four points, A, B, C, and D, in (a). A high temperature is observed in the middle of the filament, which accelerates the dissolution process. φ denotes the radius of the filament. © [2007] IEEE. Reprinted, with permission, from Ref. [24].

and the metal in the oxide is reduced to a lower valence or even a metallic state [6]. The off switching is a result of thermal dissolution of the metallic filament under Fick diffusion. Figure 7.5 shows a reset simulation result by Russo et al. [24]. They assume that there is only one filament and it is of a cylindrical shape. The rupture of the filament starts from a local hot spot near the middle of the filament, and the hot spot eventually melts the filament. Since a partially ruptured filament leads to a higher electric field at the breaking point, the dissolution is accelerating itself, leading to abrupt off switching [24].

A CF in unipolar switching TiO_2 has been observed by in situ TEM by Kwon et al. [25]. The device has a structure of Pt/TiO_2/Pt. In the forming process, a blown-off region is observed, which is probably due to a sudden evolution of compressed oxygen gas released from the oxide during electroforming [26]. According to the in situ TEM observation, as a voltage is applied to the device, some part of the TiO_2 film is changed to Ti_4O_7, a Magnéli phase that is conductive at room temperature. The conversion between such Magnéli phase and the pristine TiO_2 phase are responsible for the observed unipolar resistive switching in the TiO_2 thin film.

7.3.4 Interface Switching Memristor

The main difference between the interface-type memristive switching mechanism and the three aforementioned mechanisms is that it does not form a localized filament to connect the top and bottom electrodes. Instead, the switching takes place uniformly across the entire device area at or near the interface between the dielectric layer and one of the electrodes. A forming process is usually not needed for such switching.

A proposed explanation of interface switching is the modulation of electronic transport by field-induced redistribution of oxygen vacancies. The redistribution of oxygen vacancies changes the band structure and in turn the height of the energy barrier for electron transport, which can induce resistance change in the stack. Meyer et al. studied interface-type switching based on a stack consisting of a tunnel oxide (TO) layer and a conductive metal oxide (CMO) [27]. The switching mechanism and a simulation of the oxygen vacancy drifting are shown in Fig. 7.6. Due to a high electric field caused by the application of an external voltage, oxygen ions migrate from the CMO into the TO. The negatively charged oxygen ions lead to a high potential barrier, and the electron tunneling probability is reduced as a result. The device is thus programmed into a high resistance state. By changing the polarity of the voltage, oxygen vacancies are driven back from the TO to the CMO, which restores the tunneling barrier and the device switches back to its LRS.

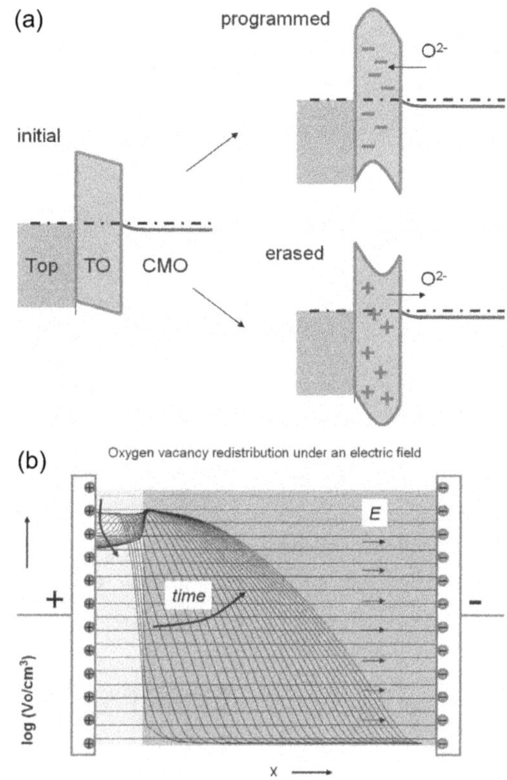

Figure 7.6 (a) Sketched band structure of the initial device, the programmed device, and the erased device. (b) Simulated time evolution of oxygen vacancy distribution. © [2008] IEEE. Reprinted, with permission, from Ref. [27].

It should be noted that more than one switching mechanism can be at play in a certain device. For instance, when an electric field drives metal ions or oxygen vacancies, a thermochemical reaction can also take place. This is the reason why different operation modes sometimes coexist in the same material system under different operation conditions [28]. Although unipolar switching usually shows a higher on/off conductance ratio compared to bipolar switching, it typically consumes more energy and generates more heat, which usually deteriorates the device's reliability.

Table 7.1 Electrical performance of different types of memristors

		Electrochemical metallization bridge	**Metal oxide: Bipolar filament**	**Metal oxide: Unipolar filament**	**Metal oxide: Bipolar interface effects**
Storage mechanism		Electrochemical filament formation	Valence change filament formation	Thermochemical effect filament formation	Change in interface
Feature size F	Best projected	<5 nm	1 nm	<30 nm	<10 nm
	Demonstrated	<10 nm (CuGeSe)	<3 nm (HfO_x)	30 nm	40 nm
Cell area (2D)	Best projected	$4F^2$	$4F^2$	$4F^2$	$4F^2$
	Demonstrated	$4F^2$	$4F^2$	$4F^2$	$4F^2$
Read current	Best projected	Not available	Not available	Not available	Not available
	Demonstrated	Not available	~ 100 pA/~10 nA	1 nA/20 nA	Not available
W/E time	Best projected	<1 ns	<100 ps	Not available	10 ns
	Demonstrated	<1 ns	100 ps	10 ns (W), 5 ns (E)	<100 ns
Retention time	Best projected	>10 years	$>10^5$ years @ 85 C	>10 years	>10 years
	Demonstrated	1000 h @ 200 C or 1 h @ 400 C	55.55 h @ 300 C	1008 h @ 150 C	~3 h @ 200 C
Write cycles	Best projected	$>10^{10}$	10^{12}	$>10^7$	$>10^7$
	Demonstrated	10^{10}	10^{12}	10^7	10^7
Write operating voltage (V)	Best projected	<0.15 V	<1 V	2 V	Not available
	Demonstrated	<0.2 V	1–3 V	1 V (Write)/1.5 V (Erase)	2
Read operating voltage (V)	Best projected	<0.02 V	0.1 V	0.3 V	0.1
	Demonstrated	0.02 V	0.1–0.2 V	0.3 V	0.5
Write energy (J/bit)	Best projected	<200 fJ	<200 fJ	<100 pJ	1 fJ
	Demonstrated	200 fJ	230 fJ	100 pJ	1.23 fJ

Source: [29].

The performance of different types of memristors is summarized in Table 7.1 [29].

7.4 Critical Characteristics

7.4.1 Switching Speed

The reading and writing speed of a memory device is a critical factor determining its overall performance. Complementary metal-oxide semiconductor static random-access memory (CMOS SRAM) can work at the GHz level, but it has a large area footprint. Traditional Flash memories are much slower due to the charging process of the floating gate over a large electronic barrier via electron tunneling. Among various emerging memory devices, memristors have shown less than 100 ps switching [17], which is likely more than sufficient for most of the foreseeable applications in the near future.

Although the mechanism that determines the switching speed of a memristor is debatable, researchers are trying to investigate the effect of certain known factors. One of them is ion mobility. As discussed in Section 7.3, most resistive switching is related to migration of either metal ions or oxygen (or nitrogen) ions. Ion migration can be described by Eq. 7.4. Under an electric field, charged ions are driven by an electrical force. If the electrical force is large enough, the ions can overcome the energy barrier and move to the next energy favorable neighboring site. The ion drift velocity can also be significantly enhanced under a high electric field. As a result, both material selection and applied voltage will determine the ion transport process. At the same time, temperature also has a complicated effect on switching not only in ion migration but also in the formation of chemical bonds. While it is generally observed that a higher voltage can switch the device exponentially faster [9], increasing the voltage will have some negative consequences with regard to power consumption and device reliability. Material-oriented approaches to develop a fast device are also valuable.

Encouraging speed performances have been reported recently. Choi et al. reported 85 ps switching in a TiN/AlN/Pt device (Fig. 7.7) [17]. In this work, nitride vacancies are used instead of oxygen

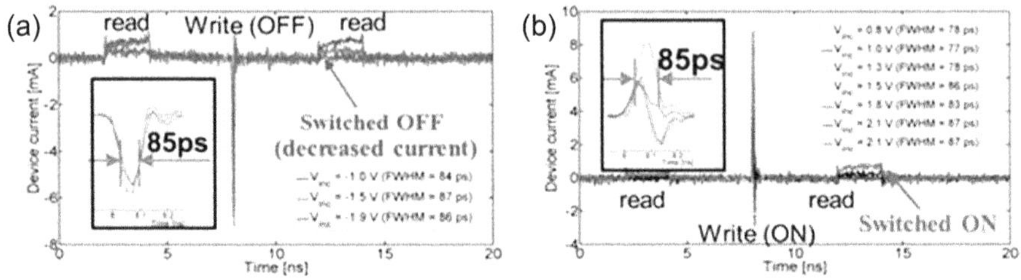

Figure 7.7 (a) 85 ps voltage pulse off switching. (b) 85 ps voltage pulse on switching. The insets to (a) and (b) are the zoom-in images of the switching pulses. A read pulse was applied before and after the switching pulse. Three attempts with an increasing voltage magnitude were used. The devices were successfully switched only if the threshold voltages were reached, that is, –1.9 V for off switching and 2.1 V for on switching. FWHM means full width at half maximum. From Ref. [17]. Copyright © 2016. Reproduced with permission of John Wiley & Sons.

vacancies as the migration species, because nitride vacancies have a valence of +3, which might be driven by a stronger electric force compared to +2 valued oxygen vacancies. In an earlier work, Pickett and Williams observed that a very small volume of switching site and joule heating–induced phase change are reasons for subnanosecond switching [30]. Conductive bridging random-access memory (CBRAM) is relatively slower compared to other types of switching mechanisms. A possible reason is that in CBRAM before a stable filament is formed, some short-lived metastable filaments are formed and dissolved. It takes some time before a stable metallic filament bridges two electrodes [12]. In addition, multiple steps are involved in CBRAM switching, including oxidation, detaching ions from one electrode, ion transport, attaching ions to the other electrode, and reduction, all of which contribute to a longer switching time. However, such switching can still take place within a few nanoseconds, which has been demonstrated experimentally [31].

7.4.2 Endurance

Endurance of a memory device denotes for how many cycles it can be programmed and erased before failure. Several factors are related to the endurance of memristors. The main ones are material stack and electrical operation schemes. In general, the reason for device degradation after repeated switching is attributed to an irreversible change of the switching material at the active region of the device, so it gets stuck in the HRS or LRS state.

Chen et al. summarized three mechanisms of endurance degradation in transition-metal-oxide (TMO)-based resistive random-access memory (RRAM) [32]. They proposed that oxygen species can react with an electrode and form a metal oxide layer, resulting in a highly resistive region, preventing the device from a successful set. Another case is when excessive oxygen vacancies make the filament too thick to be ruptured, leading to a reset failure. The third type is due to the effect of a continuously applied electric field and joule heating, causing oxygen species to migrate away from the filament site. Therefore, the available oxygen is inadequate to recombine oxygen vacancies, also resulting in a reset failure.

In CBRAM, change in the distribution of metal is usually the reason for endurance degradation. Figure 7.8 shows the material evolution after repeated set and reset [33]. The CBRAM structure is HfO_x sandwiched by Cu (active electrode) and Pt (inert electrode). The Cu electrode serves as a source of mobile Cu ions, which can form a CF. According to TEM observations and energy-dispersive X-ray spectroscopy (EDS) characterizations after different numbers of cycling, Cu accumulates on the Pt electrode after being excessively cycled. In consequence, the gap between the filament edge and the electrode becomes narrow and electrons can tunnel through, which causes the device to be stuck in the on state.

According to the 2015 International Technology Roadmap for Semiconductors, the valence-change-type memristor has the best endurance performance. An important requirement of the high endurance is a thermodynamically stable filament. For example, in TaO_x, the conductive Ta filament will not react with Ta_2O_5 even under a high temperature [34]. The HfO_x-based memristor has also exhibited a high endurance [35], which is probably due to the fact that the Hf-O system allows only two thermodynamically stable phases within the operation temperature range inside the device.

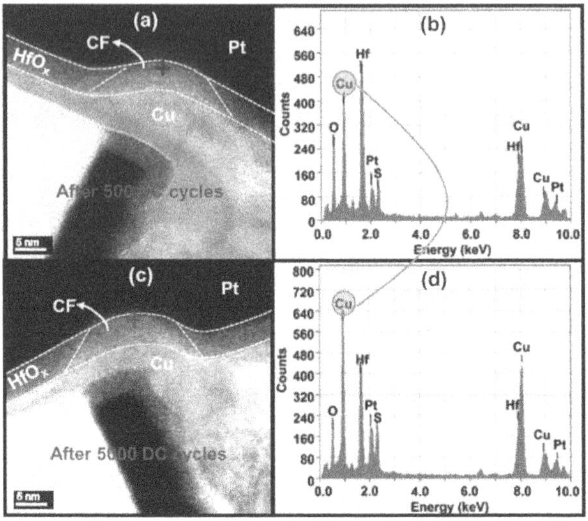

Figure 7.8 (a) TEM image of the filament region after 500 DC sweeps. (b) EDS profile of the filament region after 500 DC sweeps. (c) TEM image of the filament region after 5000 DC sweeps. (d) EDS profile of the filament region after 5000 DC sweeps. CF is short for the conductive filament. © [2015] IEEE. Reprinted, with permission, from Ref. [33].

7.4.3 Retention

Retention of a memristor device is a measure of how long the device can maintain its programmed resistance state at a certain temperature. It is of particular importance for memory applications. For nonvolatile memory, the retention is usually required to be 10 years at 85°C. Since the degradation time follows the Arrhenius dependence of temperature, researchers usually test it using high stress temperatures so that it can be performed under laboratory conditions.

Long retention has been frequently reported in filament-based memristors (electrochemical memristor, valence change memristor, and thermochemical memristor), while interface switching memristors are known to have shorter retention times [29]. A possible reason is that the switching of an interface switching device is related to charge-induced modulation of the electron barrier height.

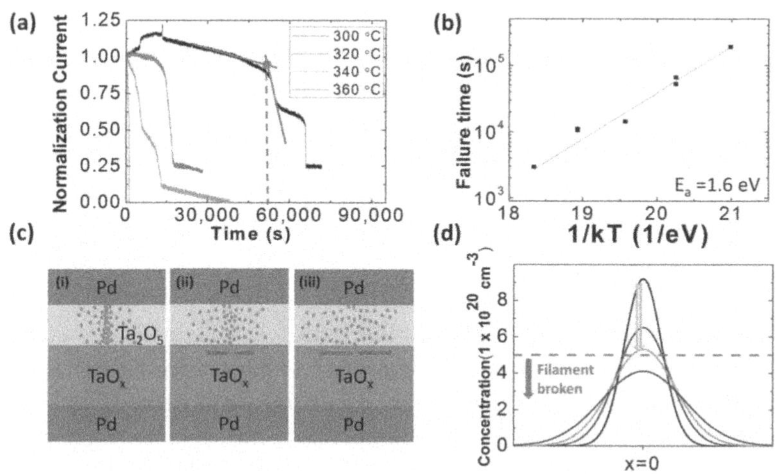

Figure 7.9 (a) Retention time of Pd(top electrode)/Ta_2O_5/TaO_x/Pd(bottom electrode) memristor at 300°C, 320°C, 340°C, and 360°C. (b) Fitted temperature dependence of the characteristic retention failure time (squares) following Arrhenius law. (c) Schematic of oxygen vacancy distribution at LRS (i), after oxygen vacancies diffused out (ii), and after the filament ruptured (iii). (d) A prediction of the vacancy concentration profile as a function of time. Reprinted from Ref. [37] with permission from AIP Publishing.

Under the same temperature it is easier for charge carriers to trap/detrap compared to the migration of ions [36].

In a filament-based memristor, the retention failure is a result of unwanted movement of oxygen or metal ions (Fig. 7.9) [37]. Oxygen vacancies or metal ions moving from the filament site to elsewhere can make the device fail to maintain its LRS. It has been observed that LRS failure usually happens before HRS failure when the testing temperature is increased, suggesting LRS retention is generally a more important issue [38].

From the material aspect, a natural way to increase the retention time in a filament-based memristor is to limit ion migration during retention tests. Improvement of retention in HfO_2-based RRAM has been achieved by introducing an additional Hf cap layer, which reduces available oxygen and slows down the recombination process [39].

However, it should be noted that reducing mobility of ions might lead to set/reset failure. Thus, a trade-off between retention and endurance needs to be achieved. Furthermore, different electrical conditions can benefit some of the performance parameters while harming others. It is an important issue to optimize both parameters according to the application requirements [40].

7.4.4 Uniformity

Uniformity of a memristor is another critical characteristic for commercial use. The uniformity of operating voltage, speed, resistance in HRS and LRS, and some other parameters will determine how well a memristor can be accommodated into a large-scale and multifunctional circuit.

Memristor devices show both device-to-device variation and cycle-to-cycle variation. Multiple factors are related to the manifestation of cycle-to-cycle variation, including randomness of ion migration, gradual changing of morphology of the filament or switching interface, and current overshoot. Device-to-device variation is mainly attributed to a nonideal fabrication process.

Researches have adopted material and electrical approaches in improving the uniformity of memristive devices. In the material-oriented method, one idea is to use the insulator doping to reduce the formation energy and the migration barrier of oxygen vacancies: the rupture or recovery of the filament is easier, under a reasonable amplitude of applied voltages [41]. Another effective approach is to insert a capping layer between the electrode and the resistive switching layer. Depending on the material stack, the capping layer may have different functions. Lv et al. inserted a GST ($Ge_2Sb_2Te_5$) layer in a Cu_xO-based memristor [42]. In the forming process, a conductive path is built in the GST layer and the filament tends to be highly localized. Yu et al. used a buffer oxide layer that has a high oxygen migration barrier and impeded the filament from easily rupturing. The buffer layer is also useful to mitigate the over-reset problem[a] and make HRS more uniform [43].

---

[a] During reset the filament is typically ruptured partially, leaving a narrow gap, but if an excessive voltage is applied the gap becomes wider.

Electrical-oriented approaches can help enhance uniformity in a different manner. In general, the voltage applied to operate the device needs to be carefully balanced. Too low a voltage may lead to a thin or even an unstable filament, while too high a voltage can cause the formation of unwanted filaments and excessive joule heating. The electrical stimuli can affect the uniformity as well. In an earlier research, better uniformity was observed in current sweeping mode than in voltage sweeping mode. The reason could be that current sweeping could prevent the formation of additional filaments once the device had been set [44].

Nonideal operating voltage is another cause of the variation in memristors. The reading process has a chance to change the state of a memristor, leading to unwanted writing/erasing. To reduce the risk of such disturbance, a writing/erasing voltage significantly higher than the reading voltage is necessary. Moreover, the reading voltage should be above the noise level and the writing/erasing voltage should not be too high, which will result in a higher power consumption and worse endurance. From the material aspects, one approach toward solving the problem is inserting a buffer layer onto the switching layer so that the voltage drop on the switching layer during reading can be limited [45].

7.5 Conventional Computing Application

Early applications of memristors focused on their capability to serve as the building blocks of resistive random-access memory (ReRAM) [43]. In 2008, Borghetti et al. demonstrated their potential for logic application [46]. The features of being nonvolatile, possessing a large on-off ratio, being fast, and having great scalability make memristors a promising candidate for future universal storage, logic operation, analog computing, and neuromorphic applications.

7.5.1 ReRAM

The state-of-the art mainstream nonvolatile storage device has been Flash. However, Flash suffers from a slow operation speed and a high voltage for the write operation. Among new types of nonvolatile

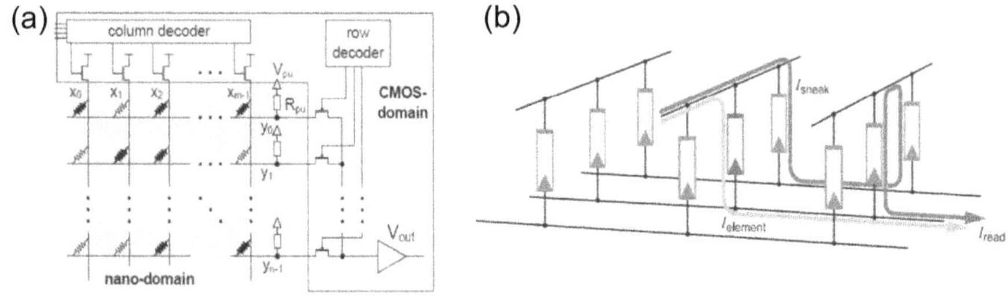

Figure 7.10 (a) Basic setup of a memristor crossbar with a peripheral CMOS circuit. © [2007] IEEE. Reprinted, with permission, from Ref. [50]. (b) Sneak current in a crossbar array. Reprinted by permission from Macmillan Publishers Ltd: *Nature Materials* (Ref. [51]), copyright (2010).

memories, ReRAM has attracted a lot of interest for its remarkable characteristics in electrical performance and scalability. Several large-scale ReRAMs have been reported recently. In 2014, Liu et al. presented a 32 Gb ReRAM chip with a 230 μs write latency [47]. In the same year Fackenthal et al. presented a 16 GB chip with only a 10 μs write latency [48].

In ReRAM, LRS and HRS of a device can represent 1 and 0, respectively. CMOS periphery circuits are used to drive reading and writing. Since a memristor has a simple metal-insulator-metal (MIM) two-terminal structure with only amorphous or polycrystalline materials, a high-density integration structure can be readily realized using a crossbar structure [49]. The crossbar structure is a rectangular or square array where devices in the same column share a common electrode and devices in the same row share the other common electrode.

Figure 7.10a shows a setup of the state-of-the art core component of a practical ReRAM. The column decoder and the row decoder receive commands from a microcontroller and select the target memristor accordingly. R_{pu} is the pull-up resistor, which can help us read out the voltage.

A potential issue in a memristor crossbar array is that writing to a certain device might cause a voltage drop in other devices, especially the devices that are sharing a common electrode with the target device. To avoid unwanted switching of those devices, a half-voltage scheme can be used. For instance, to apply a writing voltage V_{write} to the target device, only 0.5 V_{write} and −0.5 V_{write} should be applied to the selected column and row of the crossbar structure, respectively. If the switching threshold voltage is higher than 0.5 V_{write}, the partially selected or unselected devices will not be switched.

A more serious issue of the crossbar is the sneak path problem. As shown in Fig. 7.10b, it can be inferred that a large crossbar array has numerous sneak paths. If all three devices in one sneak path have been programmed into LRS, the surrounding devices cannot be properly accessible due to excessive current running through the sneak path. Even if there is no complete sneak path formed with devices all in the LRS state, as the array size becomes extremely large, the total sneak current through HRS devices will not be negligible anymore, which limits the read margin of the target device [50].

Various solutions to the sneak path problem have been proposed and investigated. In general, there are two directions. One is to use an external selector device. Transistors [52], diodes [53], threshold switching devices [54], and other devices have been studied for this purpose. Introducing a selector device can suppress the sneak path current in a reliable manner. However, introducing selectors may also complicate the structure and reduces the density.

The other direction is to equip the memristor itself with selecting capability. In general, this approach requires that the device have very large voltage-current nonlinearity, meaning the device should have a huge difference in current response under low and high voltages. Using the so-called complementary resistive switches (CRSs) is one such solution [51, 55]. A CRS device consists of two back-to-back memristive devices, where one device is in LRS and the other is in HRS and one state is defined as 0 and the opposite state is defined as 1. As a result, each cell in the array has at least one device in the pair remaining in the HRS, which prevents the formation of a sneak path composed of devices in LRS. The main drawback of this approach is the destructive reading process, which can make proper electrical operation significantly more difficult. Besides the use of CRSs, other techniques have also been developed, such as the use of highly nonlinear memristors [56] and self-rectified memristors [57].

7.5.2 Digital Application

Nonvolatile and multilevel resistive switching capabilities make memristors promising for implementing logic operations, analog computing, and reconfigurable circuits [58].

So far, researchers have developed two ways to incorporate memristors into logic circuits. One is to use memristors to build logic gates and realize a "stateful" logic computation by electrical operation [46]. The other is to let them serve as programmable switches in reconfigurable logic [59].

Metal-oxide semiconductor field-effect transistors (MOSFETs) are the most widely used device for logic operations. However, traditional metal-oxide semiconductor (MOS) technology is hitting the physical limit, and we are heading toward a more-than-Moore or post-CMOS era. Meanwhile, there is a growing need for such a device that can perform both memory and computing functions because in the traditional computer architecture paradigm too much time and energy are consumed in communication between the arithmetic logic unit (ALU) and memory. A versatile device that can fulfill this requirement together with an ultrahigh density has been explored for years.

In 2010, Borghetti et al. proposed memristor based logic and realized it through material implication (IMP), which is a Boolean logic on two variables *p* and *q* so that *p*IMP*q* is equivalent to (NOT *p*)OR*q* [46]. This work opened a new application avenue for memristors that had been limited to memory applications for a long time. Unlike traditional MOS-based logic, memristor-based logic uses resistance state rather than voltage or charge as the physical state variable. Three memristors were used to execute a NAND operation (thus all logic operation can be constructed). The basic IMP circuit is shown in Fig. 7.11. Here P and Q denote two memristor devices. If it has been set to LRS, the switch can be viewed as on, and HRS corresponds to off.

A memristor can be switched from HRS to LRS only if an applied voltage is above a certain threshold V_{th}. The HRS of the memristor is defined as 0 and LRS is defined as 1. *p* and *q* denote the state of memristor P and Q, respectively. A proper V_{SET} (which can set a memristor) and V_{COND} (which is a conditioning voltage assisting the

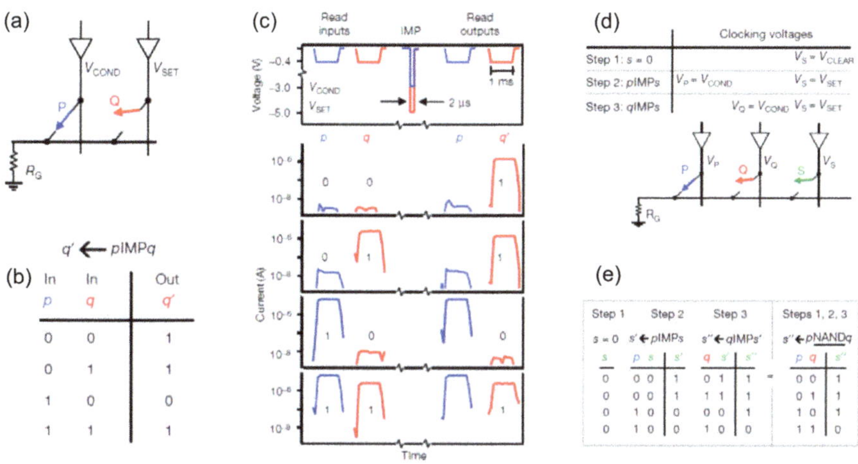

Figure 7.11 IMP and s ← *p*NAND*q* operation. (a) Schematic of a circuit where P and Q are memristive switches. (b) IMP truth table of (a). (c) Electrical measurement of pulse reading and IMP. The blue and red curves show the voltages applied and currents measured at P and Q junctions, respectively. (d) (Upper panel): Use IMP to realize NAND function in three steps. (Lower panel): Schematic of the NAND gate. (e) Truth table of three steps to perform a NAND function. Reprinted by permission from Macmillan Publishers Ltd: *Nature* (Ref. [46]), copyright (2010).

IMP) should be selected such that $V_{th} + V_{COND} > V_{SET} > V_{th}$ and an IMP pulse applied as shown in Fig. 7.11c. R_G is a standard resistor that satisfies the condition: $R_{ON} < R_G < R_{OFF}$, where R_{ON} and R_{OFF} denote the LRS and HRS, respectively. The state of memristors are read through the output current at P and Q junctions. The resulting truth table is in Fig. 7.11b. With the help of IMP, a more sophisticated logic can be realized. As shown in Figs. 7.11d and 7.11e, following three steps—(1) $s = 0$, (2) $s' \leftarrow p\mathrm{IMP}s$, and (3) $s'' \leftarrow q\mathrm{IMP}s'$—the NAND function can be realized.

After this concept had been demonstrated, more mature and larger-scale memristor-based logic circuits have been developed. The first array-level memristor logic was performed by Huang et al. in 2016 [60]. They presented a 16 by 8 crossbar array that can perform basic logic functions and some higher-level functions, like a full adder.

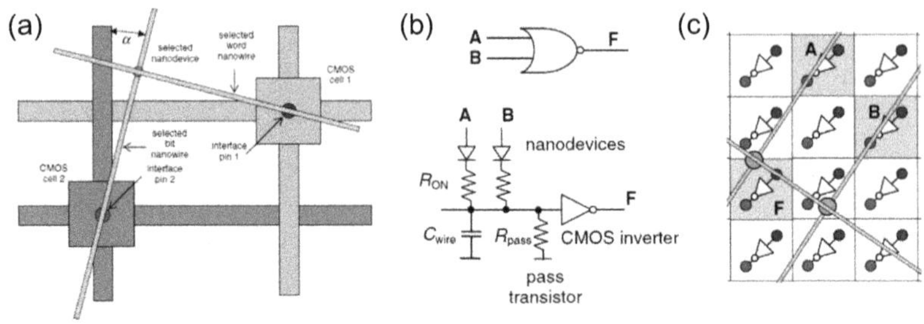

Figure 7.12 (a) The idea of addressing a nanodevice in CMOL FPGA. (b) Schematic of a CMOL NOR gate. (c) One possible configuration of (b). Reprinted from Ref. [61].

Memristors can also serve as a switch in field-programmable gate arrays (FPGAs). The idea was first proposed by Strukov and Likharev [61]. They named their design, which combined a transistor and two levels of nanowires, as CMOL (CMOS cells and two-terminal latching switches) FPGA. Nanodevices are formed at the cross point of the nanowires. The structure is shown in Fig. 7.12a. Because these nanodevices can be programmed to LRS or HRS, the structure is flexible and can be reconfigured to perform various logic functions by changing the connection patterns of the transistors. For example, Figs. 7.12b and 7.12c present a CMOL NOR gate. On the basis of the proposed idea, Xia et al. experimentally demonstrated a real memristor-CMOS hybrid integrated circuit in 2009 [59].

7.5.3 Analog Application

When a memristor is used as a switch, only its binary switching characteristics are utilized. However, due to the gradual set/reset behavior, some memristors can be tuned to multilevel resistance states [62]. This capability enables memristors to perform analog computing.

A typical application is the dot product engine (DPE). Traditional methods to multiply vectors need not only a large number of transistors but also high-efficiency software. By adopting the memristor crossbar and its multilevel switching capability, the DPE operation can be performed in one step. A small-scale experimental

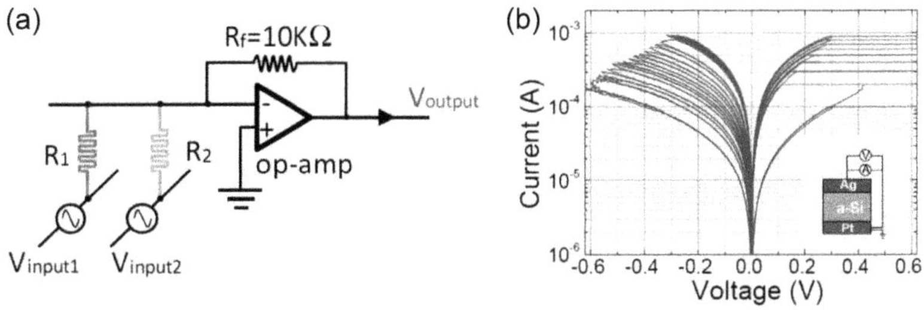

Figure 7.13 (a) Schematic of a DPE circuit. (b) I-V characteristic of the memristive device showing multilevel switching capability. (Inset): Schematic of the device structure. © [2012] IEEE. Reprinted, with permission, from Ref. [63].

demonstration of this concept was done by Gao et al. with Ag/a-Si/Pt memristive devices [63]. The electrical performance and circuit of their device are shown in Fig. 7.13. The resistance of the device can be viewed as the weight that scales its input voltage. The output voltage V_{output} is given by:

$$V_{\text{output}} = -R_{\text{f}}\Sigma V_{\text{input\_i}} \times 1/R_{\text{i}}, (7.5) \qquad (7.5)$$

where R_{f} is a reference resistor connecting the output of operational amplifiers and its inverting input and R_{i} is the resistor placed in between the input voltage V_{input} and the inverting input of the operational amplifier. If the resistance of the device is tuned to a desired value, an accurate DPE calculation can be achieved. Large-scale DPE by a memristor has been analyzed through simulation by Hu et al., along with a set of algorithms to perform its function [64], and demonstrated experimentally by Li et al. with a 128 × 64 1-transistor-1-resistor (1T1R) memristor array [65].

Similar to reconfigurable logic circuits, memristors can also be used as switches in field-programmable analog arrays (FPAAs) [66]. The traditional approach to build a FPAA is by using the switched capacitor technique [67] or the floating gate device [68]. However, both techniques have to deal with a lot of parasitic capacitance. The benefit that memristors can provide to the FPAA is that as nanodevices, memristors can be designed to have much smaller capacitances than traditional switches (like a transistor), thus providing a possible solution for large-scale reconfigurable analog circuits. The basic component of the FPAA are configurable analog blocks (CABs) and the switching matrices that route the CABs. The memristor can serve as a potentiometer in a CAB [69] or a switch in switching matrices [66].

7.6 Unconventional Neuromorphic Computing

The invention of computers with the von Neumann architecture is the main force driving human beings toward the information era. These computers quickly overtook human beings in performing sequential and repetitive tasks in speed and power aspects. However, the development of artificial intelligence calls for much more than what the current computers can offer. Furthermore, device scaling [70] has slowed down owing to quantum mechanical tunneling and power consumption issues. This has, in turn, led to an intensification of the von Neumann bottleneck as time and effort spent on moving data between memory and processor will be huge if the projected demands of the market are to be met.

Non-von-Neumann computing is a natural choice to solve the issue. The human brain provides us with an awe-inspiring computing alternative for the future. The human brain is noted for its massive parallelism. The parallel architecture links a multitude of low-power-consuming computing entities (neurons) with memory elements that can adapt (synapses). The human brain is better than modern processors at performing many tasks that involve pattern recognition and data classification.

The paradigm of computing that takes inspiration from the human brain is generally referred to as "neuromorphic computing." The term was coined by Carver Mead in the 1990s [71]. Ever since that time, researchers have been trying to implement such ideas in hardware. Such efforts were originally limited to the CMOS, which is inefficient in the long run because it is not optimized for this purpose. There has been renewed interest in this field following the advent of memristors [2]. Memristors can be scaled down to form nanometer-level arrays that are very attractive because they can be used to successfully realize the parallelism and the low power consumption of the human brain. Moreover, memristor arrays can be very dense, offering a chance to mimic hardware connectivity that can be rivalled by the connectivity observed in the brain.

There are several challenges in realizing this last computing frontier. The challenges comprise the need for peripheral circuitry that can support massively parallel access to memristor arrays [72, 73], the requirement of a suitable access device. Another obstacle that needs to be overcome is the effect of the immanent shortcomings of memristors, such as device variability, reliability issues, and finite dynamic range.

In general, synaptic weights should be analog. Hence, memristors with multiple resistance states can serve this role. In other applications, it is only required that a synapse be either connected or disconnected. Hence, binary memristors with an on/off response can also be useful.

Memristor arrays are plagued with "sneak path" current issues, as discussed earlier text. This can lead to inaccurate operation of the targeted devices. A common approach to tackle this issue is to introduce current-voltage nonlinearity into the memristors. Ideally, there is nonlinearity inherently present in the memory element. However, insufficient nonlinearity or even close-to-linear current-voltage characteristics are more commonly found in memristors. An alternative is to use a separate access device in series with the memory element [74–81].

There are several issues immanent in the memristor itself, which also need to be taken into consideration. These issues include, but are not limited to, high power consumption, excessive device variability, and a limited dynamic range, some of which have been addressed in Section 7.4.

7.6.1 Artificial Neuron Network

There are several approaches to performing brain-inspired computing. The most widely used one is the artificial neural network (ANN), which is a simplification (often oversimplified) of a biological neural network. In an ANN, the signal levels (amplitude) encode the information.

Arrays of analog resistive memory are ideal for multiply-accumulate operations critical for training and forward-inference

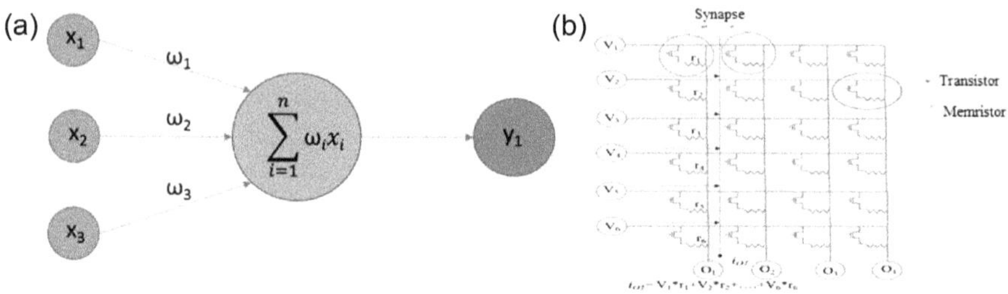

Figure 7.14 (a) A key feature of any neural network: the summation of weights with the excitation of the upward neurons emanating down. This is achieved by the product-summation rule. (b) The product-summation feature can be implemented by a vector matrix multiplication (VMM) in a memristor array, where the conductance of the memristors is equal to the weight of synapses and the voltage applied at each node is the excitation at each neuron.

of deep neural networks (DNNs) [82]. The mentioned dot product capability naturally enables emulation of the weight summation process in neural networks. Current summation (Fig. 7.14) takes places along rows or columns employing Kirchhoff's current law (KCL), and multiplication takes place at each cross point employing Ohm's law. Integration of device read currents along columns implements in parallel the sums of $\omega_{ij}\delta_j$ needed for back propagation of error terms δ_j. ω_{ij} refers to the weight (or conductance) of the memristor in the ith row and j^{th} column.

There have been several simulation and experimental demonstrations with the same idea. A 2 × 9 crossbar array was shown to perform simple pattern classification tasks employing delta learning rule [83]. The inherent nonlinearity of the memristor was determined as detrimental to the online learning rule. A read-before-write scheme was proposed, with device conductance being sensed before selection of the programming pulse [84–86]. Gradient descent learning was employed for implementing image classification (MNIST dataset) [87]. It should be noted that MNIST stands for Modified National Institute of Standards and Technology database. A combination of gradient descent and winner-take-all was implemented for compression and reconstruction of images [88]. Sanger's learning rule was used for clarification of cancer datasets [89]. The impact of parameter variation and device reliability on DNN training has also been investigated [90]. A single-layer DNN was implemented on a 12 × 12 memristor array without the need for access devices [91].

7.6.2 Spiking Neural Network

A spiking neural network (SNN) is generally considered to be any artificial implementation of synaptic functions using asynchronous spikes of identical amplitude and duration [92]. Hence, a SNN is a much closer realization of a biological neural network than a threshold logic unit (TLU)-based ANN.

One of the widely sought learning rules is spike timing–dependent plasticity (STDP). In a biological nervous system, signal propagation takes place between neurons through synapses. This can happen chemically or electrically. STDP is one of the corollaries of Hebb's learning rule [93]. The relative timing between the presynaptic action potential and the postsynaptic action potential is critical to this process. This process can be well summarized by Siegrid Lowell's quote "Cells that fire together wire together" [94].

In STDP the weight of the synapse is strengthened by causal spiking (presynaptic spike occurs before postsynaptic spike) and the strength is weakened by acausal spiking (presynaptic spike occurs after postsynaptic spike) (Fig. 7.13d). The strengthening of the synaptic weight is also called long-term potentiation (LTP). Similarly, the weakening of the synaptic strength is called long-term depression (LTD). The modulation of synaptic plasticity is widely believed to be related to processes of learning and memory [95].

There has been a lot of research done to implement STDP in memristor arrays. The basic idea is to use an array of nonvolatile devices. Memristor conductance is modified on the basis of the STDP learning rule (relative timing of the presynaptic and postsynaptic spikes). STDP can be implemented with asymmetric phase change memory (PCM) devices, though in that case two devices have to be used to implement LTP and LTD separately [96].

Ag_2S "atomic switches" were used to implement STDP learning [97]. LTP was achieved by applying several low-amplitude and narrow-width pulses in a relatively short interval. Inorganic

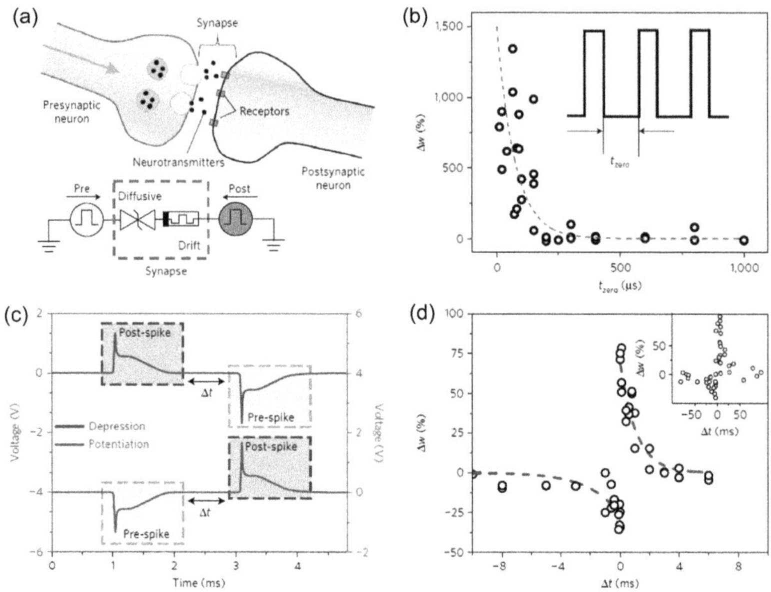

Figure 7.15 (a) Schematic of a biological synapse sandwiched between a presynaptic neuron and a postsynaptic neuron (top). An electronic analog of a synapse (bottom). (b) Short-term plasticity observed in the electronic analog. (c) Causal and anticausal spike patterns applied to the synapse analog. (d) Spike timing–dependent plasticity (STDP) observed in the analog. Reprinted by permission from Macmillan Publishers Ltd: *Nature Materials* (Ref. [100]), copyright (2016).

synapses were used to demonstrate memorization and forgetting of two simple patterns in a 7 × 7 array. STDP was also demonstrated in a 100 × 100 nm memristor connected to two CMOS-based integrate and fire neurons [34]. A low-power stochastic neuromorphic system, using CBRAM devices as binary synapses, was proposed [98]. Associative and nonassociative learning have been emulated using a single memristor device when associated with analog circuitry [99].

Short-term potentiation was realized using a SiO_xN_y:Ag diffusive memristor. In conjunction with a Pt/TaO_x/Ta/Pt drift memristor, both STDP (Fig. 7.15d) and spike rate–dependent plasticity (SRDP) (Fig. 7.15b) were realized. Spikes generated using pulsed voltage sources (Fig. 7.15c) were used as the presynaptic and postsynaptic spikes to demonstrate STDP in the diffusive memristor and drift memristor network (synaptic junction) (Fig. 7.15a). The synaptic dynamics realized here are a harbinger of biorealistic neural networks to come in the near future [100].

Engineered STDP pulses were applied to individual Si-doped HfO_2 memristors [101]. Additional simulations were performed through which learning and recognition of small black-and-white patterns were demonstrated. A memristor cellular automata network was also proposed [102]. Using binary stochastic switching synapses, a two-layer winner-take-all neural network was simulated [103]. The relationship between the application of a compliance current during forming and the synaptic behavior observed was investigated in an FeO_x-based RRAM [104]. A 1T1R synapse using HfO_2 memristors was also proposed [105]. A mixed analog-digital system combined with internal dynamics of an FeO_x-based RRAM was used to implement a recurrent neural network using recursive least-squares algorithm [106]. STDP and paired pulse facilitation were observed in a $Ta/TaO_x/TiO_2/Ti$ stack [106, 107]. Short-term plasticity was emulated in a TiO_2-based device [108]. Both short-term and long-term memory behaviors were observed in a $Pt/WO_{3-x}/Pt$ device [109]. A neuromorphic system composed of a 1 kbit cross-point array of $Pr_{0.7}Ca_{0.3}MnO_3$ (PCMO)-based RRAM stacks showed a continuously incrementing potentiation (depression) behavior under identical pulses of negative (positive) bias [110, 111]. CMOS neurons were used in conjunction with TiN/PCMO synapses to implement real-time associative memory [112]. A very high density crossbar synapse array using PCMO-based devices was also demonstrated [113]. Using signals from the brain of a live rat, fear conditioning experiments were performed [114].

A 10×10 array of PCM cells was used to demonstrate STDP-based learning of some simple patterns [115]. Individual cells were selected by using transistors. A three-layer network comprising 256 neurons was simulated, and pattern recognition was implemented by the same [116].

Neurons are the basic low-power computing entities in the mammalian nervous system. An equilibrium membrane potential is

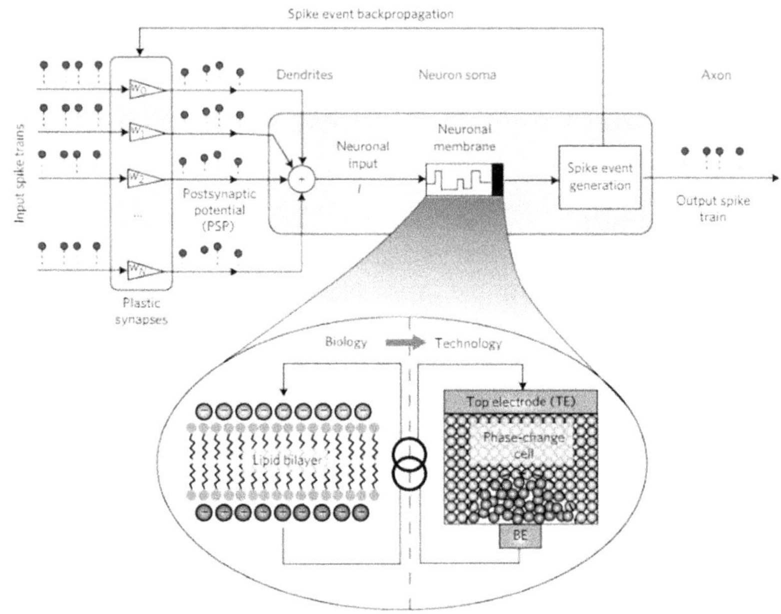

Figure 7.16 An electronic neuron analog was built using phase change memory devices whose physics was analogous to the dynamics of a biological neuron. Reprinted by permission from Macmillan Publishers Ltd: *Nature* (Ref. [117]), copyright (2016).

maintained because of the existence of a thin lipid bilayer membrane that separates the electrical charge inside the cell from that outside it. When excitatory and inhibitory postsynaptic potentials arrive through the dendrites of the neuron, the membrane potential changes and when it reaches a certain threshold, an action potential is generated ("integrate and fire"). Emulation of this functionality is critical to the development of biorealistic neuromorphic systems. A big research thrust is directed toward exploiting the physical behavior of memristors in emulating neuronal dynamics. The aim is to gain significant areal/power efficiency as well as to implement seamless integration with dense synaptic arrays.

PCM devices were first suggested for use as neurons by Ovshinsky and Pashmakov [118]. Recently, PCM device–based neurons (Fig. 7.16) were used to show that postsynaptic inputs can be integrated [117]. This experimental demonstration showed the evolution of the neuronal membrane potential by encoding through the phase configuration in the device [119]. Neuromorphic systems where both neurons and synapses have been realized using phase change devices have also been demonstrated [120]. Stochastic artificial neurons using TiO_x-based memristors have also been proposed [121]. A probabilistic winner-take-all network has been simulated by exploiting the randomness associated with the switching of a memristor [122]. A neuristor was built as an electrical analog of the Hodgkin–Huxley axon using two nanoscale Mott memristors [123]. A pattern recognition neuromorphic system was built using a PCMO-based synapse and a NbO_2-based neuron [124]. A Hopfield network was implemented using a 1 kbit array of the above system.

7.7 Summary

This chapter introduced the basic concept and typical physical mechanisms of memristors. Some important characteristics have been discussed, including speed, endurance, retention, and uniformity. High speed, high endurance, long retention, and repeatable and reproducible behavior are generally preferred for traditional memory applications. However, some of those requirements, such as speed and variability, can more or less be relaxed in certain unconventional computing applications, such as neuromorphic computing. The mainstream applications of memristor were discussed. Memristors have shown potential as both memory and computing devices. However, issues like device variability still limit their large-scale application in the market.

Further research should be carried out for a deeper understanding of the device, from physical mechanisms and microscopic pictures to circuits and architectures.

There is still plenty of room for exploration, in addition to the discussed applications, because of the rich collective behavior of the underlying electronic/ionic interactions with inherent complex nonlinear dynamics. Harnessing the dynamics in an appropriate manner can lead to breakthroughs in materials and electronics technology.

References

1. Chua, L. (1971). Memristor-the missing circuit element, *IEEE Trans. Circuit Theory*, **18**, pp. 507–519.
2. Strukov, D. B., Snider, G. S., Stewart, D. R., and Williams, R. S. (2008). The missing memristor found, *Nature*, **453**, pp. 80–83.
3. Yang, J. J., Pickett, M. D., Li, X., Ohlberg, D. A., Stewart, D. R., and Williams, R. S. (2008). Memristive switching mechanism for metal/oxide/metal nanodevices, *Nat. Nanotechnol.*, **3**, pp. 429–433.
4. Ovshinsky, S. R. (1968). Reversible electrical switching phenomena in disordered structures, *Phys. Rev. Lett.*, **21**, p. 1450.
5. Waser, R., and Aono, M. (2007). Nanoionics-based resistive switching memories, *Nat. Mater.*, **6**, pp. 833–840.
6. Waser, R., Dittmann, R., Staikov, G., and Szot, K. (2009). Redox-based resistive switching memories–nanoionic mechanisms, prospects, and challenges, *Adv. Mater.*, **21**, pp. 2632–2663.
7. Yang, Y., Gao, P., Li, L., Pan, X., Tappertzhofen, S., Choi, S., Waser, R., Valov, I., and Lu, W. D. (2014). Electrochemical dynamics of nanoscale metallic inclusions in dielectrics, *Nat. Commun.*, **5**, p. 4232.
8. Liu, Q., Sun, J., Lv, H., Long, S., Yin, K., Wan, N., Li, Y., Sun, L., and Liu, M. (2012). Real-time observation on dynamic growth/dissolution of conductive filaments in oxide-electrolyte-based ReRAM, *Adv. Mater.*, **24**, pp. 1844–1849.
9. Yang, Y., Gao, P., Gaba, S., Chang, T., Pan, X., and Lu, W. Observation of conducting filament growth in nanoscale resistive memories, *Nat. Commun.*, **3**, p. 732.
10. Xu, Z., Bando, Y., Wang, W., Bai, X., and Golberg, D. (2010). Real-time in situ HRTEM-resolved resistance switching of Ag2S nanoscale ionic conductor, *ACS Nano*, **4**, pp. 2515–2522.
11. Tian, X., Yang, S., Zeng, M., Wang, L., Wei, J., Xu, Z., Wang, W., and Bai, X. (2014). Bipolar electrochemical mechanism for mass transfer in nanoionic resistive memories, *Adv. Mater.*, **26**, pp. 3649–3654.
12. Onofrio, N., Guzman, D., and Strachan, A. (2015). Atomic origin of ultrafast resistance switching in nanoscale electrometallization cells, *Nat. Mater.*, **14**, pp. 440–446.
13. Szot, K., Speier, W., Bihlmayer, G., and Waser, R. (2006). Switching the electrical resistance of individual dislocations in single-crystalline SrTiO3, *Nat. Mater.*, **5**, pp. 312–320.

14. Kim, K. M., Kim, G. H., Song, S. J., Seok, J. Y., Lee, M. H., Yoon, J. H., and Hwang, C. S. (2010). Electrically configurable electroforming and bipolar resistive switching in Pt/TiO2/Pt structures, *Nanotechnology*, **21**, p. 305203.
15. Lee, H., Chen, P., Wu, T., Chen, Y., Wang, C., Tzeng, P., Lin, C., Chen, F., Lien, C., and Tsai, M.-J. (2008). Low power and high speed bipolar switching with a thin reactive Ti buffer layer in robust HfO2 based RRAM, in *IEEE International Electron Devices Meeting (IEDM)*, pp. 1–4.
16. Lee, M.-J., Lee, C. B., Lee, D., Lee, S. R., Chang, M., Hur, J. H., Kim, Y.-B., Kim, C.-J., Seo, D. H., and Seo, S. (2011). A fast, high-endurance and scalable non-volatile memory device made from asymmetric Ta_2O_5–x/TaO_2– x bilayer structures, *Nat. Mater.*, **10**, pp. 625–630.
17. Choi, B. J., Torrezan, A. C., Strachan, J. P., Kotula, P., Lohn, A., Marinella, M. J., Li, Z., Williams, R. S., and Yang, J. J. (2016). High-speed and low-energy nitride memristors, *Adv. Funct. Mater.*, **26**, pp. 5290–5296.
18. Knauth, P., and Tuller, H. (1999). Electrical and defect thermodynamic properties of nanocrystalline titanium dioxide, *J. Appl. Phys.*, **85**, pp. 897–902.
19. Li, C., Gao, B., Yao, Y., Guan, X., Shen, X., Wang, Y., Huang, P., Liu, L., Liu, X., and Li, J. (2017). Direct observations of nanofilament evolution in switching processes in HfO2-based resistive random access memory by in situ TEM studies, *Adv. Mater.*, **29**, p. 1602976.
20. Yao, Y., Li, C., Huo, Z., Liu, M., Zhu, C., Gu, C., Duan, X., Wang, Y., Gu, L., and Yu, R. (2013). In situ electron holography study of charge distribution in high-κ charge-trapping memory, *Nat. Commun.*, **4**, p. 2764.
21. Yasuhara, R., Fujiwara, K., Horiba, K., Kumigashira, H., Kotsugi, M., Oshima, M., and Takagi, H. (2009). Inhomogeneous chemical states in resistance-switching devices with a planar-type Pt/CuO/Pt structure, *Appl. Phys. Lett.*, **95**, p. 012110.
22. Seo, S., Lee, M., Seo, D., Jeoung, E., Suh, D.-S., Joung, Y., Yoo, I., Hwang, I., Kim, S., and Byun, I. (2004). Reproducible resistance switching in polycrystalline NiO films, *Appl. Phys. Lett.*, **85**, pp. 5655–5657.
23. Chan, M., Zhang, T., Ho, V., and Lee, P. (2008). Resistive switching effects of HfO2 high-k dielectric, *Microelectron. Eng.*, **85**, pp. 2420–2424.
24. Russo, U., Ielmini, D., Cagli, C., Lacaita, A., Spiga, S., Wiemer, C., Perego, M., and Fanciulli, M. (2007). Conductive-filament switching analysis and self-accelerated thermal dissolution model for reset in NiO-based RRAM, in *IEEE International Electron Devices Meeting (IEDM)*, pp. 775–778.

25. Kwon, D.-H., Kim, K. M., Jang, J. H., Jeon, J. M., Lee, M. H., Kim, G. H., Li, X.-S., Park, G.-S., Lee, B., and Han, S. (2010). Atomic structure of conducting nanofilaments in TiO2 resistive switching memory, *Nat. Nanotechnol.*, **5**, pp. 148–153.

26. Jeong, D. S., Schroeder, H., Breuer, U., and Waser, R. (2008). Characteristic electroforming behavior in Pt/TiO_2/Pt resistive switching cells depending on atmosphere, *J. Appl. Phys.*, **104**, p. 123716.

27. Meyer, R., Schloss, L., Brewer, J., Lambertson, R., Kinney, W., Sanchez, J., and Rinerson, D. (2008). Oxide dual-layer memory element for scalable non-volatile cross-point memory technology, in *9th Annual Non-Volatile Memory Technology Symposium (NVMTS)*, pp. 1–5.

28. Jeong, D. S., Schroeder, H., and Waser, R. (2007). Coexistence of bipolar and unipolar resistive switching behaviors in a Pt/TiO_2/ Pt stack, *Electrochem. Solid-State Lett.*, **10**, pp. G51–G53.

29. Jo, S. H., Kumar, T., Narayanan, S., and Nazarian, H. (2015). Cross-point resistive RAM based on field-assisted superlinear threshold selector, *IEEE Trans. Electron Devices*, **62**, pp. 3477–3481.

30. Pickett, M. D., and Williams, R. S. (2012). Sub-100 fJ and sub-nanosecond thermally driven threshold switching in niobium oxide crosspoint nanodevices, *Nanotechnology*, **23**, p. 215202.

31. Yang, Y. C., Pan, F., Liu, Q., Liu, M., and Zeng, F. (2009). Fully room-temperature-fabricated nonvolatile resistive memory for ultrafast and high-density memory application, *Nano Lett.*, **9**, pp. 1636–1643.

32. Chen, B., Lu, Y., Gao, B., Fu, Y., Zhang, F., Huang, P., Chen, Y., Liu, L., Liu, X., and Kang, J. (2011). Physical mechanisms of endurance degradation in TMO-RRAM, in *IEEE International Electron Devices Meeting (IEDM)*, pp. 12.3.1–12.3.4.

33. Lv, H., Xu, X., Liu, H., Luo, Q., Liu, Q., Long, S., and Liu, M. (2015). A probe of reliability issues of oxide electrolyte based CBRAM, in *IEEE International Conference on Electron Devices and Solid-State Circuits (EDSSC)*, pp. 229–232.

34. Yang, J. J., Zhang, M.-X., Strachan, J. P., Miao, F., Pickett, M. D., Kelley, R. D., Medeiros-Ribeiro, G., and Williams, R. S. (2010). High switching endurance in TaOx memristive devices, *Appl. Phys. Lett.*, **97**, p. 232102.

35. Lee, H., Chen, Y., Chen, P., Gu, P., Hsu, Y., Wang, S., Liu, W., Tsai, C., Sheu, S., and Chiang, P. (2010). Evidence and solution of over-RESET problem for HfOx based resistive memory with sub-ns switching speed and high endurance, in *IEEE International Electron Devices Meeting (IEDM)*, pp. 19.7.1–19.7.4.

36. Yoon, J. H., Kim, K. M., Song, S. J., Seok, J. Y., Yoon, K. J., Kwon, D. E., Park, T. H., Kwon, Y. J., Shao, X., and Hwang, C. S. (2015). Pt/Ta2O5/HfO2− x/Ti resistive switching memory competing with multilevel NAND flash, *Adv. Mater.*, **27**, pp. 3811–3816.

37. Choi, S., Lee, J., Kim, S., and Lu, W. D. (2014). Retention failure analysis of metal-oxide based resistive memory, *Appl. Phys. Lett.*, **105**, p. 113510.

38. Yu, S., Yin Chen, Y., Guan, X., Philip Wong, H.-S., and Kittl, J. A. (2012). A Monte Carlo study of the low resistance state retention of HfOx based resistive switching memory, *Appl. Phys. Lett.*, **100**, p. 043507.

39. Chen, Y. Y., Komura, M., Degraeve, R., Govoreanu, B., Goux, L., Fantini, A., Raghavan, N., Clima, S., Zhang, L., and Belmonte, A. (2013). Improvement of data retention in HfO2/Hf 1T1R RRAM cell under low operating current, in *IEEE International Electron Devices Meeting (IEDM)*, pp. 10.1.1–10.1.4.

40. Nail, C., Molas, G., Blaise, P., Piccolboni, G., Sklenard, B., Cagli, C., Bernard, M., Roule, A., Azzaz, M., and Vianello, E. (2016). Understanding RRAM endurance, retention and window margin trade-off using experimental results and simulations, in *IEEE International Electron Devices Meeting (IEDM)*, pp. 4.5.1–4.5.4.

41. Zhang, H., Gao, B., Sun, B., Chen, G., Zeng, L., Liu, L., Liu, X., Lu, J., Han, R., Kang, J., and Yu, B. (2010). Ionic doping effect in ZrO2 resistive switching memory, *Appl. Phys. Lett.*, **96**, p. 123502.

42. Lv, H., Wan, H., and Tang, T. (2010). Improvement of resistive switching uniformity by introducing a thin GST interface layer, *IEEE Electron Device Lett.*, **31**, pp. 978–980.

43. Yu, S., Guan, X., and Wong, H.-S. P. (2012). On the switching parameter variation of metal oxide RRAM—Part II: model corroboration and device design strategy, *IEEE Trans. Electron Devices*, **59**, pp. 1183–1188.

44. Lian, W., Lv, H., Liu, Q., Long, S., Wang, W., Wang, Y., Li, Y., Zhang, S., Dai, Y., and Chen, J. (2011). Improved resistive switching uniformity in Cu/HfO2/Pt Devices by using current sweeping mode, *IEEE Electron Device Lett.*, **32**, pp. 1053–1055.

45. Chen, Y., Lee, H., Chen, P., Gu, P., Chen, C., Lin, W., Liu, W., Hsu, Y., Sheu, S., and Chiang, P. (2009). Highly scalable hafnium oxide memory with improvements of resistive distribution and read disturb immunity, in *IEEE International Electron Devices Meeting (IEDM)*, pp. 1–4.

46. Borghetti, J., Snider, G. S., Kuekes, P. J., Yang, J. J., Stewart, D. R., and Williams, R. S. (2010). 'Memristive' switches enable 'stateful' logic operations via material implication, *Nature*, **464**, pp. 873–876.
47. Liu, T.-y., Yan, T. H., Scheuerlein, R., Chen, Y., Lee, J. K., Balakrishnan, G., Yee, G., Zhang, H., Yap, A., and Ouyang, J. (2014). A 130.7-mm2 2-layer 32-Gb ReRAM memory device in 24-nm technology, *IEEE J. Solid-State Circuits*, **49**, pp. 140–153.
48. Fackenthal, R., Kitagawa, M., Otsuka, W., Prall, K., Mills, D., Tsutsui, K., Javanifard, J., Tedrow, K., Tsushima, T., and Shibahara, Y. (2014). 19.7 A 16Gb ReRAM with 200MB/s write and 1GB/s read in 27nm technology, in *IEEE International Solid-State Circuits Conference Digest of Technical Papers (ISSCC)*, pp. 338–339.
49. Kim, K.-H., Gaba, S., Wheeler, D., Cruz-Albrecht, J. M., Hussain, T., Srinivasa, N., and Lu, W. (2011). A functional hybrid memristor crossbar-array/CMOS system for data storage and neuromorphic applications, *Nano Lett.*, **12**, pp. 389–395.
50. Flocke, A., and Noll, T. G. (2007). Fundamental analysis of resistive nano-crossbars for the use in hybrid Nano/CMOS-memory, in *33rd European Solid State Circuits Conference (ESSCIRC)*, pp. 328–331.
51. Linn, E., Rosezin, R., Kügeler, C., and Waser, R. (2010). Complementary resistive switches for passive nanocrossbar memories, *Nat. Mater.*, **9**, pp. 403–406.
52. Zangeneh, M., and Joshi, A. (2014). Design and optimization of nonvolatile multibit 1T1R resistive RAM, *IEEE Trans. Very Large Scale Integr. VLSI Syst.*, **22**, pp. 1815–1828.
53. Huang, J.-J., Hou, T.-H., Hsu, C.-W., Tseng, Y.-M., Chang, W.-H., Jang, W.-Y., and Lin, C.-H. (2012). Flexible one diode–one resistor crossbar resistive-switching memory, *Jpn. J. Appl. Phys.*, **51**, p. 04DD09.
54. Midya, R., Wang, Z., Zhang, J., Savel'ev, S. E., Li, C., Rao, M., Jang, M. H., Joshi, S., Jiang, H., Lin, P., Norris, K., Ge, N., Wu, Q., Barnell, M., Li, Z., Xin, H. L., Williams, R. S., Xia, Q., and Yang, J. J. (2017). Anatomy of Ag/Hafnia-based selectors with 10^{10} nonlinearity, *Adv. Mater.*, **29**, p. 1604457.
55. Yang, J. J., Borghetti, J., Murphy, D., Stewart, D. R., and Williams, R. S. (2009). A family of electronically reconfigurable nanodevices, *Adv. Mater.*, **21**, pp. 3754–3758.
56. Yang, J. J., Zhang, M., Pickett, M. D., Miao, F., Strachan, J. P., Li, W.-D., Yi, W., Ohlberg, D. A., Choi, B. J., and Wu, W. (2012). Engineering nonlinearity

into memristors for passive crossbar applications, *Appl. Phys. Lett.*, **100**, p. 113501.

57. Tran, X., Gao, B., Kang, J., Wu, X., Wu, L., Fang, Z., Wang, Z., Pey, K., Yeo, Y., and Du, A. (2011). Self-rectifying and forming-free unipolar HfOx based-high performance RRAM built by fab-avaialbe materials, in *IEEE International Electron Devices Meeting (IEDM)*, pp. 31.2.1–31.2.4.
58. Yang, J. J., Strukov, D. B., and Stewart, D. R. (2013). Memristive devices for computing, *Nat. Nanotechnol.*, **8**, pp. 13–24.
59. Xia, Q., Robinett, W., Cumbie, M. W., Banerjee, N., Cardinali, T. J., Yang, J. J., Wu, W., Li, X., Tong, W. M., and Strukov, D. B. (2009). Memristor—CMOS hybrid integrated circuits for reconfigurable logic, *Nano Lett.*, **9**, pp. 3640–3645.
60. Huang, P., Kang, J., Zhao, Y., Chen, S., Han, R., Zhou, Z., Chen, Z., Ma, W., Li, M., and Liu, L. (2016). Reconfigurable nonvolatile logic operations in resistance switching crossbar array for large-scale circuits, *Adv. Mater.*, **28**, pp. 9758–9764.
61. Strukov, D. B., and Likharev, K. K. (2005). CMOL FPGA: a reconfigurable architecture for hybrid digital circuits with two-terminal nanodevices, *Nanotechnology*, **16**, p. 888.
62. Miao, F., Yi, W., Goldfarb, I., Yang, J. J., Zhang, M.-X., Pickett, M. D., Strachan, J. P., Medeiros-Ribeiro, G., and Williams, R. S. (2012). Continuous electrical tuning of the chemical composition of TaOx-based memristors, *ACS Nano*, **6**, pp. 2312–2318.
63. Gao, L., Alibart, F., and Strukov, D. B. (2012). Analog-input analog-weight dot-product operation with Ag/a-Si/Pt memristive devices, in *IEEE/IFIP 20th International Conference on VLSI and System-on-Chip (VLSI-SoC)*, pp. 88–93.
64. Hu, M., Strachan, J. P., Li, Z., Grafals, E. M., Davila, N., Graves, C., Lam, S., Ge, N., Yang, J. J., and Williams, R. S. (2016). Dot-product engine for neuromorphic computing: programming 1T1M crossbar to accelerate matrix-vector multiplication, in *53nd ACM/EDAC/IEEE Design Automation Conference (DAC)*, pp. 1–6.
65. Li, C., Hu, M., Li, Y., Jiang, H., Ge, N., Montgomery, E., Zhang, J., Song, W., Dávila, N., Graves, C. E., Li, Z., Strachan, J. P., Lin, P., Wang, Z., Barnell, M., Wu, Q., Stanley Williams, R., Joshua Yang, J., and Xia, Q. (2018). Analog signal and image processing with large memristor crossbars, *Nat. Electron.*, **1**, pp. 52–59.

66. Laiho, M., Hasler, J. O., Zhou, J., Du, C., Lu, W., Lehtonen, E., and Poikonen, J. H. (2015). FPAA/memristor hybrid computing infrastructure, *IEEE Trans. Circuits Syst. I: Regul. Pap.*, **62**, pp. 906–915.

67. Kutuk, H., and Kang, S.-M. (1996). A field-programmable analog array (FPAA) using switched-capacitor techniques, in *IEEE International Symposium on Circuits and Systems (ISCAS'96), Connecting the World*, pp. 41–44.

68. Basu, A., Brink, S., Schlottmann, C., Ramakrishnan, S., Petre, C., Koziol, S., Baskaya, F., Twigg, C. M., and Hasler, P. (2010). A floating-gate-based field-programmable analog array, *IEEE J. Solid-State Circuits*, **45**, pp. 1781–1794.

69. Pershin, Y. V., and Di Ventra, M. (2010). Practical approach to programmable analog circuits with memristors, *IEEE Trans. Circuits Syst. I: Regul. Pap.*, **57**, pp. 1857–1864.

70. Dennard, R. H., Gaensslen, F. H., Rideout, V. L., Bassous, E., and LeBlanc, A. R. (1974). Design of ion-implanted MOSFET's with very small physical dimensions, *IEEE J. Solid-State Circuits*, **9**, pp. 256–268.

71. Mead, C. (1990). Neuromorphic electronic systems, *Proc. IEEE*, **78**, pp. 1629–1636.

72. Strukov, D., Merrikh-Bayat, F., Prezioso, M., Guo, X., Hoskins, B., and Likharev, K. (2015). Memory technologies for neural networks, in *2015 IEEE International Memory Workshop (IMW)*, pp. 1–4.

73. Eryilmaz, S. B., Kuzum, D., Yu, S., and Wong, H. S. P. (2015). Device and system level design considerations for analog-non-volatile-memory based neuromorphic architectures, in *IEEE International Electron Devices Meeting (IEDM) Technical Digest*, pp. 4.1.1–4.1.4.

74. Midya, R., Wang, Z., Zhang, J., Savel'ev, S. E., Li, C., Rao, M., Jang, M. H., Joshi, S., Jiang, H., Lin, P., Norris, K., Ge, N., Wu, Q., Barnell, M., Li, Z., Xin, H. L., Williams, R. S., Xia, Q., and Yang, J. J. (2017). Anatomy of Ag/Hafnia-based selectors with 10^{10} nonlinearity, *Adv. Mater.*, **29**, p. 1604457.

75. Sung Hyun, J., Kumar, T., Narayanan, S., Lu, W. D., and Nazarian, H. (2014). 3D-stackable crossbar resistive memory based on Field Assisted Superlinear Threshold (FAST) selector, in *IEEE International Electron Devices Meeting (IEDM) Technical Digest*, pp. 6.7.1–6.7.4.

76. Rohit, S. S., Geoffrey, W. B., Kumar, V., Bryan, J., Alvaro, P., Pritish, N., Charles, T. R., Robert, M. S., Donald, S. B., Karthik, V. R., Matthew, B., Eric, J., Philip, M. R., Teya, T., Andrew, J. K., Bülent, K., and Kailash, G. (2014). MIEC (mixed-ionic-electronic-conduction)-based access devices for

non-volatile crossbar memory arrays, *Semicond. Sci. Technol.*, **29**, p. 104005.

77. Kim, S., Kim, Y. B., Kim, K. M., Kim, S. J., Lee, S. R., Chang, M., Cho, E., Lee, M. J., Lee, D., Kim, C. J., Chung, U. I., and Yoo, I. K. (2013). Performance of threshold switching in chalcogenide glass for 3D stackable selector, in *2013 Symposium on VLSI Technology*, pp. T240–T241.
78. Lee, M. J., Lee, D., Kim, H., Choi, H. S., Park, J. B., Kim, H. G., Cha, Y. K., Chung, U. I., Yoo, I. K., and Kim, K. (2012). Highly-scalable threshold switching select device based on chaclogenide glasses for 3D nanoscaled memory arrays, in *IEEE International Electron Devices Meeting (IEDM) Technical Digest*, pp. 2.6.1–2.6.3.
79. Cha, E., Woo, J., Lee, D., Lee, S., Song, J., Koo, Y., Lee, J., Park, C. G., Yang, M. Y., Kamiya, K., Shiraishi, K., Magyari-Köpe, B., Nishi, Y., and Hwang, H. (2013). Nanoscale (~10 nm) 3D vertical ReRAM and NbO2 threshold selector with TiN electrode, in *IEEE International Electron Devices Meeting (IEDM) Technical Digest*, pp. 10.5.1–10.5.4.
80. Yang, H., Li, M., He, W., Jiang, Y., Lim, K. G., Song, W., Zhuo, V. Y. Q., Tan, C. C., Chua, E. K., Wang, W., Yang, Y., and Ji, R. (2015). Novel selector for high density non-volatile memory with ultra-low holding voltage and 107 on/off ratio, in *2015 Symposium on VLSI Technology (VLSI Technology)*, pp. T130–T131.
81. Luo, Q., Xu, X., Liu, H., Lv, H., Gong, T., Long, S., Liu, Q., Sun, H., Banerjee, W., Li, L., Lu, N., and Liu, M. (2015). Cu BEOL compatible selector with high selectivity ($>10^7$), extremely low off-current (~pA) and high endurance ($>10^{10}$), in *IEEE International Electron Devices Meeting (IEDM) Technical Digest*, pp. 10.4.1–10.4.4.
82. LeCun, Y., Bengio, Y., and Hinton, G. (2015). Deep learning, *Nature*, **521**, pp. 436–444.
83. Alibart, F., Zamanidoost, E., and Strukov, D. B. (2013). Pattern classification by memristive crossbar circuits using ex situ and in situ training, *Nat. Commun.*, **4**, p. 2072.
84. Chen, P.-Y., Lin, B., Wang, I.-T., Hou, T.-H., Ye, J., Vrudhula, S., Seo, J.-s., Cao, Y., and Yu, S. (2015). Mitigating effects of non-ideal synaptic device characteristics for on-chip learning, in *IEEE/ACM International Conference on Computer-Aided Design (ICCAD)*, pp. 194–199.
85. Jang, J.-W., Park, S., Burr, G. W., Hwang, H., and Jeong, Y.-H. (2015). Optimization of conductance change in Pr 1–x Ca x MnO 3-based synaptic devices for neuromorphic systems, *IEEE Electron Device Lett.*, **36**, pp. 457–459.

86. Alibart, F., Gao, L., Hoskins, B. D., and Strukov, D. B. (2012). High precision tuning of state for memristive devices by adaptable variation-tolerant algorithm, *Nanotechnology*, **23**, p. 075201.
87. Nair, M. V., and Dudek, P. (2015). Gradient-descent-based learning in memristive crossbar arrays, in *International Joint Conference on Neural Networks (IJCNN)*, pp. 1–7.
88. Sheridan, P. M., Du, C., and Lu, W. D. (2016). Feature extraction using memristor networks, *IEEE Trans. Neural Networks Learn. Syst.*, **27**, pp. 2327–2336.
89. Choi, S., Sheridan, P., and Lu, W. D. (2015). Data clustering using memristor networks, *Sci. Rep.*, **5**, p. 10492.
90. Zhao, W., Zhang, Y., Klein, J.-O., Querlioz, D., Chabi, D., Ravelosona, D., Chappert, C., Portal, J. M., Bocquet, M., and Aziza, H. (2012). Crossbar architecture based on 2R complementary resistive switching memory cell, in *Proceedings of the 2012 IEEE/ACM International Symposium on Nanoscale Architectures*, pp. 85–92.
91. Prezioso, M., Merrikh-Bayat, F., Hoskins, B., Adam, G., Likharev, K. K., and Strukov, D. B. (2015). Training and operation of an integrated neuromorphic network based on metal-oxide memristors, *Nature*, **521**, pp. 61–64.
92. Grüning, A., and Bohte, S. M. (2014). Spiking neural networks: principles and challenges, in *European Symposium on Artificial Neural Networks (ESANN)*, pp. 1–10.
93. Hebb, D. (1949). *The Organization of Behavior*, Wiely, New York.
94. Lowel, S., and Singer, W. (1992). Selection of intrinsic horizontal connections in the visual cortex by correlated neuronal activity, *Science*, **255**, p. 209.
95. Morrison, A., Diesmann, M., and Gerstner, W. (2008). Phenomenological models of synaptic plasticity based on spike timing, *Biol. Cybern.*, **98**, pp. 459–478.
96. Suri, M., Bichler, O., Querlioz, D., Cueto, O., Perniola, L., Sousa, V., Vuillaume, D., Gamrat, C., and DeSalvo, B. (2011). Phase change memory as synapse for ultra-dense neuromorphic systems: application to complex visual pattern extraction, in *IEEE International Electron Devices Meeting (IEDM)*, pp. 4.4.1–4.4.4.
97. Ohno, T., Hasegawa, T., Nayak, A., Tsuruoka, T., Gimzewski, J. K., and Aono, M. (2011). Sensory and short-term memory formations observed in a Ag2S gap-type atomic switch, *Appl. Phys. Lett.*, **99**, p. 203108.

98. Suri, M., Bichler, O., Querlioz, D., Palma, G., Vianello, E., Vuillaume, D., Gamrat, C., and DeSalvo, B. (2012). CBRAM devices as binary synapses for low-power stochastic neuromorphic systems: auditory (cochlea) and visual (retina) cognitive processing applications, in *IEEE International Electron Devices Meeting (IEDM)*, pp. 10.3.1–10.3.4.
99. Ziegler, M., Soni, R., Patelczyk, T., Ignatov, M., Bartsch, T., Meuffels, P., and Kohlstedt, H. (2012). An electronic version of Pavlov's dog, *Adv. Funct. Mater.*, **22**, pp. 2744–2749.
100. Wang, Z., Joshi, S., Savel'ev, S. E., Jiang, H., Midya, R., Lin, P., Hu, M., Ge, N., Strachan, J. P., and Li, Z. (2017). Memristors with diffusive dynamics as synaptic emulators for neuromorphic computing, *Nat. Mater.*, **16**, pp. 101–108.
101. Wang, Z., Ambrogio, S., Balatti, S., and Ielmini, D. (2015). A 2-transistor/1-resistor artificial synapse capable of communication and stochastic learning in neuromorphic systems, *Front. Neurosci.*, **8**, p. 438.
102. Itoh, M., and Chua, L. O. (2008). Memristor oscillators, *Int. J. Bifurcation Chaos*, **18**, pp. 3183–3206.
103. Yu, S., Gao, B., Fang, Z., Yu, H., Kang, J., and Wong, H.-S. P. (2013). Stochastic learning in oxide binary synaptic device for neuromorphic computing, *Front. Neurosci.*, **7**, p. 186.
104. Wang, C., He, W., Tong, Y., and Zhao, R. (2016). Investigation and manipulation of different analog behaviors of memristor as electronic synapse for neuromorphic applications, *Sci. Rep.*, **6**, p. 22970.
105. Ambrogio, S., Balatti, S., Nardi, S., Facchinetti, S., and Ielmini, D. (2013). Spike-timing dependent plasticity in a transistor-selected resistive switching memory, *Nanotechnology*, **24**, p. 384012.
106. Deng, L., Li, G., Deng, N., Wang, D., Zhang, Z., He, W., Li, H., Pei, J., and Shi, L. (2015). Complex learning in bio-plausible memristive networks, *Sci. Rep.*, **5**, p. 10684.
107. Wang, I.-T., Lin, Y.-C., Wang, Y.-F., Hsu, C.-W., and Hou, T.-H. (2014). 3D synaptic architecture with ultralow sub-10 fJ energy per spike for neuromorphic computation, in *IEEE International Electron Devices Meeting (IEDM)*, pp. 28.5.1–28.5.4.
108. Lim, H., Kim, I., Kim, J.-S., Hwang, C. S., and Jeong, D. S. (2013). Short-term memory of TiO2-based electrochemical capacitors: empirical analysis with adoption of a sliding threshold, *Nanotechnology*, **24**, p. 384005.

109. Yang, R., Terabe, K., Yao, Y., Tsuruoka, T., Hasegawa, T., Gimzewski, J. K., and Aono, M. (2013). Synaptic plasticity and memory functions achieved in a WO3− x-based nanoionics device by using the principle of atomic switch operation, *Nanotechnology*, **24**, p. 384003.

110. Park, S., Kim, H., Choo, M., Noh, J., Sheri, A., Jung, S., Seo, K., Park, J., Kim, S., and Lee, W. (2012). RRAM-based synapse for neuromorphic system with pattern recognition function, in *IEEE International Electron Devices Meeting (IEDM)*, pp. 10.2.1–10.2.4.

111. Park, S., Noh, J., Choo, M.-L., Sheri, A. M., Chang, M., Kim, Y.-B., Kim, C. J., Jeon, M., Lee, B.-G., and Lee, B. H. (2013). Nanoscale RRAM-based synaptic electronics: toward a neuromorphic computing device, *Nanotechnology*, **24**, p. 384009.

112. Moon, K., Park, S., Jang, J., Lee, D., Woo, J., Cha, E., Lee, S., Park, J., Song, J., and Koo, Y. (2014). Hardware implementation of associative memory characteristics with analogue-type resistive-switching device, *Nanotechnology*, **25**, p. 495204.

113. Park, S., Chu, M., Kim, J., Noh, J., Jeon, M., Lee, B. H., Hwang, H., Lee, B., and Lee, B.-G. (2015). Electronic system with memristive synapses for pattern recognition, *Sci. Rep.*, **5**, p. 10123.

114. Lee, D., Park, J., Moon, K., Jang, J., Park, S., Chu, M., Kim, J., Noh, J., Jeon, M., and Lee, B. H. (2015). Oxide based nanoscale analog synapse device for neural signal recognition system, in *IEEE International Electron Devices Meeting (IEDM)*, pp. 4.7.1–4.7.4.

115. Eryilmaz, S. B., Kuzum, D., Jeyasingh, R. G., Kim, S., BrightSky, M., Lam, C., and Wong, H.-S. P. (2013). Experimental demonstration of array-level learning with phase change synaptic devices, in *IEEE International Electron Devices Meeting (IEDM)*, pp. 25.5.1–25.5.4.

116. Ambrogio, S., Ciocchini, N., Laudato, M., Milo, V., Pirovano, A., Fantini, P., and Ielmini, D. (2016). Unsupervised learning by spike timing dependent plasticity in phase change memory (PCM) synapses, *Front. Neurosci.*, **10**, p. 56.

117. Tuma, T., Pantazi, A., Le Gallo, M., Sebastian, A., and Eleftheriou, E. (2016). Stochastic phase-change neurons, *Nat. Nanotechnol.*, **11**, pp. 693–699.

118. Ovshinsky, S. R., and Pashmakov, B. (2003). Innovation providing new multiple functions in phase-change materials to achieve cognitive computing, *MRS Proc.*, 803, p. HH1.1.

119. Tuma, T., Le Gallo, M., Sebastian, A., and Eleftheriou, E. (2016). Detecting correlations using phase-change neurons and synapses, *IEEE Electron Device Lett.*, **37**, pp. 1238–1241.
120. Pantazi, A., Woźniak, S., Tuma, T., and Eleftheriou, E. (2016). All-memristive neuromorphic computing with level-tuned neurons, *Nanotechnology*, **27**, p. 355205.
121. Al-Shedivat, M., Naous, R., Cauwenberghs, G., and Salama, K. N. (2015). Memristors empower spiking neurons with stochasticity, *IEEE J. Emerging Sel. Top. Circuits Syst.*, **5**, pp. 242–253.
122. Nessler, B., Pfeiffer, M., Buesing, L., and Maass, W. (2013). Bayesian computation emerges in generic cortical microcircuits through spike-timing-dependent plasticity, *PLoS Comput. Biol.*, **9**, p. e1003037.
123. Pickett, M. D., Medeiros-Ribeiro, G., and Williams, R. S. (2013). A scalable neuristor built with Mott memristors, *Nat. Mater.*, **12**, pp. 114–117.
124. Moon, K., Cha, E., Park, J., Gi, S., Chu, M., Baek, K., Lee, B., Oh, S., and Hwang, H. (2015). High density neuromorphic system with Mo/Pr0.7Ca0.3MnO3 synapse and NbO2 IMT oscillator neuron, in *IEEE International Electron Devices Meeting (IEDM)*, pp. 17.6.1–17.6.4.

Chapter 8

Resistive Memories for Spike-Based Neuromorphic Circuits

E. Vianello,[a] O. Bichler,[b] B. De Salvo,[a] and L. Perniola[a]
[a]*CEA-LETI, Minatec Campus, Grenoble, France*
[b]*CEA-LIST, Saclay, France*
elisa.vianello@cea.fr

In the last decade machine learning algorithms have proven unprecedented performance to solve many real-world detection and classification tasks, for example, in image or speech recognition. Despite these advances, there are still some deficits. First, these algorithms require significant memory access, thus ruling out an implementation using standard platforms (e.g., GPUs, CPUs, and FPGAs) for embedded applications. Second, most machine learning algorithms need to be trained with huge datasets (*supervised learning*). Resistive memories (RRAM) have demonstrated to be a promising candidate for overcoming both these constraints. RRAM arrays can act as a dot product accelerator, which is one of the main building blocks in neuromorphic computing systems. This approach could provide improvements in power consumption and speed with respect to the GPU- or CPU-based networks. We present a possible hardware implementation of a spike-based convolutional neural network for visual pattern recognition. Thanks to the use of RRAM

Emerging Devices for Low-Power and High-Performance Nanosystems: Physics, Novel Functions, and Data Processing
Edited by Simon Deleonibus

ISBN 978-981-4800-11-2 (Hardcover), 978-0-429-45873-6 (eBook)
www.panstanford.com

synapses to implement the kernel, the convolution operations are performed directly in memory, allowing one to reduce the latency per image recognition with respect to software implementations on a GPU. Moreover, RRAM devices are promising candidates for emulating synaptic plasticity, the capability of synapses to enhance or diminish their connectivity between neurons, which is widely believed to be the basis for learning and memory in the brain. Neural systems exhibit various types of plasticity. In this work we propose an architecture that implements both spike timing–dependent plasticity (STDP) (a type of long-term plasticity) and synaptic adaptation (a type of short-term plasticity). We show the benefits of utilizing both kinds of plasticity on a real application, visual pattern extraction in real time. Long-term STDP allows the neural network to learn patterns without a training dataset, and short-term synaptic adaptation makes the learning process very robust against background noise in the input data.

8.1 Introduction

Extracting useful information from sensor data (such as video, audio, and biological data) is a key requirement for future mobile and Internet of Things (IoT) applications. Mining data from raw sensors is challenging: systems capable of handling large volumes of noisy and incomplete real-life data are required. The most promising approaches for coping with this challenge are brain-inspired machine learning algorithms (e.g., deep learning [1]). Advances in this field have demonstrated its powerfulness in various tasks, including image classification, speech recognition, natural language processing, and bioinformatics. The two most popular deep learning architectures are fully connected neural networks (FCNNs, Fig. 8.1) and convolutional neural networks (CONVs, Fig. 8.2) [2, 3]. A FCNN is composed of fully connected layers, each of which contains a collection of neurons and synapses. The neurons of a given layer are connected to every neuron of the previous layer by a large number of synapses. Raw data (e.g., video, audio, and biological data) initialize the values of the first layer (the input layer). The output layer corresponds to the inference classes (each

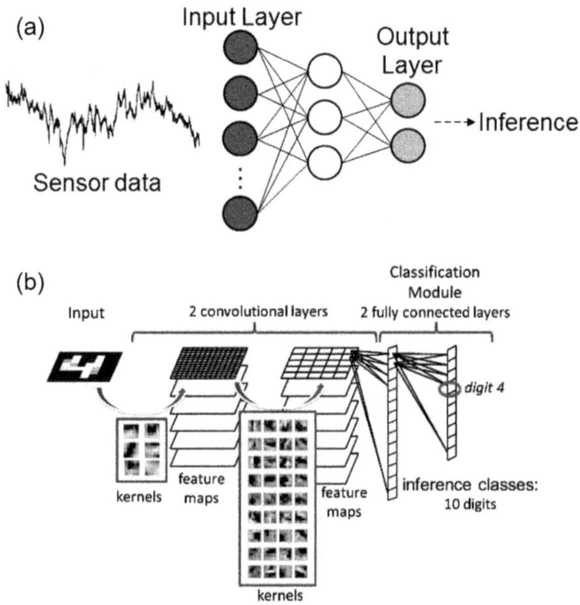

Figure 8.1 (a) Example of a two-layer fully connected neural network (FCNN). (b) Example of a convolutional neural network (CONV) architecture for handwritten digits recognition (Modified National Institute of Standards and Technology [MNIST] database) [15].

output neuron is associated to a class of objects, e.g., dog, cat, or car). A CONV is composed of one or more convolutional layers, pooling or subsampling layers, and fully connected output layers. In a convolutional layer a small set of synapses (constituting a kernel) is shared among different neurons to connect layer N and $N-1$ through a convolutional operation. Each kernel corresponds to a feature that has to be localized in the input image. The output of the convolutional layers (feature maps) contains information about the locations where the kernel features are present in the input. The fully connected layer (classification module) is applied to complete the classification. Inference in a CONV is similar to that of a FCNN. The data initialize the first layer, and the algorithm moves forward layer by layer. The output layer corresponds to the inference classes as for the FCNN.

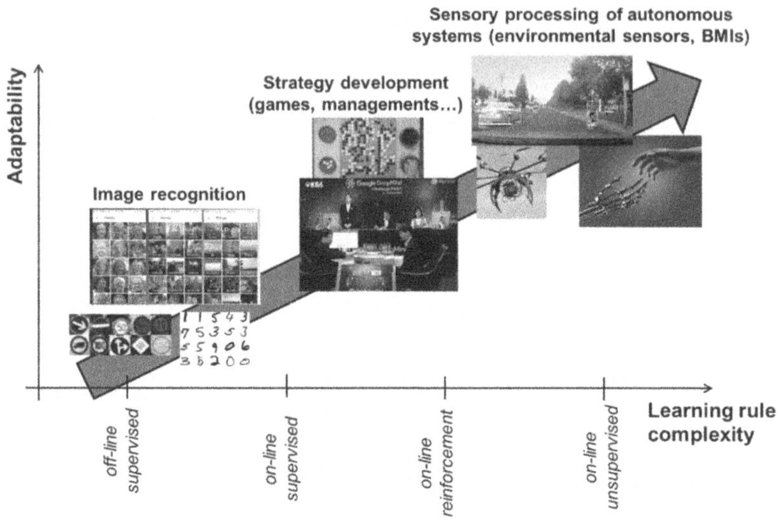

Figure 8.2 Classification of the learning schemes as a function of their adaptability (i.e., their capability to perform lifelong learning) and their complexity.

Despite the benefits of deep learning with IoT and mobile hardware, its adoption faces significant barriers. In contrast to more traditional programming, these algorithms require minimal computing but significant memory access intensity. It has been demonstrated that a CONV with the AlexNet network topology (i.e., an object recognition model that supports more than 1000 object classes [2]) accesses nearly 300 MB parameters from the main memory while running [4]. 300 MB is roughly a third of the main memory available in standard hardware platforms for smartphones and IoT like Qualcom Snapdragon 800 and Intel Edison. The main memory access intensity (von Neumann bottleneck) rules out an implementation of the deep learning algorithms using standard platforms (graphics processing units [GPUs], central processing units [CPUs], etc.) for embedded applications. Today software implementations of machine learning algorithms in mobile devices (e.g., speech recognition, object identification, photo search, etc.) are cloud assisted. This introduces negative effects: first, the latency and throughput depend on the network quality; second, the bit transmission through the cloud is energetically expensive; and third, it exposes users to privacy dangers as data are processed off-device by a third party.

To avoid access to the cloud for embedded systems it is foreseeable to research custom neuromorphic chips, which bring memory closer to the processor, thus overcoming the von Neumann bottleneck. One promising solution is the architecture that leverages the distributed computing in the neuron nodes and localized storage in the synaptic networks. Nowadays, most neurocomputing approaches use standard complementary metal-oxide semiconductor (CMOS) technologies. Examples include Heidelberg's HICANN [5], IBM's TrueNorth [6], and Zurich's ROLLS [7] chips. The synaptic weights are stored in analog or digital devices such as capacitors [7], 4 bit static random-access memory (SRAM) [5, 8], 8 T SRAM [9], and 1 bit SRAM [10].

The fabrication of large RRAM arrays has been recognized as another promising path to build synaptic connections in energy-efficient neuromorphic circuits [11–15]. In neural networks, the base operation is the multiply and accumulate one (MAC), between the input data and their associated synaptic weights (RRAM conductance values) in a given receptive field (i.e., the full input matrix in a FCNN or a submatrix with the kernel size in a CONV). In a conventional von Neumann architecture, the MAC operation is carried out using digital multipliers, adders, and registers. The operands are stored in system memory, and at every clock cycle they are retrieved for computation. In the RRAM array the multiply operation is performed at every memory point by Ohm's law, with current summation along rows or columns. In this way the MAC operation is performed in parallel at the location of data, by avoiding the time and the power consumption of moving data. Therefore, this approach could provide significant improvements in power and speed compared to GPU-based architectures.

The pattern detection and classification in neural networks are the result of a learning phase; the synaptic connections among neurons (memories) are created, modified, and preserved according to a learning model. The computation schemes to define the RRAM synaptic weights can be divided into three classes:

- Supervised learning: The neural network learns from training examples, both the inputs and outputs are available, and an external teacher tells each output neuron its desired response to a given input signal (backpropagation, extreme learning machine, etc.).
- Reinforcement learning: There is no direct access to the correct output. However, some measure of the quality of an output is available.
- Unsupervised learning: There are no training data and no external teacher (see Fig. 8.2).

Machine learning algorithms (such as backpropagation, extreme learning machine, and support vector machine) have achieved unprecedented performance to solve many real-world learning tasks, for example, in image or speech recognition [1]. However, there are some critical deficits yet to be resolved. First, they need to be trained with huge datasets (supervised learning) and their performances are usually impressive for solving a very specific task but drop largely for slightly changed demands. Second, deep learning networks are still widely lacking the ability for lifelong learning. Once it has learned, it is almost impossible to train the same algorithm or network on a different task without it having to completely relearn all parameters. The implementation of RRAM-based spiking neural networks with bioinspired learning rules, like the spike timing–dependent plasticity (STDP), is a first effort to overcome these limits. STDP shows promising opportunities to achieve on-chip unsupervised learning [11, 12]. The online unsupervised learning could enable the design of future sensory processing of robotics and autonomous systems.

In this work, we will present a RRAM synaptic compound composed of several binary RRAM cells connected in parallel, thereby providing synaptic analog behavior. First, the proposed synapse has been used for the implementation of CONVs for visual pattern recognition with off-line supervised learning. Second, the same RRAM-based synaptic compound coupled with unsupervised learning has been proposed for a visual-pattern extraction application based on a FCNN. Two different kinds of spike-based plasticity rules (long-term plasticity STDP and short-term synaptic

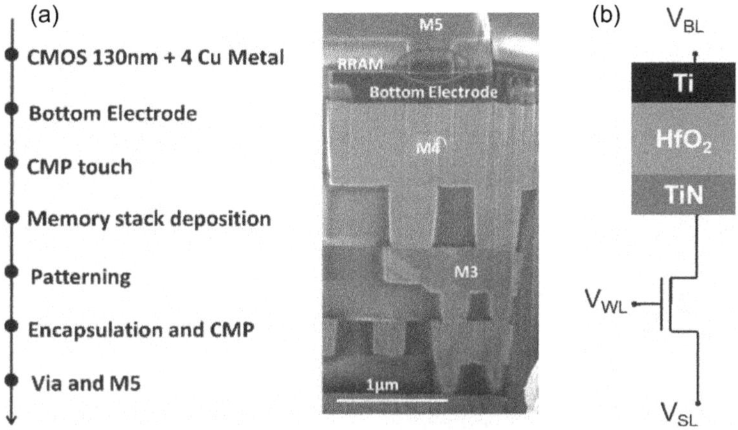

Figure 8.3 (a) Description of the integration flow and SEM image of a CMOS stack including the RRAM cell between M4 and M5. (b) Schematic view of the 1T1R cell configuration.

adaptation) have been proposed to enable on-chip unsupervised learning.

8.2 Array Test Vehicle and Key Features of RRAM

8.2.1 Key Features of RRAM

A 4 kb RRAM array was fabricated using a 130 nm CMOS front end (Fig. 8.3a) [16]. The memory element integration starts on top of the fourth metal layer (Cu). The RRAM active stack is composed of a 10-nm-thick atomic layer deposition (ALD) HfO_2 resistive switching layer sandwiched between physical chemical deposition (PVD) TiN and PVD Ti(10 nm)/TiN bottom and top electrodes. The OxRAM cell diameter is 300 nm. The array is logically organized into 64 rows (word line [WL]) and 64 columns (bit line [BL]). The cell configuration is 1 transistor 1 resistor (1T1R) (Fig. 8.3b). The WL is connected to the nMOS gate, setting the programming current, I_{set}. Forming and set operations are performed by applying a positive voltage pulse on the BL that is connected to the RRAM top electrode, whereas reset is performed by applying a voltage pulse on the

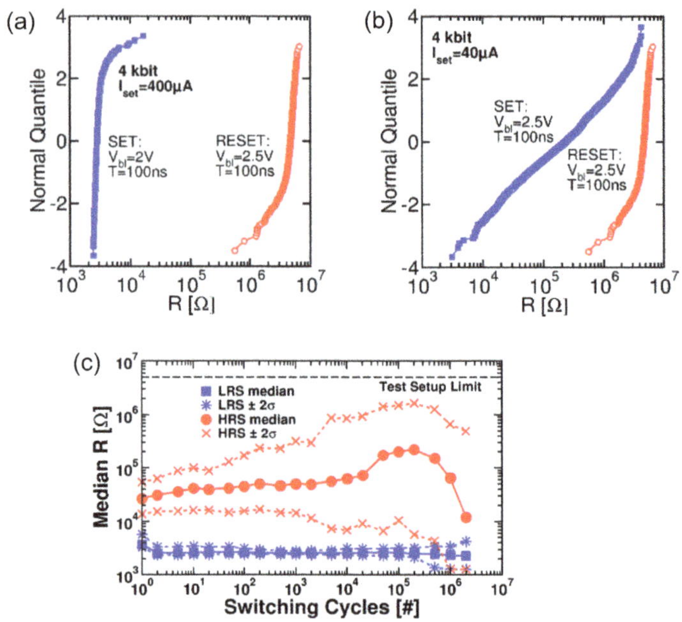

Figure 8.4 (a) 4 kbit resistance distributions for strong ($I_{set} \approx 400$ µA) and (b) weak ($I_{set} \approx 40$ µA) programming currents. 5 MΩ is the resistance measurement limit. (b) Endurance corresponding to the same programming conditions presented in (a).

source line (SL), which is in contact with the transistor side (bottom electrode).

RRAM devices switch between two distinct resistance states, a low resistance state (LRS) and a high resistance state (HRS), when set and reset conditions are applied, respectively. Figures 8.4a and 8.4b present the 4 kbit LRS and HRS distributions for a programming current $I_{set} \approx 400$ µA and $I_{set} \approx 40$ µA, respectively. The higher programming current strongly reduces the low-resistance-state variability (black curve), thus improving the window margin. Figure 8.4c demonstrates the programming stability of the low and high resistance states during endurance for $I_{set} \approx 400$ µA.

When a low BL voltage is applied during set immediately after a reset, a probabilistic switching of the device may appear and the set operation fails (LRS > 10 kΩ) in several cycles as the set programming conditions are not strong enough to switch the device

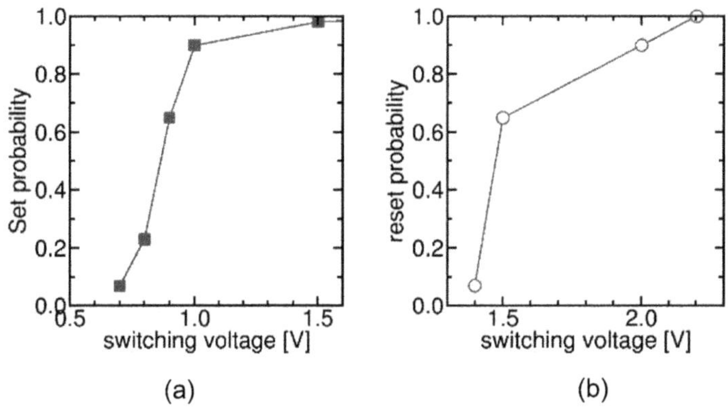

Figure 8.5 Set (a) and reset (b) switching probabilities extracted from a 4 kbit array (switching being considered successful if LRS < 10 kΩ and HRS > 100 kΩ) using low BL and SL voltages during set and reset, respectively. $I_{set} = 400$ μA.

in those cycles. A similar behavior has been observed when a low SL voltage is applied during reset after a set operation (HRS < 100 kΩ). Figures 8.5a and 8.5b show the percentage of switched cells as a function of the applied BL and SL voltages, respectively. Set and reset conditions can be identified to control the probability to switch the memory state.

8.2.2 Synaptic Compound Based on Multiple Binary RRAM Devices

The ideal RRAM synapse presents an analog conductance modulation under identical pulses in both programming directions: the conductance gradually increases (decreases) when a train of set (reset) pulses is applied. However, in most RRAM devices, both set and reset operations are noncumulative. As shown in Fig. 8.4a, HfO_2-based RRAM are intrinsically binary devices: they switch between two distinct states, LRS and HRS. It is possible to provide multilevel memory by carefully choosing the programming current or the reset voltage [17], while applying the same programming pulse multiple times is useless in terms of conductance variation. On the other hand, the use of only two resistance levels per synapse, with respect to the multilevel approach, can be insufficient to

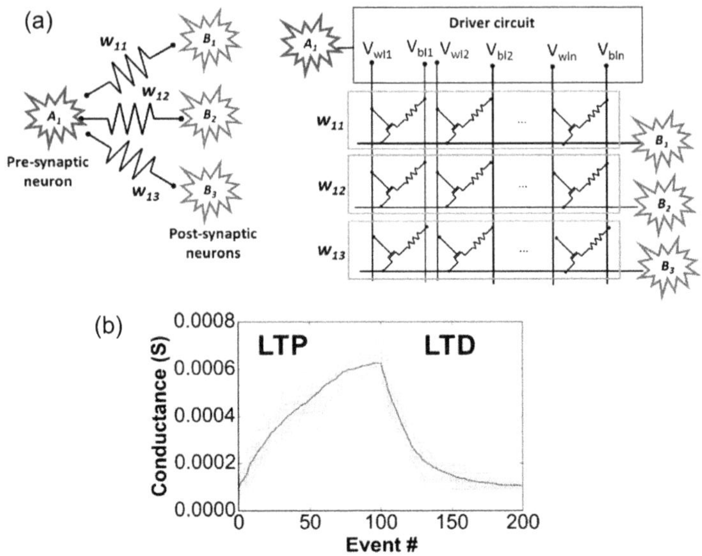

Figure 8.6 (a) Schematic of an RRAM synapse into a 1T1R array; devices in the same row build one equivalent synapse (nTnR structure). A driver circuit is used to individually program RRAM devices and propagate spikes to the next neuron layer. (b) Conductance evolution corresponding to 100 consecutive set (LTP) and reset (LTD) events with $P_{SET} = 0.02$ and $P_{REST} = 0.04$. Gray lines are representative of 25 synapses composed of 20 RRAMs each.

achieve good performances in neuromorphic systems designed for some complex applications, as for example image recognition [18]. To achieve a multilevel/analog synaptic element we proposed a synaptic compound composed of n multiple binary RRAM cells operating in parallel (Fig. 8.6a) [19]. Since parallel conductances sum up, the conductance of the equivalent synapse ranges from the sum of the n conductance values in the HRS to the sum of all the n conductance values in the LRS. A probabilistic programming scheme can be adopted to emulate analog conductance modulation under identical pulses set (or reset) [19–24]. Figure 8.6b shows the evolution of the conductance corresponding to 100 set pulses (long-term potentiation [LTP]) followed by 100 reset pulses (long-term depression [LTD]) for a synapse composed of $n = 25$ RRAM devices connected in parallel (Fig. 8.6a), with set switching probability

$P_{SET} = 0.02$ and reset switching probability $P_{REST} = 0.04$. The switching probability can be governed by the RRAM itself (internal switching probability): all the cells are programmed simultaneously using low set and reset voltages tuned to control the probability to switch the cells (Fig. 8.5). Another possibility consists in using an external pseudorandom number generator (PRNG) circuit block (extrinsic switching probability). In this case strong programming conditions that do not show intrinsic stochasticity (i.e., a switching probability equal to 1) are adopted and at each programming event only a few cells of the synaptic compound (defined by the PRNG) are sequentially programmed. The extrinsic switching probability allows a finer tuning of the switching probability, at the expense of a small increase in the switching complexity, with respect to the internal switching probability.

The weakest point of the proposed approach is the silicon area consumption for each synapse that is proportional to the number of cells needed to obtain the multilevel behavior. Vertical RRAM technology has been proposed as a possible solution to gain area by realizing one pillar per synapse [25].

8.3 Implementation of Spike-Based Neural Networks Using RRAM Arrays

8.3.1 Convolutional Neural Network for Image Classification with Supervised Learning

Software implementations of CONVs were applied with great success in applications such as traffic sign recognition [26]; the analysis of biological images [27]; and the detection of faces, complex text, pedestrians on the streets, and human bodies in natural images [28–30]. A major recent practical success of software implementations of CONVs is the face recognition software proposed by Facebook [31]. The power consumption to perform convolution operations is computationally expensive in the CONV implementation on CPUs and GPUs. This hinders its integration in portable devices. In recent years, dedicated systems-on-a-chip (SOC) solutions and field-programmable gate array (FPGA) platforms have been used to implement these networks for increasing performances while decreasing their power consumption. A hardware implementation of CNNs based on RRAM devices can further improve the power efficiency. We proposed the use of RRAM synapses to store the kernel features [32, 33].

The proposed CONV architecture is composed of a feature extraction module (made of two cascaded convolutional layers) and a classification module (made of two fully connected layers), shown in Fig. 8.1b. While in the fully connected classification module the neurons of a given layer are connected to every neuron of the previous layer by a large number of synapses, in convolutional layers a small set of synapses (constituting a kernel) are shared among different neurons to connect layers N and $N - 1$. A convolutional layer is composed of several (output) feature maps, each of them being connected to some or all the feature maps of the previous layer through a convolutional kernel. The kernel corresponds to a feature that has to be localized in the input image. Each output feature map contains information about the locations where the kernel features are present in the input map. The extraction module transforms the input image into a simpler set of feature maps. The classification module connects the obtained set of feature maps to the output neuron layer. Each output neuron is associated to a category: an output neuron spikes when the image presented to the input of the network belongs to its category. Two different visual pattern classification applications were demonstrated: the Modified National Institute of Standards and Technology (MNIST) handwritten digit database and the German Traffic Sign Recognition Benchmark (GTSRB) database. For both applications, in the first convolutional layer each output feature map is connected to the input image through a convolutional kernel. The input image is connected to 16 (32) output feature maps through 16 (32) kernels for the MNIST (GTSRB) application. For the second convolutional layer we adopted a noncomplete connection scheme; each output feature map is connected to only some of the input features maps through a convolutional kernel [34]. For MNIST 16 input feature maps are connected to 24 output feature maps through 90 kernels, while for the complex GTSRB 32 input feature maps are connected to 48 output feature maps through 186 kernels. To produce smaller output volume data it is possible to increase the stride with which the kernel is sided on the input data or to add a max-pooling layer [34]. For the MNIST and GTSRB the stride is 2 and 1, respectively. In the case of the GTSRB network the down-sampling operation is performed by adding a max-pooling layer.

For the hardware implementation we proposed the use of RRAM-based synapses to store the kernel features [15, 35]. To implement a synapse in hardware, n RRAM devices connected in parallel are used as shown in Fig. 8.6a. The kernel features, and the corresponding synaptic weights, are defined off-line using the supervised back-propagation learning algorithm. The calculated synaptic weight values are then discretized and written in the RRAM-based synaptic compound. To define the optimal value of the synaptic weight resolution, the number (n) of RRAM devices needed per synapse, simulations have been performed on both MNIST and GTSRB databases. Figure 8.7 shows the simulation results in terms of recognition success as a function of the value of RRAM cells per synapse, n. The RRAM characteristics have been modeled on the data of Fig. 8.4a. The recognition success improves as n increases; for n higher than 12 the maximum network performance greater than 99% is reached. The blue curve reports the recognition success for the GTSRB database. As this application poses a more complex task, a higher number of cells per synapse, here $n = 20$ is required. More complex application tasks are more demanding in terms of the number of cells per synapse. The estimated size of the RRAM array needed to implement the kernels of the proposed convolutional neural networks (CNNs) is 600 kb for MNIST and 1 Mb for GTSRB, respectively.

8.3.2 Fully Connected Neural Network with Unsupervised Learning for Complex Visual Extraction in Real Time

In the brain, memory is stored in synapses, which are the connections between neurons. Synapses not only transmit information from one neuron to another but also modify their strength in response to their local experience. This adaptation is known as synaptic plasticity, and it is considered to be the biological substrate

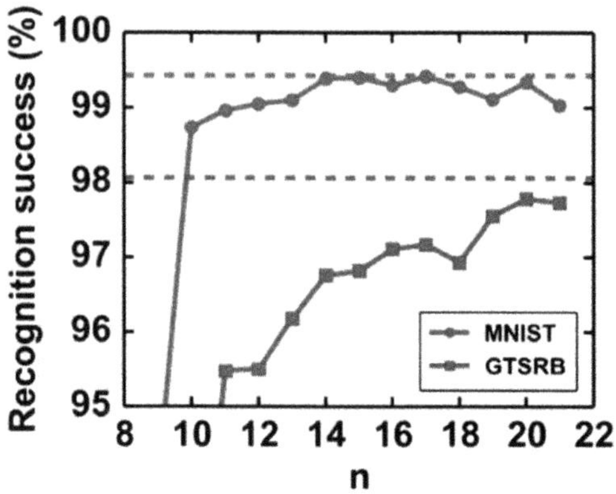

Figure 8.7 Recognition success as a function of the number n of parallel RRAM devices used to implement an equivalent synapse (Fig. 8.6a). The RRAM characteristics have been modeled on the data of Fig. 8.4a.

for learning and memory of the brain. Recent developments in the neuroscience community evidence that biological synapses exhibit different kinds of plasticity rules. We will focus on the coimplementation of both short-term synaptic adaptation and long-term STDP [36, 37]. Long-term STDP produces stable modifications of the synapses according to the timing of pre- and postsynaptic spike events. Along with long-term STDP, cortical synapses undergo short-term synaptic adaptation, which is the dynamic modulation of the synaptic strength as a function of input stimulations (presynaptic activity) and it shows retrievable dynamics in short time scales [37].

We proposed the implementation of the short-term synaptic adaptation and long-term STDP rules using two separate nTnR synaptic compounds, $y_i(t)$ and w_{ij}, respectively. w_{ij} obeys a stochastic and simplified STDP learning rule as described in Ref. [22]. When an output neuron is active (i.e., fires), if the input neuron has been active immediately before, each RRAM cell of the synaptic compound has a given probability to switch into the low resistance state (potentiation); if not, each RRAM cell of the synaptic compound has a given probability to switch to the high resistance state (depression). $y_i(t)$ undergoes the short-term synaptic adaptation rule according to the model presented in Refs. [22–24]. The synaptic weight decreases in inverse proportion to the presynaptic spiking frequency. Each presynaptic event corresponds to a reset operation (conductance decrease). The synaptic weight gradually recovers the initial high conductance value in the absence of input spikes; to this purpose a series of set operations is applied after every time slot $\Delta T =$ 1 ms. The degree of depression and the recovery time are modulated by tuning the set and reset probabilities, respectively. This strategy allows the use of the same nonvolatile memory technology to implement both short-term adaptation and long-term STDP.

We proposed the implementation of long-term STDP and short-term synaptic adaptation using two separate synaptic compounds. However, in biology there is only one synaptic connection between two neurons, which obeys both types of plasticity mechanisms. The strength of the synaptic connection (or synaptic conductance) between the input neuron, A_i, and the output neuron, B_j, is $w_{ij}*y_i(t)$, where w_{ij} is the strength modulated according to the long-term STDP rule and $y_i(t)$ is a scaling factor that describes the short-term synaptic adaptation rule. Therefore, we proposed the circuit of Fig. 8.8 to connect the long-term STDP and the short-term synaptic adaptation synaptic compounds. The conductance multiplication during the reading operation is performed by means of a buffer that modulates the read voltage for w_{ij} as a function of the conductance value $y_i(t)$, that is, the lower read voltage as shown in Fig. 8.8. When an input neuron receives an incoming event, it generates a small voltage pulse (V_{in}) that propagates to the output neuron through $y_i(t)$, the buffer, and w_{ij}. The resulting postsynaptic current (I_{out}) is proportional to $y_i(t)*w_{ij}$.

Long-term STDP and short-term synaptic adaptation can manifest simultaneously, and it is complicated in practical cases to make a clear distinction between them. In this work, by means of system-level simulations of a real application, we tried to shed new light on the role of these two different types of plasticity. We performed simulations of a real application for complex visual extraction with our special-purpose event-based N2D2 simulator tool [38, 39]. The neuron circuits are modeled with behavioral equations as in Ref. [39]. Both long-term STDP and short-term synaptic adaptation synapses are modeled by fitting data of Fig. 8.2a with a log-normal distribution, to consider the experimental spread

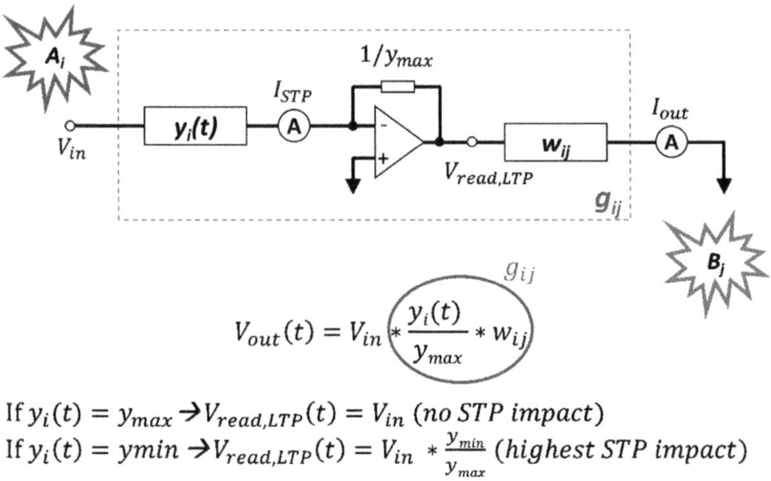

Figure 8.8 Circuit proposed to reproduce both long-term STDP and short-term synaptic adaptation rules. The conductance multiplication during the read operation is performed by means of a buffer that modulates the read voltage for the long-term STDP synapse w_{ij} as a function of the short-term synaptic adaptation conductance value $y_i(t)$.

in the conductance parameters. We simulated a FCNN to process highly noisy temporally encoded video data. A video of cars passing on a six-lane freeway is recorded in the address event representation (AER) format [40]. AER sensors take inspiration from biology; each pixel sends an event (spike) when light changes by a given relative amount (increment or decrement). The event-driven representation can have a strong reduction in terms of computational power, time, and energy with respect to frame-based representation, where each frame is passed through the entire processing chain. The simulated architecture is composed of 60 neurons in the first layer and 10 neurons in the second one (Fig. 8.9a). The bottom-most layer represents a 128×128 pixel AER vision sensor. Every input pixel requires two synapses, to send the increase and decrease illumination events. The total number of long-term synapses in this system is thus $2 \times 128 \times 128 \times 60 + 60 \times 10 = 1{,}966{,}680$, while the total number of short-term synapses is $2 \times 128 \times 128 + 60 = 32{,}768$. A similar network has been implemented in Refs. [41] and [14], exploiting multilevel phase-change memory and

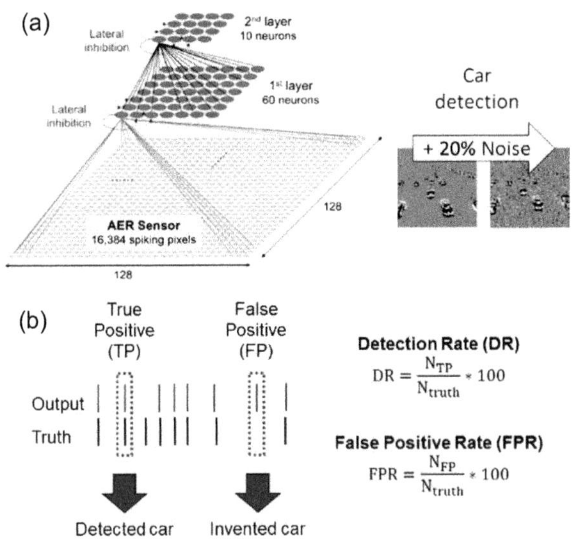

Figure 8.9 (a) Two-layer SNN used for car detection and example of unsupervised detection of cars in different traffic lanes. (b) The reference activity (truth) is compared to the activity of the corresponding output neurons (output) to calculate the number of missed events and false positive events.

binary conductive bridge RRAM synapses to capture the long-term plasticity rule only. After unsupervised learning, each output neuron becomes sensitive to one traffic lane. In Fig. 8.9b, the reference activity (truth) is compared to the output neuron activity of the corresponding traffic lane (output). False positive (invented cars) and true positive (detected cars) are put in evidence. A false positive event occurs when the output neuron spikes even if there is no car in the corresponding line, while the true positive event is a detected car. The output neuron spikes as a consequence of a car passing through the corresponding line. The metrics to evaluate the network performances (the detection rate [DR] and the false positive rates [FPRs]) are presented. The false positive estimation is fundamental for highly noisy input data, due to the increased spiking activity of the input neurons in the presence of input noise. Figure 8.10a reports the DR obtained as a function of the number of RRAM cells per long-term STDP synapse, that is, the number of resistance levels

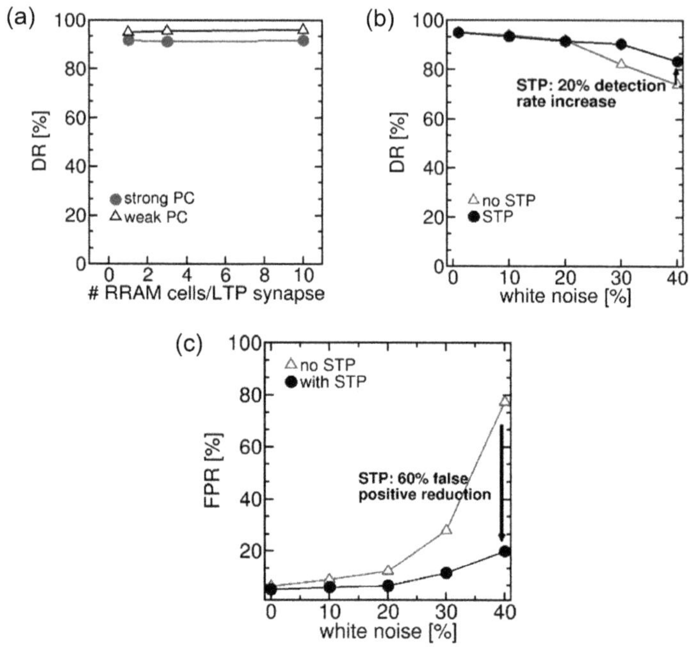

Figure 8.10 (a) Detection rate as a function of the number of cells per long-term potentiation synapse. Only long-term STDP is considered. Results have been obtained on a bench of 20 simulations (no added noise). The parameters used to implement the long-term STDP are $P_{set} = 0.133$ and $P_{reset} = 0.250$. Detection rate (b) and false positive rate (c) as a function of the noise level artificially introduced in the address event representation (AER) input data. The introduction of the short-term synaptic adaptation (black circles) allows one to improve the detection rate by about 20% and to decrease the false positive rate by 60%.

that can be achieved. Since it is a detection (not a classification) problem, the network performances do not improve on increasing the number of cells (levels) per synapse; therefore, we adopt one RRAM per long-term STDP synapse in the following. The synapses have been modeled by fitting data of Figs. 8.2a and 8.2b, strong and weak programming conditions, respectively. Even if the low programming current strongly reduces the window margin, it does not degrade the network performances.

To study the impact of the short-term plasticity rule, we artificially introduced white noise into the already noisy data recorded from the AER sensor. Simulations with added noise (up to 40% of the total amount of spikes in the sequence are random) are presented in Figs. 8.10b and 8.10c. The noise introduction strongly increases the false positive events. The network is not able to distinguish between real input stimuli and noise. For a noise level of 40% the system does not work and the FPR is about 80%. Short-term synaptic adaptation is fundamental to maintaining the functionality of the network for the highest noise levels. Thanks to short-term synaptic adaptation the network learns to stop responding to noise stimulus, thus allowing a 60% reduction of the FPR for a noise level of 40%. Therefore, short-term potentiation (STP) implementation enables the network to be used in highly noisy environments (light reflections, fog, rain, etc.).

8.4 Conclusions

This chapter showed how 1T1R RRAM arrays can be exploited to implement spike-based neuromorphic circuits. A RRAM synaptic compound composed of several binary RRAM cells connected in parallel, thereby providing synaptic analog behavior, is presented. First, the proposed synapse has been used for the implementation of a CONV for visual pattern classification. The synaptic weights are defined off-line with supervised back-propagation learning algorithm (one-time programming). The proposed network achieves recognition rates of 99% for the MNIST database and 97% for the GTSRB database using 11 and 20 RRAM devices per synapse, respectively. Second, the same RRAM-based synaptic compound associated with a bioinspired methodology for programming has been used to perform on-chip unsupervised learning in FCNNs. Two different kinds of spike-based plasticity rules (short-term synaptic adaptation and long-term STDP) have been proposed to enable on-chip unsupervised learning. Long-term STDP allows the neural network to learn patterns without training data set (unsupervised learning). Short-term synaptic adaptation allows one to improve accuracy in the presence of significant background noise in the input data (reduction of the false positive events generated by white noise in the input data). A visual-pattern extraction application is presented. Thanks to the short-term synaptic adaptation implementation, despite highly noisy input data (noise level of 40%), the demonstrated application exhibits very high performances (DR of 83% and FPR lower than 20%), while systems without STP would no longer deliver valid results (FPR $\approx$ 80%).

References

1. LeCun, Y., Bengio, Y., and Hinton, G. (2015). Deep learning, *Nature*, **521**, pp. 436–444.
2. Krizhevsky, A., Sutskever, I., and Hinton, G. E. (2012). ImageNet classification with deep convolutional neural networks, in *Advances in Neural Information Processing Systems (NIPS)*, pp. 1097–1105.
3. Goodfellow, I., Bengio, Y., and Courville, A. (eds.) (2016). *Deep Learning*, MIT Press, Cambridge.
4. Lane, N. D., Bhattacharya, S., Georgiev, P., Forlivesi, C., and Kawsar, F. (2015). An early resource characterization of deep learning on wearables, smartphones and internet-of-things devices, *HotMobile*, pp. 117–122.
5. Schemmel, J., Bruderle, D., Grubl, A., Hock, M., Meier, K., and Millner, S. (2010). A wafer-scale neuromorphic hardware system for large-scale neural modeling, in *IEEE International Symposium on Circuits and Systems (ISCAS)*, pp. 1947–1950.
6. Merolla, P. A., Arthur, J. V., Alvarez-Icaza, R., Cassidy, A. S., Sawada, J., Akopyan, F., Jackson, B. L., Imam, N., Guo, C., Nakamura, Y., Brezzo, B., Vo, I., Esser, S. K., Appuswamy, R., Taba, B., Amir, A., Flickner, M. D., Risk, W. P., Manohar, R., and Modha, D. S. (2014). A million spiking-neuron integrated circuit with a scalable communication network and interface, *Science*, **345**(6197), pp. 668–673.
7. Qiao, N., Mostafa, H., Corradi, F., Osswald, M., Stefanini, F., Sumislawska, D., and Indiveri, G. (2015). A reconfigurable on-line learning spiking neuromorphic processor comprising 256 neurons and 128K synapses, *Front. Neurosci.*, **9**, p. 141.
8. Meier, K. (2015). A mixed-signal universal neuromorphic computing system, in *IEEE International Electron Devices Meeting (IEDM)*, pp. 4.6.1–4.6.4.

9. Seo, J., Brezzo, B., Liu, Y., Parker, B. D., Esser, S. K., Montoye, R. K., Rajendran, B., Tierno, J. A., Chang, L., Modha, D. S., and Friedman, D. J. (2011). A 45nm CMOS neuromorphic chip with a scalable architecture for learning in networks of spiking neurons, in *IEEE Custom Integrated Circuits Conference (CICC)*, San Jose, CA, USA, pp. 1–4.
10. Akopyan, F., Sawada, J., Cassidy, A., Alvarez-Icaza, R., Arthur, J., Merolla, P., Imam, N., Nakamura, Y., Datta, P., Nam, G. J., Taba, B., Beakes, M., Brezzo, B., Kuang, J. B., Manohar, R., Risk, W. P., Jackson, B., and Modha, D. S. (2015). TrueNorth: design and tool flow of a 65 mW 1 million neuron programmable neurosynaptic chip, *IEEE Trans. Comput. Aided Des. Integr. Circuits Syst.*, **34**(10), p. 1537.
11. Serrano-Gotarredona, T., Masquelier, T., Prodromakis, T., Indiveri, G., and Linares-Barranco, B. (2013). STDP and STDP variations with memristors for spiking neuromorphic learning systems, *Front. Neurosci.*, **7**, p. 2.
12. Querlioz, D., Bichler, O., Vincent, A. F., and Gamrat, C. (2015). Bioinspired programming of memory devices for implementing an inference engine, *Proc. IEEE*, **103**(8), p. 1398.
13. Burr, G. W., Shelby, R. M., di Nolfo, C., Jang, J., Shenoy, R., Narayanan, P., Virwani, K., Giacometti, E., Kurdi, B., and Hwang, H. (2014). Experimental demonstration and tolerancing of a large-scale neural network (165,000 synapses), using phase-change memory as the synaptic weight element, *IEEE International Electron Devices Meeting*, pp. 29.5.1–29.5.4.
14. Suri, M., Querlioz, D., Bichler, O., Palma, G., Vianello, E., Vuillaume, D., Gamrat, C., and De Salvo, B. (2013). Bio-inspired stochastic computing using binary CBRAM synapses, *IEEE Trans. Electron Device*, **60**(7), p. 2402.
15. Garbin, D., Bichler, O., Vianello, E., Rafhay, Q., Gamrat, C., Perniola, L., Ghibaudo, G., and DeSalvo, B. (2014). Variability-tolerant convolutional neural network for pattern recognition applications based on OxRAM synapses, *IEEE International Electron Devices Meeting*, pp. 28.4.1–28.4.4.
16. Grossi, A., Nowak, E., Zambelli, C., Pellissier, C., Bernasconi, S., Cibrario, G., El Hajjam, K., Crochemore, R., Nodin, J. F., Olivo, P., and Perniola, L. (2016). Fundamental variability limits of filament-based RRAM, *IEEE International Electron Devices Meeting*, pp. 4.7.1–4.7.4.
17. Vianello, E., Thomas, O., Molas, G., Turkyilmaz, O., Jovanović, N., Garbin, D., Palma, G., Alayan, M., Nguyen, C., Coignus, J., Giraud, B., Benoist, T.,

Reyboz, M., Toffoli, A., Charpin, C., Clermidy, F., and Perniola, L. (2014). Resistive memories for ultra-low-power embedded computing design, *IEEE International Electron Devices Meeting*, pp. 6.3.1–6.3.4.

18. Garbin, D., Vianello, E., Bichler, O., Azzaz, M., Rafhay, Q., Candelier, P., Gamrat, C., Ghibaudo, G., DeSalvo, B., and Perniola, L. (2015). On the impact of OxRRAM-based synapses variability on convolutional neural networks performance, in *IEEE/ACM International Symposium on Nanoscale Architectures (NANOARCH)*, pp. 193–198.
19. Garbin, D., Vianello, E., Bichler, O., Rafhay, Q., Gamrat, C., Ghibaudo, G., De Salvo, B., and Perniola, L. (2015). HfO2-based RRAM devices as synapses for convolutional neural networks, *IEEE Trans. Electron Device*, **68**(8), p. 2494.
20. Suri, M., and Parmar, V. (2015). Exploiting intrinsic variability of filamentary resistive memory for extreme learning machine architectures, *IEEE Trans. Nanotechnol.*, **14**(6), p. 963.
21. Bill, J., and Legenstein, R. (2014). A compound memristive synapse model for statistical learning through STDP in spiking neural networks, *Front. Neurosci.*, **8**, p. 412.
22. Werner, T., Vianello, E., Bichler, O., Grossi, A., Nowak, E., Nodin, J.-F., Yvert, B., De Salvo, B., and Perniola, L. (2016). Experimental demonstration of short and long term synaptic plasticity using OxRAM multi k-bit arrays for reliable detection in highly noisy input data, in *IEEE International Electron Devices Meeting (IEDM)*, pp. 16.6.1–16.6.4.
23. Vianello, E., Werner, T., Bichler, O., Grossi, A., Nowak, E., Yvert, B., De Salvo, B., and Perniola, L. (2017). Bio-inspired programming of resistive memory devices for implementing spiking neural networks, in *Proceedings of the on Great Lakes Symposium on VLSI 2017 (GLSVLSI)*, Banff, Alberta, Canada, pp. 393–398.
24. Vianello, E., Werner, T., Bichler, O., Valentian, A., Molas, G., Yvert, B., De Salvo, B., and Perniola, L. (2017). Resistive memories for spike-based neuromorphic circuits, in *IEEE International Memory Workshop (IMW)*, Monterey, CA, USA.
25. Piccolboni, G., Molas, G., Portal, J. M., Coquand, R., Bocquet, M., Garbin, D., Vianello, E., Carabasse, C., Delaye, V., Pellissier, C., Magis, T., Cagli, C., Gely, M., Cueto, O., Deleruyelle, D., Ghibaudo, G., De Salvo, B., and Perniola, L. (2015). Investigation of the potentialities of vertical resistive RAM (VRRAM) for neuromorphic applications, in *IEEE International Electron Devices Meeting (IEDM)*, pp. 17.2.1–17.2.4.

26. Ciresan, D., Meier, U., Masci, J., and Schmidhuber, J. (2012). Multi–column deep neural network for traffic sign classification, *Neural Networks*, **32**, pp. 333–338.
27. Ning, F., Delhomme, D., LeCun, Y., Piano, F., Bottou, L., and Barbano, P. E. (2005). Toward automatic phenotyping of developing embryos from videos, *IEEE Trans. Image Process.*, **14**, pp. 1360–1371.
28. Sermanet, P., Kavukcuoglu, K., Chintala, S., and LeCun, Y. (2013). Pedestrian detection with unsupervised multi–stage feature learning, in *Proceedings of the 2013 IEEE Conference on Computer Vision and Pattern Recognition (CVPR)*, pp. 3626–3633.
29. Vaillant, R., Monrocq, C., and LeCun, Y. (1994). A convolutional neural network hand tracker, *IEEE Proc. on Vision, Image, and Signal Processing*, **141**, pp. 245–250.
30. Garcia, C., and Delakis, M. (2004). Convolutional face finder: a neural architecture for fast and robust face detection, *IEEE Trans. Pattern Anal. Mach. Intell.*, **26**, pp. 1408–1423.
31. Taigman, Y., Yang, M., Ranzato, M., and Wolf, L. (2014). Deepface: closing the gap to human-level performance in face verification, in *Proceedings of the 2013 IEEE Conference on Computer Vision and Pattern Recognition (CVPR)*, pp. 1701–1708.
32. Garbin, D., Bichler, O., Vianello, E., Rafhay, Q., Gamrat, C., Perniola, L., Ghibaudo, G., and DeSalvo, B. (2014). Variability–tolerant convolutional neural network for pattern recognition applications based on RRAM synapses, in *IEEE International Electron Devices Meeting (IEDM)*, pp. 28.4.1–28.4.4.
33. Garbin, D., Vianello, E., Bichler, O., Azzaz, M., Rafhay, Q., Candelier, P., Gamrat, C., Ghibaudo, G., DeSalvo, B., and Perniola, L. (2015). On the impact of RRAM–based synapses variability on convolutional neural networks performance, in *IEEE/ACM International Symposium on Nanoscale Architectures (NANOARCH)*, pp. 193–198.
34. LeCun, Y., Bottou, L., Bengio, Y., and Haffner, P. (1998). Gradient-based learning applied to document recognition, *Proc. IEEE*, **86**, pp. 2278–2324.
35. Vianello, E., Garbin, D., Bichler, O., Piccolboni, G., Molas, G., De Salvo, B., and Perniola, L. (2017). Multiple binary OxRAMs as synapses for convolutional neural networks, in *Advances in Neuromorphic Hardware Exploiting Emerging Nanoscale Devices* (Suri, M., eds.), Cognitive Systems Monographs, vol. 31, Springer, New Delhi, pp. 109–127.

36. Bi, G. Q., and Poo, M. M. (1998). Synaptic modifications in cultured hippocampal neurons: dependence on spike timing, synaptic strength, and postsynaptic cell type, *J. Neurosci.*, **18**(24), pp. 10464–10472.
37. Tsodyks, M.-V., and Markram, H. (1997). The neural code between neocortical pyramidal neurons depends on neurotransmitter release-probability, *Proc. Natl. Acad. Sci. U. S. A.*, **94**(2), pp. 719–723.
38. Bichler, O., Querlioz, D., Thorpe, S., Bourgoin, J., and Gamrat, C. (2011). Unsupervised features extraction from asynchronous silicon retina through spike-timing-dependent plasticity, in *The 2011 International Joint Conference on Neural Networks (IJCNN)*, pp. 859–866.
39. Bichler, O., Querlioz, D., Thorpe, S. J., Bourgoind, J.-P., and Gamrat, C. (2012). Extraction of temporally correlated features from dynamic vision sensors with spike-timing-dependent plasticity, *Neural Networks*, **32**, pp. 339–348.
40. Lichtsteiner, P., Posch, C., and Delbruck, T. (2008). A 128 × 128 120 dB 15 μs latency asynchronous temporal contrast vision sensor, *IEEE J. Solid-State Circuits*, **43**(2), pp. 566–576.
41. Suri, M., Bichler, O., Querlioz, D., Cueto, O., Perniola, L., Sousa, V., Vuillaume, D., Gamrat, C., and DeSalvo, B. (2011). Phase change memory as synapse for ultra-dense neuromorphic systems: application to complex visual pattern extraction, in *International Electron Devices Meeting*, pp. 4.4.1–4.4.4.

Chapter 9

Nanomagnet Logic: Routes to Enhanced Dot-Dot Coupling

H. Dey,[a] F. Shah,[a] G. Csaba,[b] W. Porod,[a] and G. H. Bernstein[a]

[a]*University of Notre Dame, Notre Dame (IN), USA*
[b]*Pazmany University, Budapest, Hungary*
gbernste@nd.edu

Nanomagnet logic (NML) is a novel scheme for performing Boolean logic by encoding digital bits as directions of magnetization in nanomagnets that are on the size scale of a few hundred nanometers. Prior research on NML has relied solely on field coupling among nanomagnets, and any directionality was built in using shape anisotropy. Here we discuss two important improvements to the NML scheme. The first method is the addition of unidirectional anisotropy in ferromagnets achieved by exploiting exchange bias between layers of each nanomagnet that alternate directions. The second is the use of exchange coupling among nanomagnets mediated by a common underlying magnetic film.

Emerging Devices for Low-Power and High-Performance Nanosystems: Physics, Novel Functions, and Data Processing
Edited by Simon Deleonibus

ISBN 978-981-4800-11-2 (Hardcover), 978-0-429-45873-6 (eBook)
www.panstanford.com

9.1 Introduction

As complementary metal-oxide semiconductor (CMOS) device features are shrinking [1], the end of Moore's law is finally within sight. The benefits of "more-Moore" have slowed, and attention has turned to "more-than-Moore" and "beyond-Moore" technologies. The term "more-Moore" implies a continued attempt at scaling CMOS, and "more-than-Moore" implies adding value to electronic systems through improved displays, packaging, sensors and actuators, power devices, RF devices, passive devices, etc. "Beyond-Moore" implies replacing CMOS with a different kind of "switch," be that based on quantum, magnetic, tunneling, molecular, or some other physical phenomenon [2]. In this chapter, we very briefly review one potential beyond-Moore technology known as quantum-dot cellular automata (QCA), and quickly move to the related magnetic computing technology called nanomagnet logic (NML).

The notion of QCA was first proposed in 1993 [3–6]. The original embodiment of QCA exploits quantum mechanical tunneling of single electrons to implement a type of classical cellular automata architecture [7]. The classical cellular automaton is a finite-state machine of interacting entities arranged as cells. Each cell can have a finite number of states, and the state of an individual cell depends on those of its neighboring cells. In QCA, these cells are realized by quantum dots, that is, highly localized regions that hold a discrete number of electrons. These quantum dots, having predictable interactions with their neighbors, can be made to perform Boolean logic operations. So far, two major QCA-related schemes, namely electronic and magnetic QCA, have been demonstrated. In this chapter we focus on logic devices based on magnetic QCA, referred to here as "nanomagnet logic."

9.2 Electronic and Magnetic QCA

The basic building block of the electronic QCA scheme is an ensemble of four quantum dots placed at the corners of a square.

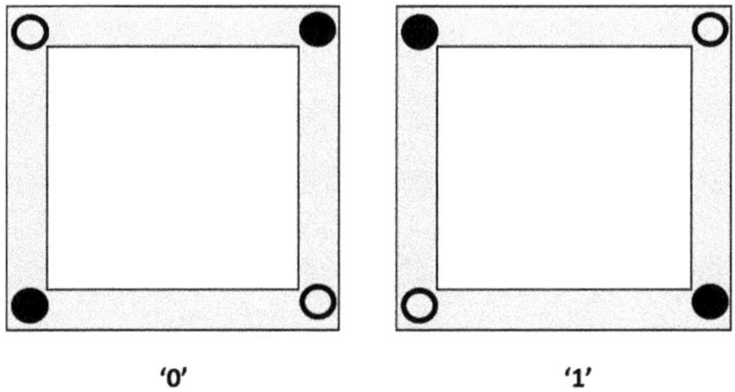

Figure 9.1 Basic QCA cell showing two possible binary states.

The actual physical arrangement requires only that electrons in each quantum dot affect the probability of occupancy of a nearby site. The dots are isolated from each other by a thin insulator, and electrons placed in one dot can get transferred to another one in the same cell by tunneling. Figure 9.1 illustrates the basic building block of electronic QCA with two possible binary states. The cells interact with each other through coulombic interactions. On the basis of this intercell interaction, inverters, binary wires, and majority logic gates can be implemented, as illustrated in Fig. 9.2 [8].

In 1997, Orlov et al. [9] experimentally demonstrated the electronic QCA scheme using metal dots. Semiconductor quantum dots have also been utilized to demonstrate the electronic QCA scheme [10–12]. However, due to the very small, yet finite, capacitance of the dots, all of these electronic QCA systems work only at very low temperatures (mK range). That said, room-temperature QCA-like operation has been demonstrated using a nanomagnetic implementation.

Magnetic materials have a huge advantage over single electrons confined to quantum dots. In the electronic QCA scheme, the energies of confinement are very small compared to any reasonable kilotesla, even at temperatures as low as 4 K. On the other hand, nanomagnets that are large enough to be beyond the superparamagnetic limit retain their magnetization states due to the relatively large energies contained in the magnetization of the ensemble of magnetic dipoles. These energies, on the order of 100

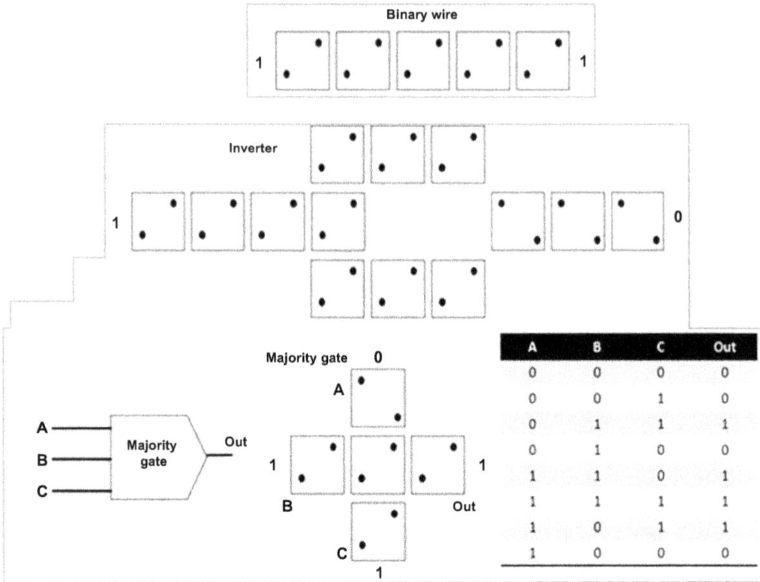

A	B	C	Out
0	0	0	0
0	0	1	0
0	1	1	1
0	1	0	0
1	1	0	1
1	1	1	1
1	0	1	1
1	0	0	0

Figure 9.2 Implementation of basic logic elements using electronic QCA [8].

Figure 9.3 Illustration of bistability in NML, where the arrow represents the magnetization direction of the single-domain "dot."

to 200 kT, are large enough to withstand thermal effects, yet small enough to offer important energy advantages compared to CMOS.

The magnetic version of QCA, known as NML or magnetic quantum-dot cellular automata (MQCA), is the room-temperature realization of the original QCA scheme. The basic building block of NML is a nanoscale, single-domain, magnetic dot. Bistability in the magnetic dots is achieved by built-in shape anisotropy. The two opposite magnetization directions of these bistable magnets denote logic ones and zeros. Figure 9.3 illustrates the bistability in NML.

One of the primary concerns of magnetic logic and storage technology is that the scaling of these devices is limited by the superparamagnetic limit imposed by thermal fluctuations. At sufficiently small dimensions (sub-100 nm), magnetic structures or particles prefer to be in single-domain state [13] in which all the magnetic moments in a sample behave as a giant macrospin and respond collectively to external excitations. Typically, magnetization of such a small nanomagnet can be represented by "magnetization-up" or "magnetization-down" states (Fig. 9.3). These two stable ground states are separated by an energy barrier that is defined by the anisotropy energy of the nanomagnet [14]. At room temperature, there is a finite probability that a nanomagnet can flip from one of its stable states to the other due to thermal fluctuations. The probability of flipping is a negative exponential function of the ratio between the energy barrier and thermal energy (kT). If the energy barrier is on the same order as the thermal energy, the nanomagnet can randomly flip [15]. This is the biggest challenge faced by the magnetic hard disk drive (HDD) industry at this time. The areal density of a HDD is limited by the bit size [16], and if the bit size is reduced to below the superparamagnetic limit, the bits can randomly flip, destroying the control over them needed to represent and retain information.

However, this is also a very serious problem for devices that are made of an assembly of small nanomagnets that interact with each other by magnetic dipole (magnetostatic) coupling, that is, stray fields emanating from the individual nanomagnets of the assembly. In such devices, stronger dipole coupling between the neighboring elements is extremely important for correct operation of the device. NML devices, for example, the one developed at Notre Dame [8], also rely on dipole field coupling between neighboring nanomagnets. Significant work has already been done in order to understand the energetics of dipole-coupled systems like NML. In such systems, the two magnetization states are separated by a well-defined energy barrier [14, 17]. These two well-defined states are shown in Fig. 9.4 on the easy-axis hysteresis loop (left), and the energy barrier is also shown (right) [18].

Two nanomagnets that are close enough together for strong dipole coupling will arrange themselves in either an antiferromagnetic (AFM) or a ferromagnetic (FM) configuration. Utilizing these coupling configurations, and by engineering the shape and spatial separations of an assembly of nanomagnets, more complex

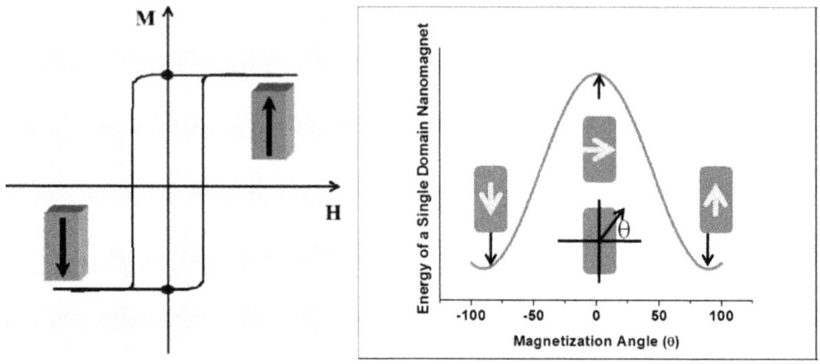

Figure 9.4 Two stable states of a single-domain nanomagnet; (left) hysteresis loop along the easy axis showing magnetization, **M**, as a function of applied field, **H**, and (right) the energy barrier between the two states [8, 18].

systems, such as logic blocks and data chains, can be built [19–21], as discussed below. Energetics and switching of these nanomagnets can be tailored by changing the shape and aspect ratio of the nanomagnets, and thus programmable operation and dense integration are possible [22–24]. The strength of the dipole coupling between two closely spaced nanomagnets is critical for correct functionality of these systems, and this energy landscape can be disturbed, mostly due to thermal fluctuations and lithographic variations, which can cause unwanted reversal of the nanomagnets and lead to errors [15, 25–27].

Thermal fluctuations can be integrated into the micromagnetic model for simulating such systems. Figure 9.5 shows the simulated switching distribution [15] of a single $60 \times 90 \times 10$ nm^3 nanomagnet in the presence of random thermal fluctuations. It is a Gaussian distribution with a mean switching field of 61 mT, but the width of the distribution is 10 mT, meaning that the switching field can be as high as 66 mT or as low as 56 mT. This is a large variation, and if the coupling field between two neighboring nanomagnets is comparable to this variation, there can be random reversal of the nanomagnets. In addition to that, process variations at sub-100 nm dimensions can also cause significant modification of the switching field. Therefore, it is important to have stronger coupling between

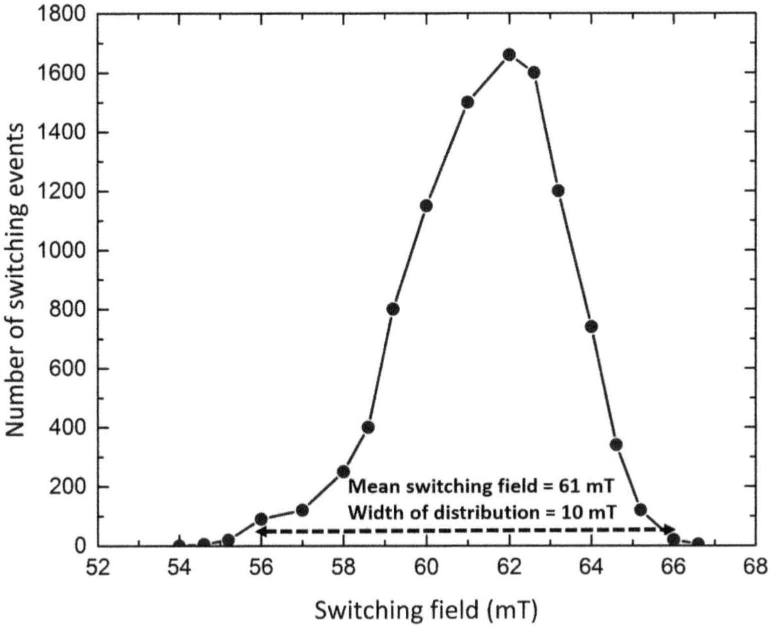

Figure 9.5 Wide switching-field distribution of a 60 × 90 × 10 nm^3 single-domain nanomagnet at room temperature in the presence of random thermal noise [15].

adjacent elements in order to make them switch in a desirable way and to be able to retain information reliably for a very long time.

9.3 Nanomagnet Logic

In this section, we discuss how Boolean logic operation and data propagation in NML are based on nearest-neighbor magnetostatic coupling, but later we add the effects of exchange coupling (EC) to an underlying layer. Figure 9.6 illustrates data propagation and logic operation in NML. Logic operation is realized by a majority gate, in which the center nanomagnet, which we call the "compute" magnet, is simultaneously influenced by the three "input" nanomagnets. The final state of the compute magnet is passed to the "output magnet," which responds to the influence of the majority of the inputs.

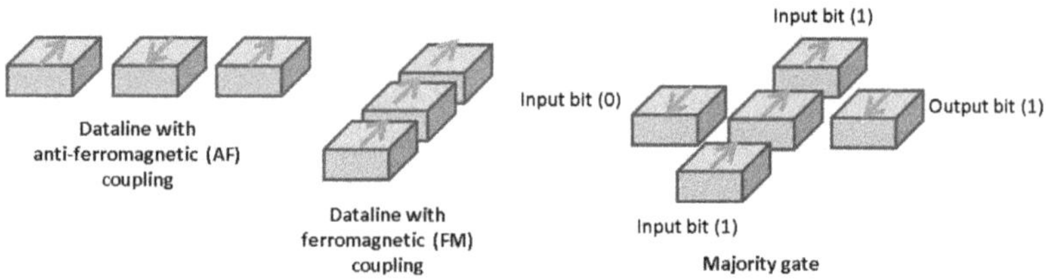

Figure 9.6 Illustration of data propagation and logic operation in NML. Here, interactions between magnets is due only to dipole coupling, and the magnets respond antiferromagnetically or ferromagnetically, depending on their relative positions, as shown in the figure.

The output magnet settles into the inverted state of the compute magnet due to AFM coupling. Two-input Boolean logic operations are realized using a majority gate by fixing the state of one of the inputs.

The fringing field exerted by a nanomagnet alone is not sufficient to flip its neighboring nanomagnet from one stable state to another. To overcome this energy barrier, an external field, called the "clock-field" or "nulling-field," is applied along the short axis, which is commonly called the "hard axis," of the magnets to bring them to their highest energy states. The long axis is called the "easy axis." It is important to understand that due to the self-energy demagnetizing field, it takes more energy to magnetize a rectangular magnet along its hard axis than along its easy axis (hence the nomenclature). Because alignment along the easy axis is neither a "1" nor a "0," we call that the "null state" of the magnets. The dipolar coupling of the fringing field of an adjacent magnet tends to nudge an adjacent magnet slightly into an AFM direction (for magnets adjacently aligned); upon removal of the nulling field, the magnets settle in either of their low-energy states to minimize the overall system energy, wherein that final state is the one preferred by the original adjacent magnet.

As discussed above, NML can suffer from errors that are dominated by premature switching of the magnets due to lack of stability in the null state in the presence of thermal noise [28]. Conventional NML circuits can be operated in two different modes [27]. In the soliton-operation mode, the clock field is removed

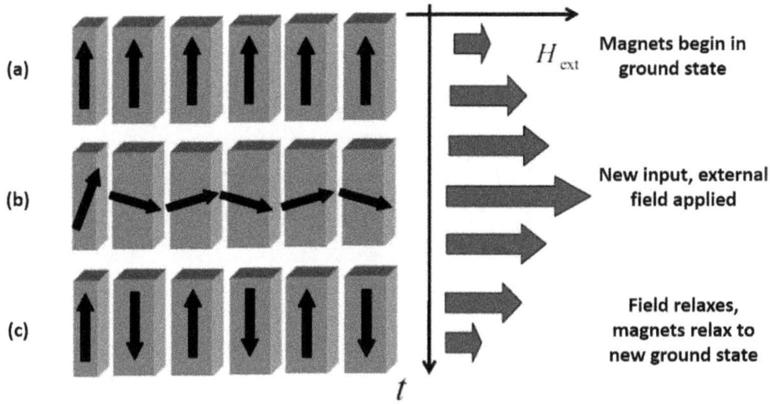

Figure 9.7 Illustration of the clocking process in NML. (a) Initial state of an NML data line with an input on the left. (b) An increasing clock field is applied to null the magnets, and the input is set to the upward magnetization direction. In response to the input, the line of magnets interacts antiferromagnetically with a tendency toward the correct final state. (c) The clock field is removed, and the line of nanomagnets relaxes under the influence of the input nanomagnet to the final state [14].

abruptly and the magnets are required to remain in the null state until they are set by their neighbors. As a result, the magnet furthest from the input is required to stay in the null state the longest. Note that the longer the magnet stays in the null state, the higher is the probability that it switches prematurely due to thermal noise [29]. Alternatively, in the adiabatic-operation mode, every magnet imposes an easy-axis bias to its neighbor during the gradual removal of the clock field, as shown in Fig. 9.7 [30]. As the clock field is reduced to zero, the magnets settle according to the bias from their neighbors. In this scheme, each successive magnet enacts a weaker bias to its neighbor. Therefore, the further the magnet is from the input, the higher is the probability that the thermal noise overrides the neighbor's bias and causes premature switching. As a result, regardless of the operating mode, premature switching is more likely in magnets further from the inputs of a long NML chain [31]. Stability of the magnets in the null state could be enhanced by introducing biaxial anisotropy that exploits

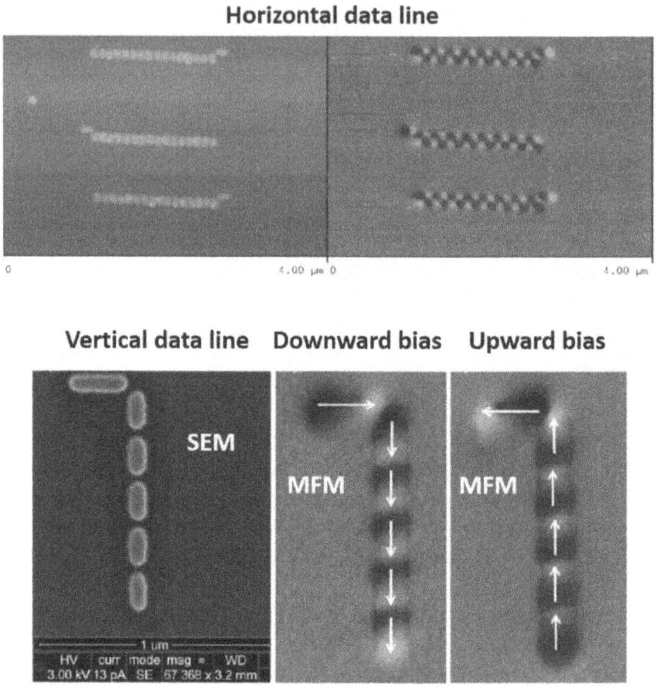

Figure 9.8 Experimental demonstration of NML data lines [8, 19]. (Top): Atomic force microscope and MFM images of a horizontal (antiferromagnetically coupled) data line. An MFM image is formed by the dipole coupling of a magnetic coating on an atomic force microscope tip with the two poles of the nanomagnets, where one pole attracts and the other pole repulses the tip; the contrast in the color indicates the two poles of a magnet. (Bottom): SEM and MFM images of a vertical (ferromagnetically coupled) data line.

the crystalline properties of single-crystal magnets [28] or the configurational anisotropy property of shape-engineered magnets [32]. However, single-crystal nanomagnets with biaxial anisotropy are hard to realize and yet to be demonstrated experimentally. Lambson et al. [32] used concave-shaped magnets to introduce a metastable state along the hard axis. However, optimizing the concave shape to tune the uniaxial and biaxial anisotropy in a way that allows spontaneous data propagation without errors is challenging.

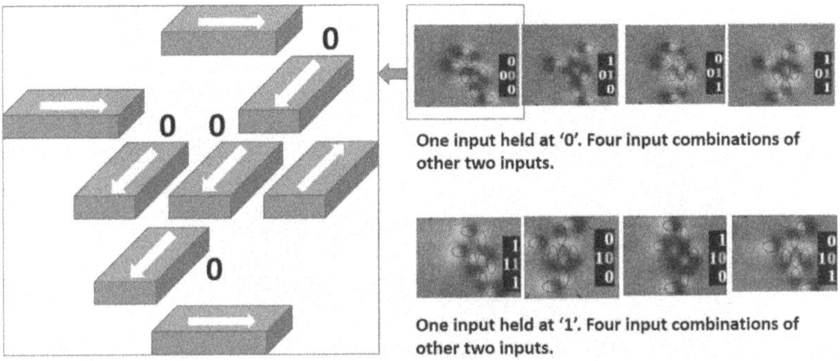

Figure 9.9 MFM images of majority gates providing an experimental demonstration of Boolean logic operation [20]. Schematic of a particular input combination is shown on the left.

9.3.1 Experimental Demonstration of NML Elements

Over the past several years, progress has been made in the experimental demonstration of basic NML elements. Imre [8] demonstrated successful data propagation in AFM and FM data lines. Figure 9.8 shows atomic force microscope images and corresponding magnetic force microscope (MFM) images of AFM and FM data lines. For FM data lines, both vertically upward data propagation and vertically downward data propagation are shown.

In 2006, Imre et al. [20] demonstrated Boolean logic operations with majority gates. Figure 9.9 shows MFM images of majority gates with all possible input combinations. A schematic of one particular input combination is shown to the left of the figure. In these demonstrations, inputs were switched using externally generated magnetic fields. Later, a two-input, shape-based logic gate was introduced to perform Boolean logic [33]. Varga et al. [21] demonstrated fanout in NML.

All of the experimental work in NML discussed so far used an external magnetic field to null the magnets. Alam et al. [34, 35] introduced an on-chip clocking scheme to null the nanomagnets. Figure 9.10 illustrates the on-chip clocking scheme [34]. They used on-chip copper wires to generate a magnetic field through local currents. They were successful in nulling individual magnets as well as magnets in data lines on top of copper clock wires.

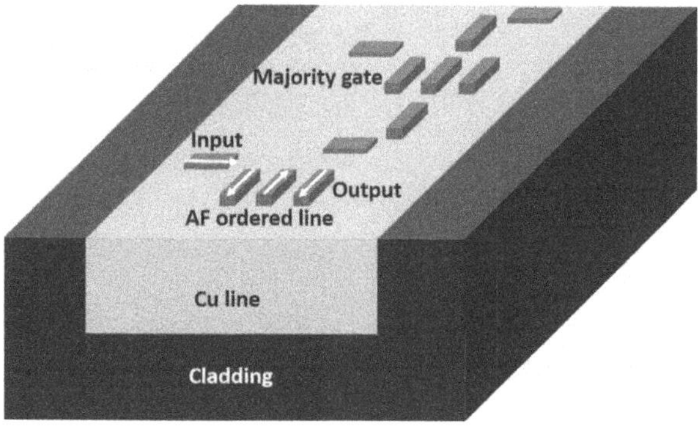

Figure 9.10 Illustration of the on-chip clocking scheme [34].

9.3.2 Advantages and Disadvantages of NML

NML has quite a few advantages over CMOS circuits. NML does not produce static power dissipation, as does CMOS. In addition, switching of nanomagnet states requires very little energy. Simulation studies have shown that the energy dissipated by the switching of 10^{10} magnets at a frequency of 10^8 Hz is only 0.1 W of power [36]. Although this does not include other overhead costs, such as the generation of on-chip clock fields, it suggests the potential for low-energy computing. Simulations also show that a 32-bit NML-based adder (including the nanomagnets, clock-wire, and CMOS-based clock-wire driver) requires less power than a conventional CMOS-based 32-bit adder [30]. Unlike most charge-based devices (e.g., CMOS), NML circuits are radiation hard. We have done our own radiation studies that show no switching after irradiating nanomagnets with 10 Mrads, which is an extremely high dose [19]. Nanomagnets are also nonvolatile, which is a major advantage over CMOS-based logic devices. NML requires less space than its CMOS-based counterpart to perform the same logic operation [2, 27]. Moreover, the operating temperature range for NML is higher than it is for CMOS-based devices, as the curie temperature of most magnetic materials is several 100°C, and magnets well above the superparamagnetic limit are still in the tens of nanometer range.

Despite all the advantages, NML has a major disadvantage: the switching frequency of NML devices is much lower than that of

CMOS devices. The predicted maximum operating frequency of NML devices is only about 1 GHz [37], which is much lower compared to state-of-the-art CMOS devices. As the clock field is reduced, the limiting factor is the speed of rotation of the magnetic dipoles. Therefore, the switching speed is due to the internal dynamics rather than external geometric factors. This drawback is ameliorated by the ability to create highly pipelined architectures with NML [30, 38], so parallel computing could make up for some of this issue. Most importantly, and the drawback that is addressed in this chapter, is the high error rate and lack of directionality observed so far in NML prototypes [39]. NML prototypes so far suffer from high error rates, low throughput, and high clock-power dissipation, all of which must be addressed to make NML an attractive post-CMOS logic technology. In this work, we address these fundamental issues.

9.4 Two Novel Approaches

Due to lack of directionality in NML, premature switching in NML can also trigger back propagation. Imre [8] simulated a pair of magnets with trapezoidal shapes to demonstrate sequential switching that can provide directionality. However, such an approach is not useful in longer data lines or logic gates. Additionally, switching of individual magnets in NML is not synchronized with the clocking, so simultaneous input bits at a logic gate cannot be ensured, resulting in possible output errors.

In previous instantiations of NML architecture, all the magnets on a single clock line were clocked simultaneously [34]. This means that even in the presence of a pipelining mechanism with a two- or three-phase clocking scheme, delay in data propagation will be determined by the time required to propagate data from one edge of the clock line to other edge of the line at the boundary of the clock line of the next phase, through all the magnets. An example of such a clocking scheme (two phase) is shown in Fig. 9.11 [40]. Such delays make NML slow (on the order of several nanoseconds), although switching of individual magnets is quite fast ($\sim$100 ps) [41]. To improve the reliability and speed of NML, we must re-evaluate its fundamental operation.

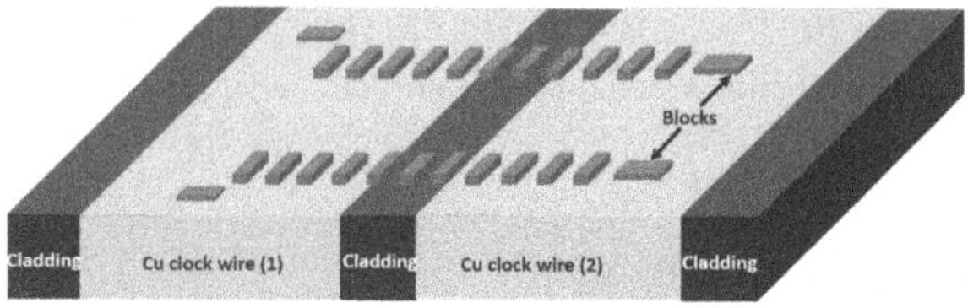

Figure 9.11 An NML pipelining mechanism with a two-phase clocking scheme.

In this chapter, we offer two novel approaches that promise a solution to all of the challenges discussed above. The first approach, discussed in the first half of this chapter, called synchronous nanomagnet logic (SNML), exploits exchange bias within layers of a nanomagnet along with the field asymmetry of shape-engineered magnets to clock magnets in such a way as to avoid premature switching and achieve directionality [42].

The second approach, discussed in the second half of this chapter, is also based on EC, but rather than using EC within a single magnet, it is used to couple separate magnets. In this interlayer exchange-coupled (IEC) system, two FM layers are separated by a thin, nonmagnetic metal layer, also called a spacer layer. Depending on the thickness of the spacer layer, the FM layers can be brought into parallel or antiparallel orientations [43], and, as will be discussed, short-range coupling of magnets by EC is shown to be much stronger than dipole field coupling, which is critical to eliminating errors in NML. This scheme not only offers stronger interaction among thin nanomagnets but also promises greater scalability and better data retention at the deep submicron level, allowing magnetic interaction to be manipulated both in the vertical and lateral directions at the same time [44].

9.5 Selective Clocking in NML

Selective clocking of subgroups of magnets within a clock zone is the primary requirement for SNML. As identified in the previous section, magnets that are clocked along with their neighbors on a clock line often fail to remain under the influence of their neighbors

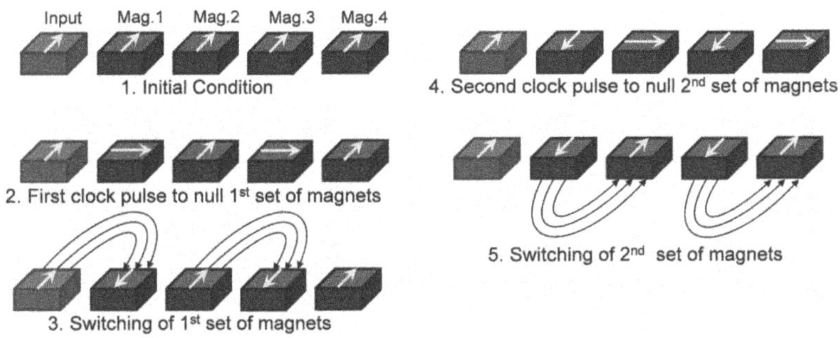

Figure 9.12 Illustration of the selective clocking in NML.

to avoid premature switching. Therefore, we devise an alternative mechanism to clock the magnets individually. It appears that if we can selectively clock alternate magnets simultaneously, we will achieve the effect of individual clocking.

Figure 9.12 illustrates the idea of selective clocking. In this section, we make the assumption that each of the nulled magnets is influenced only by its left neighbor. Later we present a technique to create asymmetric influence of the neighbors. For now, we present only the selective clocking process and assume the asymmetric field influences. In Step 1, all the data magnets and the input magnet are initialized along the up direction. In Step 2, the first set of magnets (Magnets 1 and 3) is clocked. Note that since the neighbors are not clocked, the nulled magnet is always under their influence, so premature switching is eliminated. In Step 3, the left neighbors switch the nulled magnets due to the aforementioned asymmetric field. As a result, the data bit from the input moves to Magnet 1 and the data bit from Magnet 2 moves to Magnet 3. In Step 4, the second set of magnets (Magnets 2 and 4) is clocked. Finally, in Step 5, the magnets relax according to their left neighbors. As a result, the data bit from Magnet 1 moves to Magnet 2 and the data bit from Magnet 3 moves to Magnet 4. During the aforementioned five steps, equivalent to a full clock cycle, the data bit from the input bit has moved by two magnets to Magnet 2 and the data bit stored initially in Magnet 2 has moved by two magnets to Magnet 4. Since at each half clock cycle the data bit moves by one magnet, the propagation of data can be said to

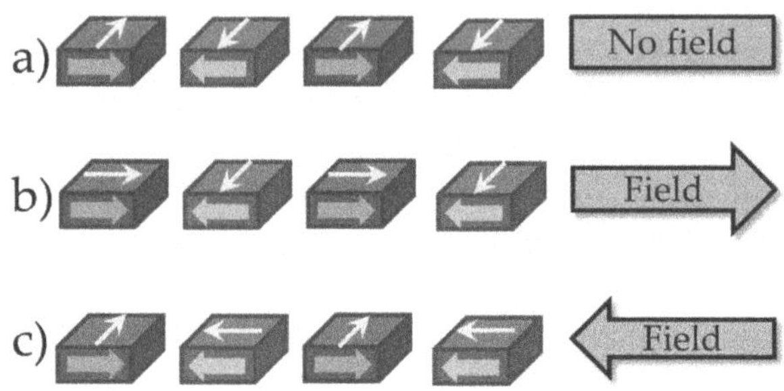

Figure 9.13 Hard-axis biased magnet for selective clocking.

be synchronous with the clock field. This SNML scheme, therefore, behaves as a shift register.

Here, we discuss how selective clocking in SNML that was assumed above is actually achieved. To implement selective clocking, we must either realize separate clock lines for the individual magnets (which is unrealistic) or adjust the properties of the individual magnets so that they interact differently from each other with the clock field. The magnets in conventional NML circuits are unbiased along both rightward and leftward clock-field directions, that is, they have no preference in their nulling behaviors along their hard axes. However, if we can cause the neighboring magnets to have a nulling preference, that is, make them biased, along opposite clock-field directions, then we can selectively clock the magnets by changing the polarity of the clock field. Figure 9.13 illustrates the idea in which neighboring magnets are biased in opposite directions to allow selective clocking. Bias directions are denoted by a wide arrow on the front edge of each magnet. When a right clock field is applied, only the magnets with the right bias are nulled (as shown in Fig. 9.13b). When a left clock field is applied, the magnets with the left bias are nulled (as shown in Fig. 9.13c). Therefore, such alternating bias in neighboring magnets will allow selective clocking. In the next section, we discuss how we can achieve such bias.

9.5.1 Unidirectional Anisotropy for Selective Clocking

Figure 9.14a shows the energy landscape of a typical rectangular magnet for NML with uniaxial shape anisotropy (along the Y axis),

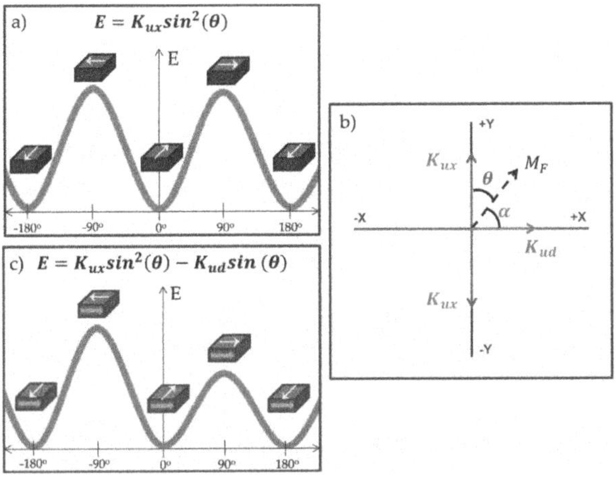

Figure 9.14 (a) Energy landscape of a magnet with uniaxial anisotropy, (b) relative orientation of the anisotropy constants with respect to magnetization, and (c) energy landscape of a magnet with combined uniaxial and unidirectional anisotropy.

which follows the equation

$$E_{\mathrm{ux}} = K_{\mathrm{ux}} \sin^2(\theta), \tag{9.1}$$

where K_{ux} is the uniaxial anisotropy constant per unit volume and θ is the magnetization direction with respect to the positive Y direction. Unidirectional anisotropy along the positive X direction adds the following energy term to the total energy

$$E_{\mathrm{ud}} = -K_{\mathrm{ud}} \cos(\alpha), \tag{9.2}$$

where K_{ud} is the unidirectional anisotropy constant per unit volume and α is the magnetization direction with respect to the positive X direction. Figure 9.14b shows uniaxial and unidirectional anisotropy orientation with respect to the magnetization direction. The total energy, combining both of the anisotropy terms, is

$$\begin{aligned} E_{\mathrm{tot}} &= E_{\mathrm{ux}} + E_{\mathrm{ud}} \\ &= K_{\mathrm{ux}} \sin^2(\theta) - K_{\mathrm{ud}} \cos(\alpha) \\ &= K_{\mathrm{ux}} \sin^2(\theta) - K_{\mathrm{ud}} \sin(\theta), \end{aligned} \tag{9.3}$$

and the energy landscape would look like Fig. 9.14c. It is evident from Fig. 9.14c that the magnet with unidirectional anisotropy along

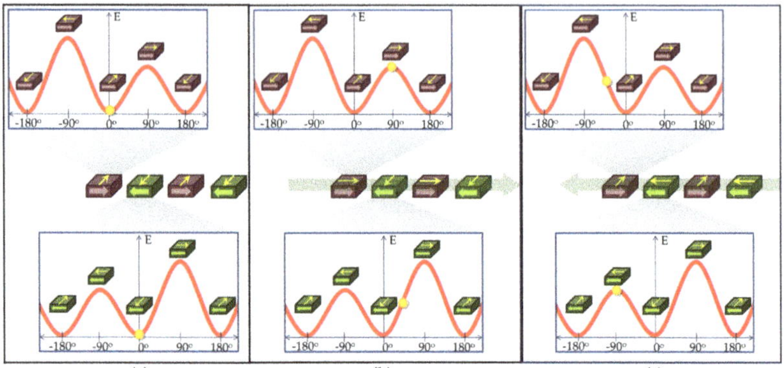

Figure 9.15 Energy state of the right- and left-biased magnets under (a) no clock field, (b) rightward clock field, and (c) leftward clock field. The yellow circle indicates the energy state of the magnet at each condition.

the positive X direction requires less Zeeman energy to be clocked along the positive X direction than to be clocked along the negative X direction. This asymmetry in barrier height can be utilized to selectively clock the magnets. If we have neighboring magnets with opposite unidirectional anisotropy, then we can selectively clock an individual magnet without clocking its neighbor.

Figure 9.15 illustrates how unidirectional anisotropy can be used to selectively clock the magnets. When no clock field is applied, magnets of both types (i.e., biased both left and right) stay at the lowest energy state as the Zeeman energy acting on the magnets is zero (indicated by the position of the yellow dot in Fig. 9.15a). Referring to Fig. 9.15b, when a rightward clock field is applied, the acting Zeeman energy on the magnets with a rightward bias (second magnet from the left) is sufficient to bring the magnet to the top of the energy barrier. However, the magnets with a leftward bias (leftmost magnet) do not reach the top of their energy barrier. As a result, only the magnets with a rightward bias become nulled (and are available to change state based on their neighbors). Similarly, when a leftward clock field is applied, the magnets with a leftward bias arrive at the top of the energy barrier and the magnets with a rightward bias do not. In this case, only the magnets with a leftward bias become nulled (Fig. 9.15c). Given that the actual direction of nulling has no effect on the final state, this scheme, then, allows each of the two sets of magnets, that is, those biased leftward and rightward, can be selectively nulled and switched in alternate half clock cycles.

So far, we have explained the idea of how unidirectional anisotropy can be used to selectively clock the magnets. In the following section we will discuss how we can practically incorporate such anisotropy into a magnet and then how to add directionality to data propagation.

9.5.2 Exchange Bias for Unidirectional Anisotropy

Unidirectional anisotropy in ferromagnets can be achieved by exploiting the exchange bias phenomenon present at FM/AFM interfaces [45]. However, for our application, adjacent magnets must have unidirectional anisotropy in opposite directions, which means that adjacent magnets must be exchange-biased in opposite directions. The exchange-bias direction depends on the direction of the field applied during the field cooling process required to introduce the exchange bias. In thermally assisted magnetic random access memories (TAMRAMs) [46], the reference layer and the storage layer are exchange-biased independently using two AFM layers of different blocking temperatures. The AFM layer for the storage-layer bias has a lower blocking temperature than that for the reference-layer bias. Therefore, there exists a temperature window within which the storage layer can be written with thermal assistance without disturbing the reference layer. Similarly, if we use two AFM materials (e.g., AFM1 and AFM2) having two different blocking temperatures for the neighboring ferromagnets, we can bias the neighboring magnets independently and along opposite directions. Guo et al. [47] demonstrated opposite exchange bias in a sample using two different antiferromagnet materials with different blocking temperatures. To achieve such a property, they annealed the samples at two different temperatures, first at a higher blocking temperature and then at a lower blocking temperature, under opposite field directions.

Figure 9.16 illustrates a possible way to introduce an opposite exchange bias in neighboring magnets. In the first step, the set of

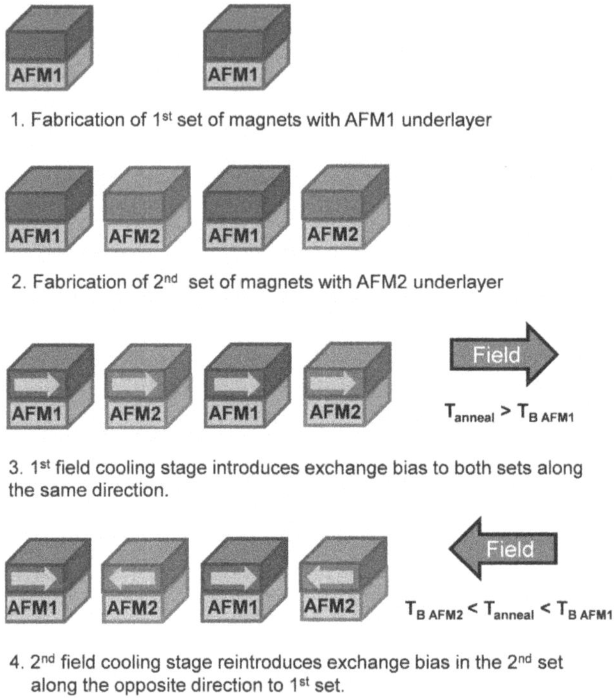

Figure 9.16 Process steps for realizing a synchronous data line with neighboring magnets exchange-biased in opposite directions.

magnets with an AFM1 underlayer will be fabricated. In the second step, a second set of magnets with an AFM2 underlayer will be fabricated with nanoscale overlay accuracy. In our laboratory at Notre Dame, 4 nm layer-to-layer accuracy has been demonstrated with nanomagnets having dimensions of around 50 × 120 nm [48].

AFM1 and AFM2 will be chosen carefully so that AFM1 has a higher blocking temperature than AFM2 (i.e., $T_{\mathrm{B\ AFM1}} > T_{\mathrm{B\ AFM2}}$). For example, IrMn ($T_\mathrm{B} \approx 230°\mathrm{C}$) and FeMn ($T_\mathrm{B} \approx 153°\mathrm{C}$) can be used as AFM1 and AFM2, respectively. In the third step, the sample will be annealed at a temperature higher than $T_{\mathrm{B\ AFM1}}$ and cooled slowly under a high field (typically 1 T). At this point, both sets of magnets will be exchange-biased in the same direction. In the fourth, and final, step, the sample will be annealed at a temperature that is in between $T_{\mathrm{B\ AMF1}}$ and $T_{\mathrm{B\ AFM2}}$ and cooled under a high field (1 T as

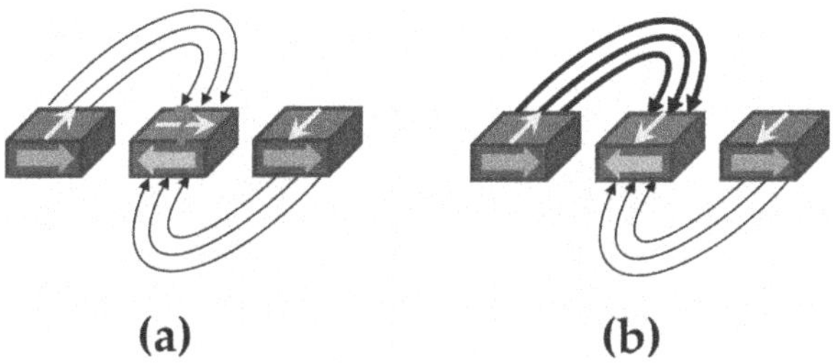

Figure 9.17 (a) The central magnet is influenced by both of the neighbors equally and assumes a random state. (b) The central magnet is influenced by the left neighbor more than the right one and follows the left magnet.

before) with the opposite polarity. At this point, the bias direction of the second set of magnets (with the AFM2 underlayer) will change along the second applied field. This will result in neighboring magnets with opposite biases.

As explained above, in SNML, only alternate magnets are clocked at any one time. In such a situation, the clocked magnets will be under the influence of both of their unclocked neighbors. In the worst-case scenario, when the two neighbors compete to switch the clocked magnet, the clocked magnet may not switch to the correct direction intended by the specified propagation direction. The situation is illustrated in Fig. 9.17a. If somehow the influence from either of the neighbors could be enhanced compared to the other, the clocked magnet would follow the neighbor with the stronger influence, and the problem would be solved. Figure 9.17b shows the result when the left magnet has a stronger influence on the center magnet than does the right one, and therefore, the center magnet switches according to the left one. Such unequal influence from the neighbors would introduce the desired directionality in NML operation. Next, we describe how shaped magnets help solve this problem.

9.5.3 Asymmetric Field Distribution

To achieve such variation in influence from the neighbors, we need magnets with asymmetric stray field distribution along the propagation axis. Such asymmetric field distribution can be achieved

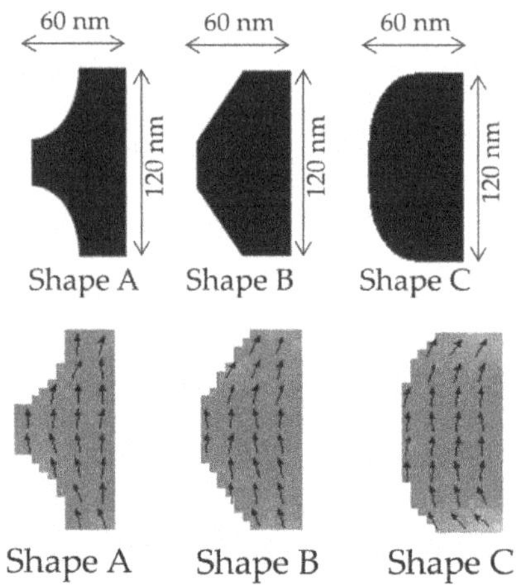

Figure 9.18 OOMMF snapshot of the shapes in the relaxed state showing canted orientation of the spins at the top and the bottom.

by optimizing the shape of the magnets. Therefore, we have considered several promising shapes to investigate their field distributions. Figure 9.18 (top) shows three different Co dots that have longer curved edges on the left side. Spins in the magnets align with the longer curved edge to minimize the demagnetization energy and result in a canted spin orientation at the top and bottom of the magnets. Figure 9.18 (bottom) shows simulation snapshots of the shapes in the relaxed state with the canted spins at the top and bottom of the magnets. Such orientation of the spins produces an asymmetric stray field distribution along the hard axes of the magnets. Figures 9.19 and 9.20 show M-H curves of the shapes along their easy and hard axes, respectively. All three exhibit typical hysteresis curves when taken along both easy and hard axes. The M-H curves confirm that the shapes fulfill the requirement for proper NML operation, that is, that they have bistable states that can be used to represent data.

Next, we calculate the average stray field from the magnets (magnetized along the easy axis) on both of their neighbors as a function of clock field. Figure 9.21 shows the magnets and the representative regions of the neighboring magnets used in the

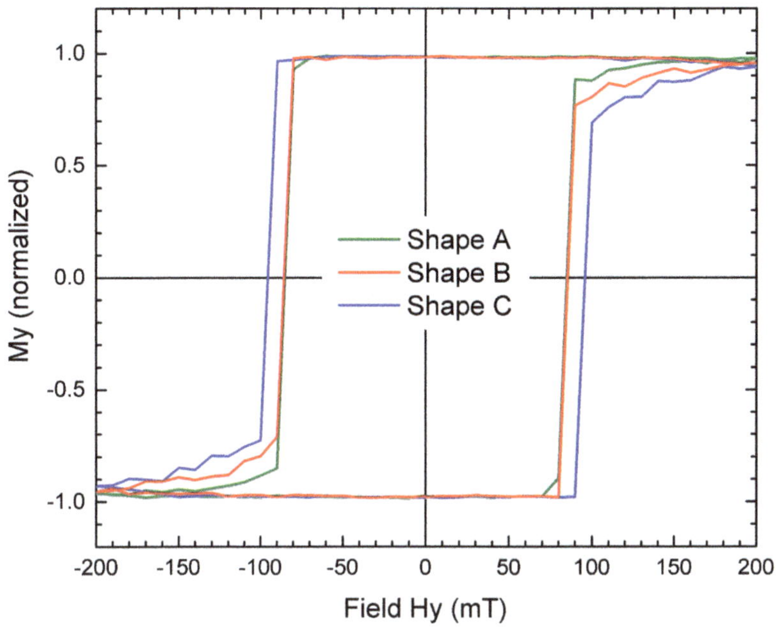

Figure 9.19 Easy-axis M-H curves of the shapes.

calculation of the average stray field. The average field is calculated within the region defined by the adjacent magnets. The field asymmetry ΔF is calculated by taking the difference between the average stray field on the right area and that on the left area.

Figure 9.22 shows the dependence of ΔF on the clock field along both directions for all three shapes. Note that shape C has the highest ΔF at a zero clock field but ΔF reduces sharply with the clock field. Shape B has a moderately high ΔF at a zero clock field, which does not decrease as sharply with the clock field. Since in SNML even the unclocked magnet will be under the influence of the clock field during the clocking step of its neighbor, it is important to choose a shape that gives the largest ΔF over a wide range of external fields. In other words, we seek a shape for which ΔF remains large during the clocking step at the highest possible clock field so that the asymmetric influence on the neighbors is most effective during clocking. Shape B best satisfies these constraints.

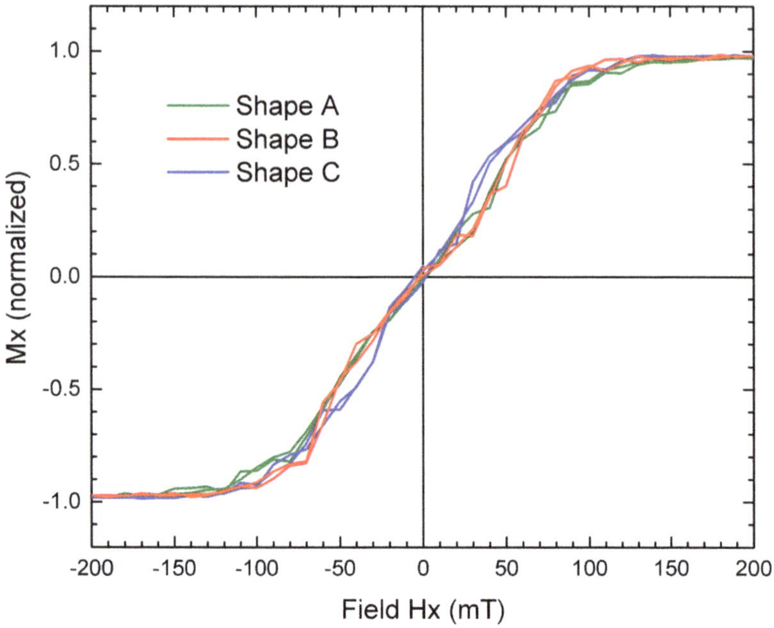

Figure 9.20 Hard axis M-H curves of the shapes.

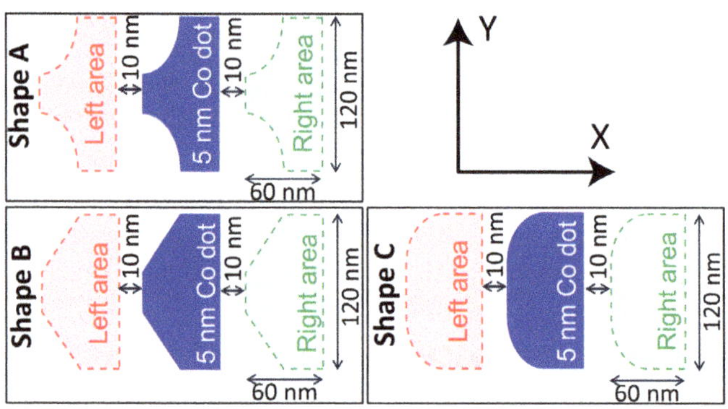

Figure 9.21 Configuration used to calculate asymmetric field for the shapes.

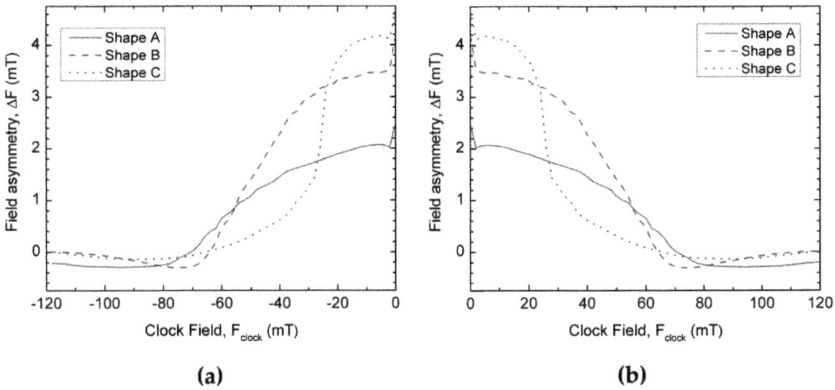

Figure 9.22 Field asymmetry (ΔF) of the shapes as a function of (a) negative and (b) positive clock fields.

9.5.4 Simulation of SNML Operation

To validate the idea of synchronous operation, we performed room-temperature micromagnetic simulations for the basic components of an NML circuit. For all the simulations, we used shape B as specified in the previous section. The damping constant (α) for all the simulations was set to 0.012 [45]. The exchange-bias field for shape B was chosen to be large enough to allow selective clocking but not so large as to saturate the magnet along its hard axis. Shape B requires an ~100 mT clock field to saturate along the hard axis. A 50 mT bias field will not saturate the magnet along either of its hard-axis directions. 50 mT of exchange bias in a 5 nm Co layer would require an exchange-bias energy (ΔE) of 0.35 erg/cm^2 at the AFM/FM interface. Such exchange-bias energy (or higher) has already been observed at various AFM/FM interfaces [49]. In the presence of such a bias field, an external field as low as 50 mT along the bias direction will clock the magnet. Since the neighboring magnets will be biased in opposite hard-axis directions, selective clocking can be achieved just by changing the clock-field direction.

Figure 9.23 shows the field asymmetry ΔF for shape B with a negative bias (Group 1) and a positive bias (Group 2) at each of the three stages of a clock cycle. At stage 1, when no clock field is applied, Groups 1 and 2 experience a total field of −/+50 mT, respectively,

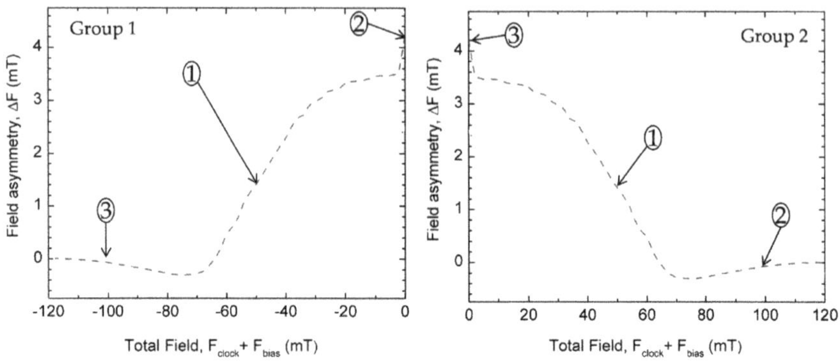

Figure 9.23 Field asymmetry (ΔF) of shape B with (a) negative and (b) positive bias fields at different stages of the clock field cycle.

due to their built-in bias fields, and have a ΔF corresponding to their bias fields, as indicated in the figure. At stage 2, when a $+50$ mT clock field is applied, Group 1 experiences a total field of 0 mT (as the $+50$ mT clock field compensates the -50 mT bias field) and has the highest ΔF. On the other hand, Group 2 experiences a total field of 100 mT (as the $+50$ mT clock field aids the $+50$ mT bias field) and has a ΔF close to zero. At stage 3, when a -50 mT clock field is applied, Group 1 experiences a total field of -100 mT (as the -50 mT clock field aids the -50 mT bias field) and has a ΔF close to zero. On the other hand, Group 2 experiences a total field of 0 mT (as the -50 mT clock field compensates the $+50$ mT bias field) and has the highest ΔF.

In summary, during the alternate clocking in synchronous operation, clocked magnets show close to zero ΔF while unclocked neighbors show the maximum ΔF and ensure directionality.

9.5.4.1 Horizontal data line

First, we simulated the synchronous operation in a horizontal data line with a fixed input on the left. The magnets were 10 nm apart, and the neighboring magnets were exchange-biased in opposite directions, with a bias field of 50 mT along their hard axes. A clock field of +/− 50 mT was applied alternatively. For each direction of clock field, alternate magnets were nulled. Figure 9.24 shows

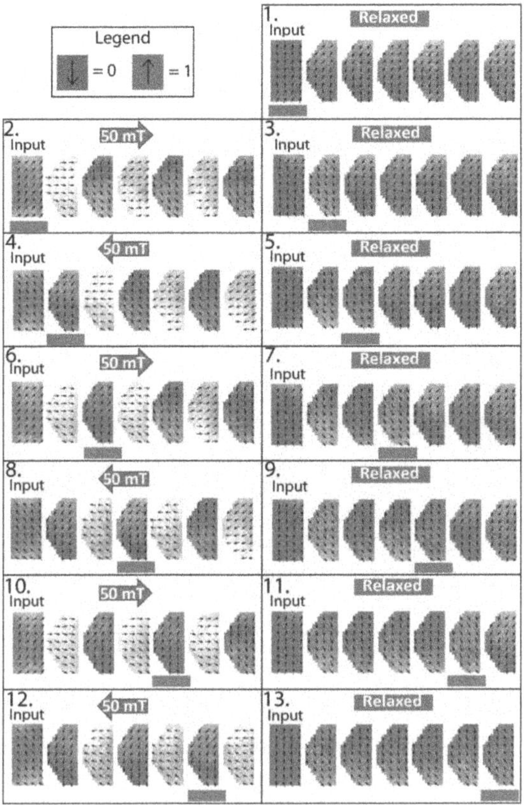

Figure 9.24 Synchronous NML operation in a horizontal data line.

the simulated data line at different stages. The wide bar under the magnets indicates the position of the data bit under inspection. During each half clock cycle, the data bit progressed by one magnet to the right, and after six consecutive half clock cycles (or three clock cycles), the data bit reached the end magnet.

9.5.4.2 Vertical data line

For the synchronous vertical (ferromagnetically coupled) data line, we set the vertical spacing at 10 nm. However, in order to exploit the asymmetric field distribution for directionality, the magnets are required to be shifted along the horizontal direction. Figure 9.25a

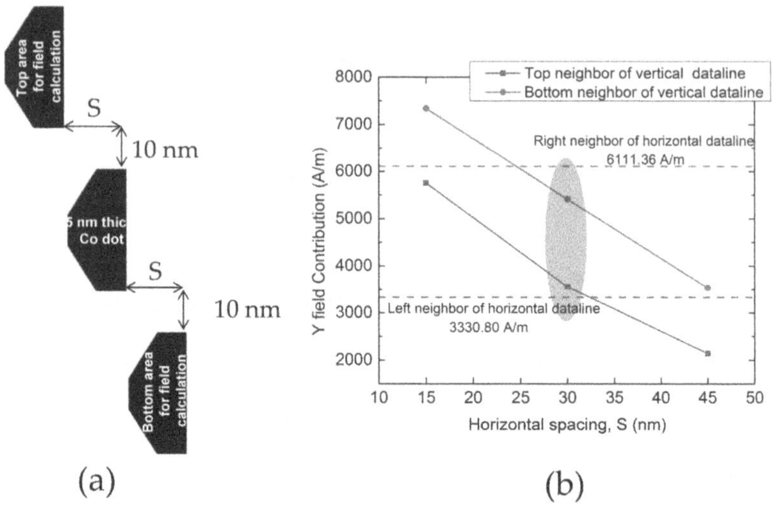

Figure 9.25 (a) Arrangement of magnets in a vertical line used for ΔF calculation. (b) Field contribution to the top and bottom magnets as a function of horizontal spacing (S) for the vertical arrangement shown in (a).

shows the arrangement of the magnets in a vertical data line for synchronous operation. The arrangement is such that the data propagate from top to bottom. We want to keep the field contribution to the top and bottom neighbors in the vertical data line the same as that to the left and right neighbors in the horizontal data line. Such similarity between the vertical and horizontal arrangement, in terms of field contribution to the neighbors, is important for applications in which both types of arrangement are required (e.g., logic gates). Figure 9.25b shows the dependency of the field contribution to the top and bottom neighbors (arranged vertically) on the horizontal spacing (S). The dotted horizontal lines represent the field contribution to the left and right neighbors in a horizontal data line. Neighbors in the vertical data line with a 30 nm horizontal spacing experience a field contribution similar to what they do in a horizontal data line. Therefore, a horizontal spacing of 30 nm was chosen for the vertical data line with a 10 nm vertical spacing.

Next, we simulated synchronized data propagation in the vertical data line with a fixed input magnet at the top left. The neighboring

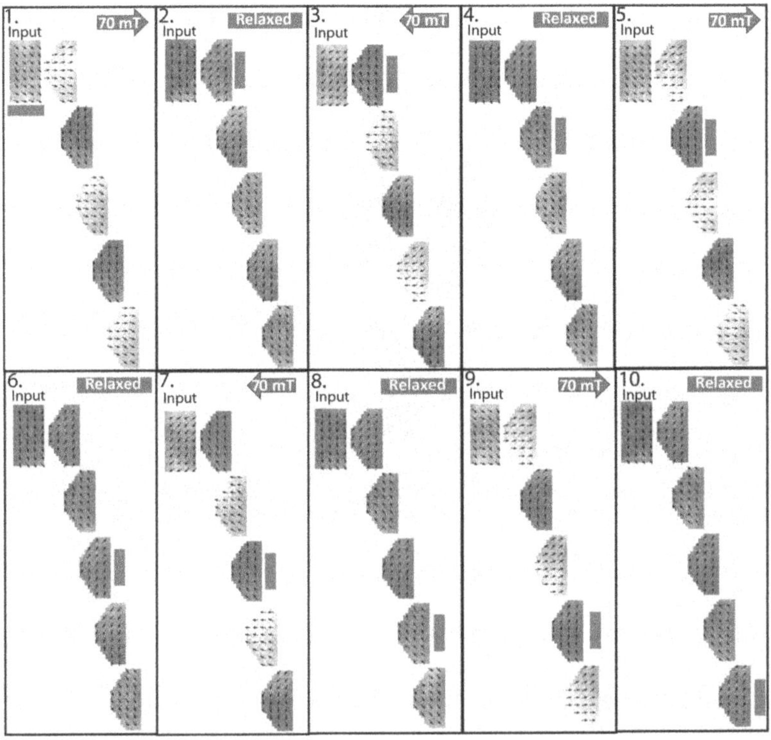

Figure 9.26 Synchronous NML operation in a vertical data line.

magnets were exchange-biased in the opposite directions, with a bias field of 50 mT along their hard axes. Magnets arranged along the vertical direction require more field to clock because of the stronger FM coupling between the magnets along the easy axis but less than that of a conventional (rectangular) FM line of magnets. Therefore, an alternating clock field of $+/-$ 70 mT was applied. Figure 9.26 shows synchronized data propagation in the vertical data line.

9.5.4.3 Majority gate

Figure 9.27 illustrates the arrangement of magnets for a synchronous majority logic operation. Note that the horizontally spaced magnets in the majority gate have 10 nm of horizontal spacing, which is the same as in the horizontal data line. Similarly, vertically

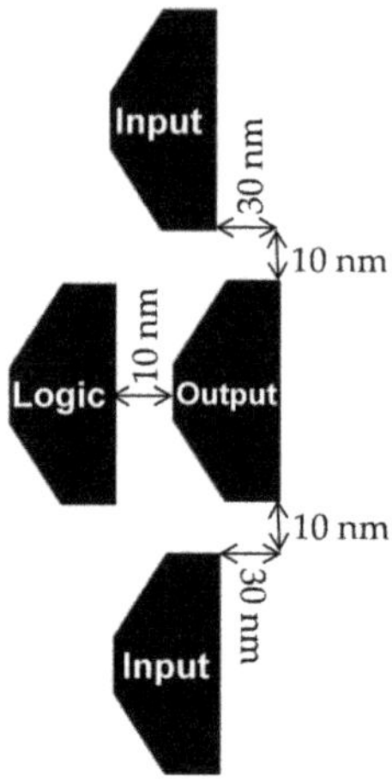

Figure 9.27 Arrangement of magnets in a majority gate for synchronous operation.

spaced magnets in the gate have 10 nm of vertical and 30 nm of horizontal spacing, which is the same as in the vertical data line. The biasing field and the clock field were set to be +/− 50 mT and +/− 70 mT, respectively.

Next, we simulated the majority gate for all possible combinations of inputs. To demonstrate the synchronous operation of the gate, we fed the input bits to the gate through data lines, synchronously from fixed inputs, instead of feeding them directly. We kept an additional magnet at the output end of the gate to demonstrate that the output of the gate can also be propagated synchronously to a data line. Figure 9.28 shows the simulation results at the final stage for all possible input combinations. The majority gate works properly for all the combinations. As the simulation result shows, with the logic bit set to "0" we obtain the OR operation and with the logic bit set to "1" we obtain the AND operation.

9.5.5 Robustness of SNML Operation

The fundamental reason for developing this novel idea of SNML operation is to mitigate the reliability issues of prior NML methods.

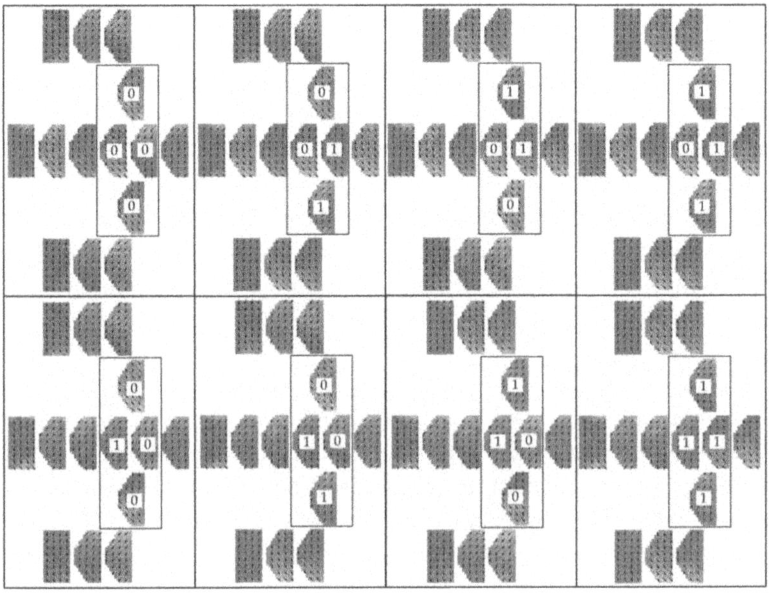

Figure 9.28 Final stage of the synchronous operation in a majority gate for all possible input combinations.

However, realizing this idea would require some approaches that may introduce additional reliability issues. Reliability of this novel synchronous approach in the presence of thermal noise must be evaluated first. In the SNML operation, neighboring magnets have different AFM layers underneath them to introduce exchange bias in opposite directions. Therefore, neighboring magnets must be fabricated separately, and there will be some finite misalignment between these two process steps. Such misalignment can raise reliability issues. In addition, optimizing both of the AFM/FM interfaces to obtain exactly equal and opposite exchange bias fields between both Group 1 and Group 2 sets of magnets will be extremely challenging. Therefore, there will be some deviation in the bias fields of the two magnet groups that can cause consistency issues. Vertical data lines require more clock field compared to horizontal data lines in SNML, as they do in conventional NML. To clock an entire system, we must clock it at a field that can null both the vertical and horizontal data lines. We must ensure that the overclocking of the horizontal lines does not jeopardize their synchronous operation.

To ascertain that SNML operation is a reliable approach, we must evaluate the effect of the aforementioned factors on it. Therefore, we performed the micromagnetic simulations by taking into account these nonideality factors. These simulations were done using the Object Oriented MicroMagnetic Framework (OOMMF) package by the National Institute of Standards & Technology (NIST) [50]. The following sections elaborate on these findings from the simulations.

9.5.5.1 Thermal noise

Although all the simulations on synchronous operation so far presented are done at room temperature, to evaluate the thermal stability of the system it is required to perform hundreds of simulations with different random seeds. Simulations for the synchronous operation are time consuming, as one must apply an alternating field successively to propagate the data. Performing such simulations hundreds of times to estimate the error rate is unrealistic. Thermal stability of the individual magnets is quite high, as we need 50 mT to clock the magnet in the presence of a 50 mT exchange-bias field. However, we must evaluate the coupling energy to assess how stable is the coupling between the magnets in the presence of thermal noise. The net coupling field acting on a magnet in synchronous NML is the difference in the field contribution from its neighbors, ΔF. This field is understandably lower than the coupling field in a conventional NML system. Stability of the coupling in the synchronous NML will depend on this field.

First, we must calculate the energy contribution from this field. The coupling energy for a given field can be calculated using

$$E_{\text{cpl}} = \mu_0 \mathbf{M}_S V\, \mathbf{m} \cdot \mathbf{H}, \tag{9.4}$$

where μ_0 is the permeability constant $4\pi \times 10^{-7}\ \mathbf{H} \cdot \mathbf{m}^{-1}$, $\mathbf{M}_S$ is the saturation magnetization of the magnet material, V is volume of the magnet, $\mathbf{m}$ is the unit vector along the magnetization, and $\mathbf{H}$ is the coupling field vector.

The ΔF for shape B is 2780.56 A/m, and the total volume of shape B is 2.925×10^{-23} m^3. Using Eq. 9.4, the coupling energy is

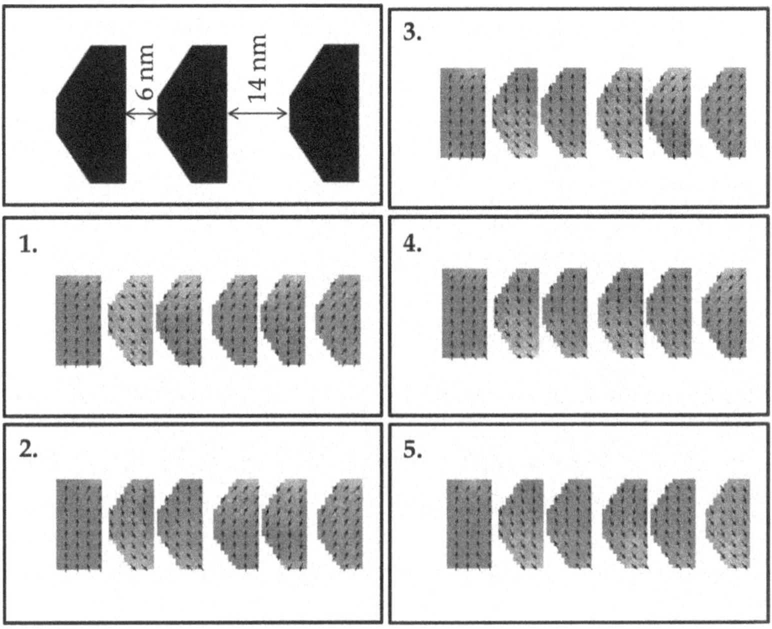

Figure 9.29 Synchronous NML operation in a horizontal data line with a 4 nm misalignment along the *x* axis.

found to be ~35 $k_B T$ at room temperature. Csaba and Porod [15] showed that the coupling error probability decreases exponentially with coupling energy

$$P_{\text{cpl. err.}} = \exp^{-\frac{E_{\text{cpl}}}{k_B T}}. \tag{9.5}$$

Therefore, the probability of coupling error for a coupling energy of 35 $k_B T$ is 6.3×10^{-16}, which is consistent with the standard CMOS error probability of 10^{-17} [51]. Even with the reduced coupling field, the error probability in the synchronous operation remains within the acceptable limit.

9.5.5.2 Misalignment

On the basis of our experience of achieving sub-4-nm overlay accuracy, it is reasonable to assume that the misalignment between the neighboring magnets fabricated in two separate process steps will not be more than 4 nm. Therefore, we simulated a five-magnet-long data line with alternate magnets displaced by 4 nm. The rest of the simulation parameters were the same as those used above. Figure 9.29 shows the arrangement of the misaligned magnets and the step-by-step simulation results. Despite the 4 nm misalignment, the data line showed perfect synchronous operation. Therefore, it can be inferred that synchronous NML can perform reliably in the presence of minor misalignment.

9.5.5.3 Unequal exchange bias

Exchange bias field is a strong function of the FM/AFM interface quality and the texture, which depend on the deposition conditions and the AFM/FM materials. Tuning of all the conditions to have a pair of AFM/FM interfaces that gives equal and opposite exchange-bias fields will be difficult. It is likely that we would see some deviation in exchange-bias fields between the AFM/FM pair, so SNML must be robust enough to tolerate such deviations. To test the robustness of the synchronous operation over unequal bias fields, we performed micromagnetic simulations where neighboring magnets had 40 mT and 60 mT of oppositely pointing bias fields instead of having the oppositely-pointing equal bias fields of 50 mT. Figure 9.30 confirms that such variation in the bias field does not cause any error in synchronous operation of NML. This robustness is provided by the built-in bias fields, as explained in Fig. 9.23.

9.5.5.4 Overclocking

The required clock field in NML varies with the relative orientation of the magnets. As discussed in Fig. 9.26, the vertical data line requires a 70 mT clock field, whereas the horizontal data line requires only a 50 mT clock field. We will have both kinds of data lines in a complete NML system, so we must clock the system at the highest required clock field, which is 70 mT. Therefore, we must confirm that a 70 mT clock field does not introduce errors in the horizontal data line. Figure 9.31 shows synchronous data propagation in a horizontal data line with 70 mT of alternating clock

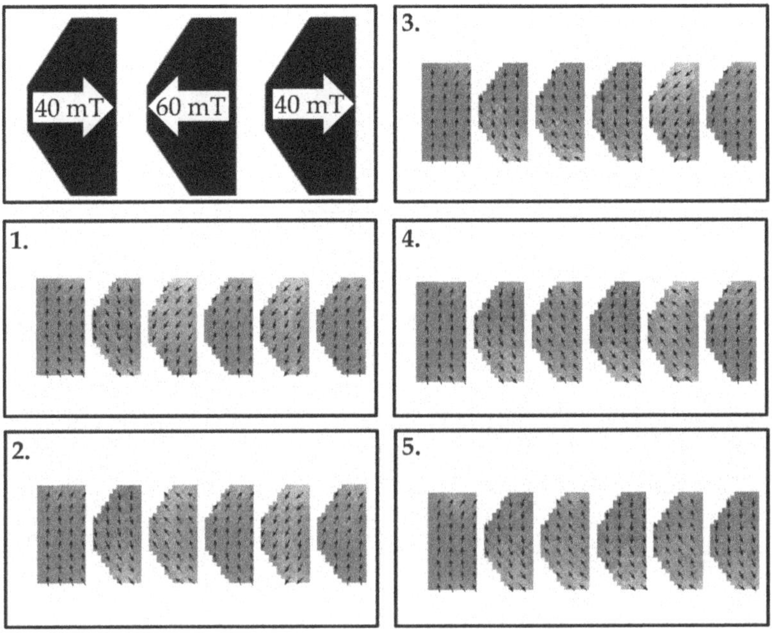

Figure 9.30 Synchronous NML operation in a horizontal data line with 20 mT mismatch in exchange bias.

field. The result confirms that such overclocking does not hinder the synchronous operation in the horizontal data line.

9.5.6 Implications of SNML

Conventional NML showed great promise owing to its low switching energy, nonvolatility, radiation hardness, and compact design. However, due to the lack of reliability, directionality, and speed, its advantages over conventional CMOS-based logic were in question.

SNML inherits all the advantages of the conventional NML operation yet promises to eliminate the reliability and directionality issues and improves the speed of operation by exploiting pipelined stages. The switching behavior of nulled magnets in conventional NML can be stochastic in nature in the presence of the thermal noise. Such a stochastic system is incompatible with Boolean logic operations. In SNML, nulled magnets are under the influence of their

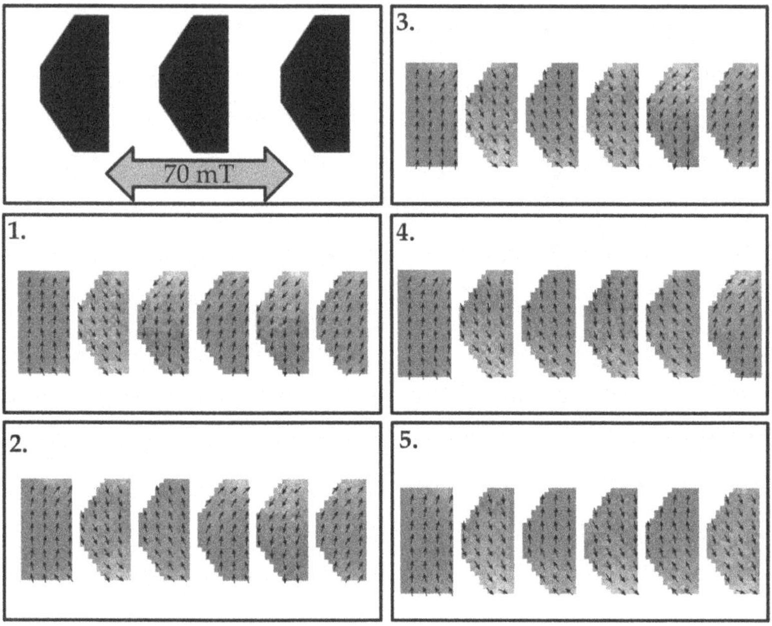

Figure 9.31 Synchronous NML operation in a horizontal data line with 70 mT of clock field.

unclocked neighbors at all times, which suppresses the stochastic switching behavior of the nulled magnets.

In conventional NML, directionality is controlled at the clock-line level using a two- or three-phase clock scheme. There is no mechanism to control the directionality within a single clock line in conventional NML. On the other hand, SNML controls directionality at the single-magnet level, eliminating back propagation. In SNML, data propagates in a shift-register-like fashion, where two neighboring magnets act as a single register. In SNML, data shifts by one magnet per half clock cycle and two magnets (or a register) per clock cycle. In other words, SNML basically realizes a synchronous pipeline mechanism of two-magnet-long stages. Such a pipeline mechanism allows data processing in SNML to be as fast as half the switching speed of a single magnet. This is a significant improvement in terms of throughput over conventional NML. Data propagation in conventional NML is governed by a two- or three-phase unipolar clocking scheme. Such a clocking scheme requires a complex arrangement of clock wires, which is difficult to realize. On the other hand, the entire SNML is based on a single-phase bipolar clocking scheme, which requires only a single clock wire.

9.5.7 Summary of SNML

We have shown that by utilizing exchange bias and shape engineering we can achieve error-free, directional, and synchronous data propagation and logic operation in NML circuits. This new approach also allows faster data processing and a simpler clocking scheme. Here, we have considered only three shapes that have an asymmetric field distribution along the hard axis. Magnets with higher asymmetric field distribution can be designed through a careful and systematic approach of shape engineering. We simulated synchronous operation of the three basic elements (i.e., horizontal line, vertical line, and majority gate) of NML circuits. Complicated NML elements such as full adders are also based on these basic elements. Since the fundamental operation in SNML is limited to the adjacent magnets only, a larger circuit does not create any added complexity in the operation. Therefore, SNML should also offer error-free operation in full-scale logic circuits. We also discussed possible reliability issues related to this novel approach. However, from the simulation results under nonideal conditions, it is evident that SNML is a robust and reliable approach toward the practical implementation of NML. We believe that SNML could make a significant impact in nanomagnet logic research by opening paths for novel architectural design and performance enhancement techniques.

9.6 Exchange Coupling between Laterally Adjacent Nanomagnets

In this section, we describe an approach to enhanced dot-dot coupling for NML by using EC between laterally adjacent nanomagnets. Some of the complex structures that function based on in-plane dipole field coupling have already been discussed above [19,

22, 24]. In addition to those, devices that realize logic functions based on out-of-plane stray fields have also been demonstrated [52–54]. To enhance the functionality of these devices and to realize schemes that are more complex, strong magnetic coupling between the neighboring elements is required. In this section, an interlayer exchange coupling (IEC)-based coupled-nanomagnet system is discussed in detail, which, if engineered in patterned structures, may open up a new avenue where stronger short-range interactions are desired and allow magnetic interactions in three dimensions. This interaction can be exploited in magnetoelectronic devices that rely on magnetic (dipole) coupling [20, 55] and where the coupling strength between neighboring elements plays a significant role in correct functioning of these devices in the presence of thermal fluctuations and process variations. EC offers some additional benefits compared to dipole coupling. Exchange interaction is short range, and therefore, there are no long-range parasitic stray-field couplings. Moreover, exchange-coupled magnets can be scaled to smaller volumes and will still be immune to thermal fluctuations. For example, EC provides additional anisotropy in high-density magnetic storage media and ensures better thermal stability around the superparamagnetic limit [56].

9.6.1 The Proposed Coupling Scheme

In the previous sections, SNML was presented and the role of EC was only to bias the preferred nulling direction of the nanomagnets. The EC played no role in the actual coupling among the magnets, which was entirely of dipole-field origins. Here, the coupling is mostly by EC and to a lesser extent by stray fields. To our knowledge, logic devices based on exchange-coupled multilayers in which the layers couple vertically with each other have not been explored, although applications of EC and the associated physics can be seen in magnetic data storage, sensing, and high-speed applications. One of the notable experimental demonstrations of information processing using exchange-coupled multilayers is the magnetic shift register in perpendicularly magnetized exchange-coupled superlattice structures [57].

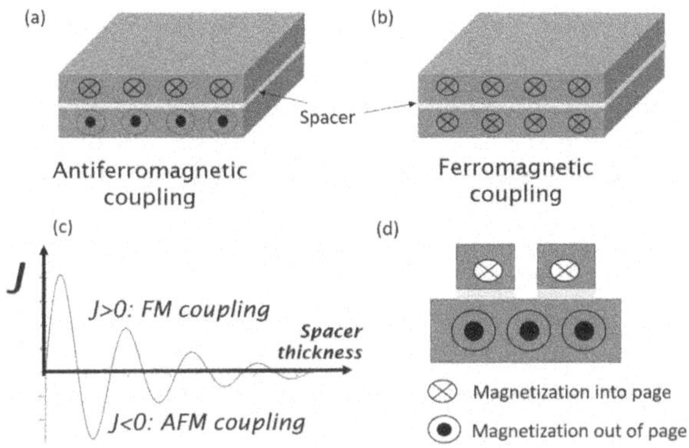

Figure 9.32 Interlayer exchange-coupled (IEC) magnetic multilayers; (a) antiferromagnetic ground state, (b) ferromagnetic ground state, (c) oscillatory variation of interlayer exchange coupling with spacer thickness, and (d) the proposed system (images not to scale).

Figures 9.32a and 9.32b show the two possible ground states of such a system. In 9.32a, the two magnetic layers (top and bottom) are pointing in antiparallel directions, and in 9.32b, they are pointing in parallel directions. In between the magnetic layers, there is a nonmagnetic spacer layer (shown yellow in the pictures). By varying the thickness of this layer, one can achieve either antiparallel or parallel alignment between the magnetic layers. The IEC strength between the magnetic layers is quantified by J, known as IEC strength or IEC constant. J varies in an oscillatory fashion with variation of the spacer thickness, which is schematically shown in 9.32c. In the figure, the magnitude of J represents the IEC strength and the nature of the coupling is determined by the sign of J. Note that $J > 0$ for FM coupling and $J < 0$ for AFM coupling. With a very slight change in the spacer thickness, the nature of the EC oscillates between AFM and FM [43]. Therefore, we expect that exquisitely fine control of layer thicknesses is needed to demonstrate this coupling scheme, although that has already been done in practice [58].

Given $J < 0$, the ground state of this system would support antiparallel alignment between the two layers. In the proposed scheme (shown in Fig. 9.32d), the top layer is patterned into two closely spaced single-domain nanomagnets, each individually antiferromagnetically exchange-coupled to the common bottom FM layer. If realized, these two nanomagnets will be brought into strong lateral interaction, which will force them to point along the same direction (against their dipole interactions), while the bottom FM layer will point along the direction opposite to both nanomagnets.

In this work, it has been shown that this lateral coupling between the two nanomagnets, mediated by the bottom layer, is much stronger than the typical dipole/stray field interaction, if they are placed in a conventional side-by-side configuration in the absence of a bottom layer. We have demonstrated the effect experimentally [44] and confirmed it with the aid of micromagnetic simulations. Exploiting this scheme as a building block, logic blocks that are more complex can be realized that are thermally robust and more immune to process variations. In addition to these very desirable functionalities, the system is inherently suitable for magnetoelectric interfaces, as its spin-valve-like building block exhibits giant magnetoresistance (GMR).

This type of ferromagnet/spacer/ferromagnet multilayer is also found in a magnetoresistive random access memory (MRAM) device [59]. In an MRAM device, one ferromagnet layer is called the "free layer" and the other the "fixed layer." Among several ways of pinning the magnetization of the fixed layer [60], one way is to strongly antiferromagnetically exchange-couple it with another FM layer, called the "pinning layer." In this case, the fixed layer/spacer/pinning layer block is also called a "synthetic antiferromagnet (SAF)" system [61]. This kind of system is also employed in perpendicular recording media [62]. The proposed coupling scheme may also find applications in non-Boolean computing, where arrays of interacting magnetic nanoelements can be used to find solutions to complex energy minimization problems (e.g., feature extraction and recognition) [63].

A schematic of the proposed coupling scheme is shown in Fig. 9.33a, along with a cross section of the experimentally used stack (Fig. 9.33b). Here, lateral FM coupling between the two rectangular nanomagnets is enforced by the vertical AFM EC mediated by the FM film under both of them. A top view of the device is also shown in Fig. 9.33c.

Interlayer EC, or RKKY (named after Ruderman, Kittel, Kasuya, and Yosida) coupling, between the two CoFe layers separated by a

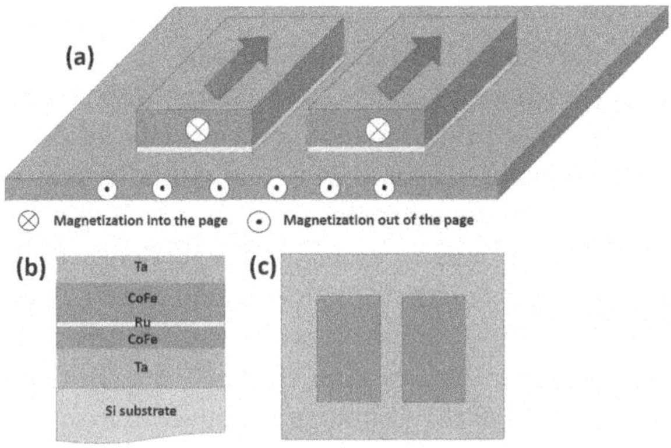

Figure 9.33 Lateral coupling between two nanomagnets antiferromagnetically exchange-coupled to a bottom layer; (a) coupling scheme, (b) implemented stack, and (c) top view of the device [44].

Ru spacer layer can be either AFM or FM in nature, depending on the thickness of the Ru layer [43]. The ground state of the system occurs when both magnets are antiparallel to the underlying layer and all the magnets are in a single-domain state. Effectively, the EC ferromagnetically couples the two laterally adjacent magnets relative to each other. This coupling mechanism works against the magnetic dipole coupling, which would prefer an antiparallel alignment of the nanomagnets [14]. The magnetic ground state and energy landscape of this system are jointly determined by shape anisotropy, dipole coupling, and EC. This scheme has some similarity to the spin transfer torque–based majority logic gate (STMG) proposed by Nikonov et al. [64], which exploits the interaction between magnetic nanopillars through a free layer underneath these pillars.

9.6.2 Interlayer Exchange Coupling Effect in SAF

Exchange interactions in a material define how neighboring spins interact over a short range, and either line up ferromagnetically (in FM materials) or antiferromagnetically. It is a quantum mechanical phenomenon that dictates the nature of short-range interaction between neighboring spins depending on how their electronic wave functions overlap. On the basis of this interaction, there is an energy difference between the parallel and the antiparallel states of two neighboring spins. This energy difference is called the exchange energy [65]. Depending on the sign of the exchange energy J, two neighboring spins will align either ferromagnetically ($J > 0$) or antiferromagnetically ($J < 0$). In a FM material, the spins align in parallel due to exchange interaction over a distance called the exchange length [66]. Ruderman, Kittel, Kasuya, and Yosida showed that this interaction can be mediated by nonmagnetic metal atoms between the two spin sites [67–69]. In this case, the conduction electrons of the nonmagnetic impurity carry oscillatory spin polarization. This interaction depends on the band structure of the nonmagnetic metal and on the distance over which the interaction is mediated [70].

This concept can be extended to a system like SAF that consists of two FM layers separated by a nonmagnetic spacer layer [71]. One can effectively tailor the nature and magnitude of the interactions between these two FM layers by varying the spacer thickness. The two FM layers exhibit two stable ground states depending on the spacer thickness, as shown above. By patterning the top layer into two closely spaced nanomagnets, one can exploit this interlayer coupling in order to laterally couple the nanomagnets [71].

In 1986, Majkrzak et al. [72] reported AFM alignment between two Gd layers in a Gd/Y/Gd superlattice structure. The spacer was 10 monolayers of yttrium (Y) favoring AFM EC between the two gadolinium (Gd) layers. Almost simultaneously, a group led by Grunberg [73] discovered AFM oscillatory EC in a Fe/Cr/Fe superlattice. Further investigation followed that varied the thickness and material of the spacer in an attempt to observe the nature and strength of this coupling. Heinrich et al. [74] showed that the interlayer coupling changes from FM to AFM between two Fe layers as the spacer (Cu) thickness is varied from 9 to 10 monolayers in an epitaxially grown Fe/Cu/Fe system. The most complete work that fully characterized an interlayer exchange-coupled multilayer system was done by Parkin et al. [43, 58, 75].

9.6.3 Micromagnetic Energy Calculations of the Proposed Coupling Scheme

This section provides the details of the magnetization evolution and energy landscape of the system shown in Fig. 9.33. Micromagnetic simulations using the OOMMF code were used for this pur'pose [50]. The "two-surface-exchange" module in OOMMF was used for these simulations at $T = 0$ K, setting the AFM interlayer EC constant to $J = -1 \times 10^{-3}$ J/m^2. This is the strongest AFM EC we routinely achieve between the magnetic layers in our experiment. The following parameters were chosen for simulation: mesh size, $5 \times 5 \times 1$ nm^3; saturation magnetization ($\mathbf{M}_s$) of CoFe, 14×10^5 A/m; exchange constant (A), 30×10^{-12} J/m^2; and damping constant (α) $= 0.5$ (at 0 K). Figure 9.34 illustrates the magnetization evolution of the simulated system with and without the presence of EC. The simulated system is schematically shown in the inset. The top two nanomagnets are each $120 \times 60 \times 10$ nm^3, and the bottom CoFe film under the magnets is 3 nm thick. Spacing between the two nanomagnets was chosen to be 20 nm.

In simulation, the short axis of the magnets is chosen to be along the X direction and the long axis is chosen to be along the Y direction. Magnetization of the left nanomagnet is held "upward," along the Y direction. An external magnetic field is applied along the X direction at the beginning of the simulation ($\mathbf{B}_x = 1$ T) and is gradually reduced to zero ($\mathbf{B}_x = 0$) in 2 mT steps. The direction of the field sweep (right to left) is shown by the green arrow in the figure. As $\mathbf{B}_x$ varies, $\mathbf{M}_y$, the Y component of magnetization of the right nanomagnet also varies. The blue curve represents the magnetization when there is strong EC between the layers ($J = -1 \times 10^{-3}$ J/m^2) and the black curve represents the evolution when there is no EC ($J = 0$). Magnetizations of the nanomagnets for both cases at $\mathbf{B}_x = 1$ T are shown schematically on the right in the figure and from OOMMF simulations. For both $J = -1 \times 10^{-3}$ J/m^2 and $J = 0$, the right nanomagnet is biased along the X direction, given the left nanomagnet is always biased to the Y direction. As $\mathbf{B}_x$ is reduced from 1 T, for $J = 0$, $\mathbf{M}_y$ is still biased along the X direction up to $\mathbf{B}_x \approx 300$ mT. In this case, only dipole coupling exists between the two nanomagnets, and beyond $\mathbf{B}_x \approx 300$ mT, the right nanomagnet

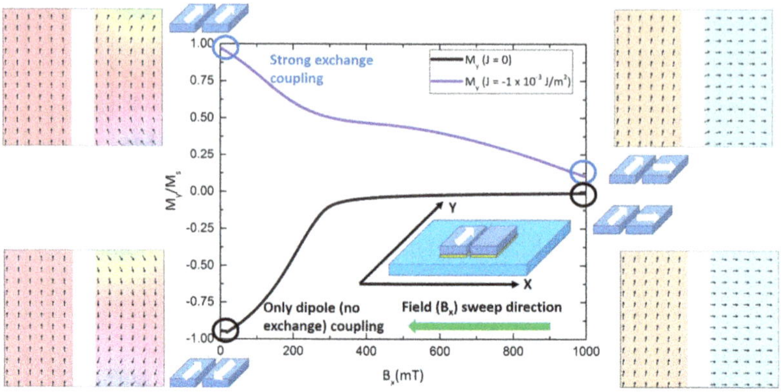

Figure 9.34 Magnetization evolution of the right nanomagnet from a hard-axis-biased state ($\mathbf{B}_x = 1$ T) to the relaxed state ($\mathbf{B}_x = 0$); $J = -1 \times 10^{-3}$ J/m^2 favors exchange coupling, and $J = 0$ favors dipole coupling. Inset shows the simulated geometry. Magnetizations of the nanomagnets at $\mathbf{B}_x = 1$ T (right) and $\mathbf{B}_x = 0$ (left) are shown for both $J = 0$ (bottom) and $J = -1 \times 10^{-3}$ J/m^2 (top).

exhibits a sharp transition of $\mathbf{M}_y$ from X to Y direction. Eventually, at $\mathbf{B}_x = 0$ T, the dipole coupling sets $\mathbf{M}_y = -\mathbf{M}_S$. This situation is indicated schematically and from OOMMF simulations in the bottom left side of the figure. However, for $J = -1\times 10^{-3}$ J/m^2, the evolution of M_y is more gradual as B_x is reduced. The presence of strong EC tends to keep the magnetizations of the top and bottom layers align antiferromagnetically, and eventually it overwhelms the dipole coupling by setting $\mathbf{M}_y = +\mathbf{M}_S$. In this case, the right nanomagnet turns upward, following the left (driver) magnet. Magnetizations of the nanomagnets are illustrated in the top left in the figure. The EC prefers a more gradual rotation of the magnetization, whereas dipole coupling prefers a nucleation/growth-type abrupt switching of magnetization [76].

The curves of Fig. 9.35 illustrate the energy landscape of the system during the magnetization process in Fig. 9.34. Typically, conventional dipole-interaction-based systems, such as those in Refs. [20, 23, 54], rely only on the lateral dipole coupling between the individual neighboring nanomagnet islands. OOMMF simulations were performed in order to estimate the dipole coupling energy between two adjacent nanomagnets in such a system where there is

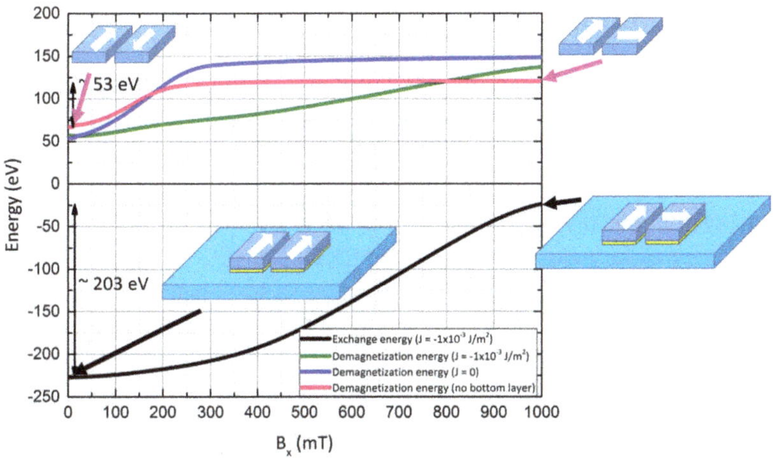

Figure 9.35 Energetics of the coupled system showing the dominance of the exchange energy term. The black curve corresponds to the change in the exchange energy. The change in the exchange energy is significantly larger than the dipole (demagnetization) terms in the system.

no FM layer underneath the magnets. The pink curve corresponds to the change in the demagnetization energy as the right nanomagnet is relaxed from $\mathbf{B}_x = 1$ T to $\mathbf{B}_x = 0$ T. Magnetization directions of the nanomagnets are schematically shown at $\mathbf{B}_x = 1$ T and $\mathbf{B}_x = 0$ T. The demagnetization energy is decreased by $E_{\text{demag}} = 53$ eV as the system goes from the high-energy state at $\mathbf{B}_x = 1$ T to the ground state at $\mathbf{B}_x = 0$, indicated by a black arrow in the figure. Then the system was modified by adding a nonmagnetic spacer layer (1 nm) and a FM layer (3 nm) under the nanomagnets. The blue curve in Fig. 9.35 corresponds to the change in the demagnetization energy in this system. Although slightly modified from the pink curve due to presence of the underlying film, the shape of the blue curve is very similar to the pink curve, since an additional flat film does not contribute to a significant change in the demagnetization energy. The black curve shows the change in the exchange energy of the right nanomagnet when the bottom layer is antiferromagnetically exchange-coupled to the nanomagnets. In this simulation, the interlayer EC constant J was set to -1×10^{-3} J/m^2 keeping everything else the same. Magnetizations of the nanomagnets in this system are also schematically shown at $\mathbf{B}_x =$ 1 T and $\mathbf{B}_x = 0$. The green curve corresponds to the change in demagnetization energy for this exchange-coupled system. The EC energy, $E_{\text{ex}} = 203$ eV, indicated by another black arrow in the figure, is almost four times larger than the demagnetization energy in a conventional dipole-coupled system (shown by the pink curve) as the system settles into its ground state to $\mathbf{B}_x = 0$ T from $\mathbf{B}_x = 1$ T, which is clear evidence of the dominance of EC energy over the dipole-coupling energy in this system.

This enhanced dot-dot coupling would be significantly beneficial to many magnetologic devices, where correct device operations are limited by errors occurring during switching. Since the probability for a nanomagnet to switch is an exponential function of the ratio of the coupling energy to the thermal noise (kT), such enhancement in coupling energy promises information transfer between magnets even in the presence of thermal fluctuations or process variations [15].

9.6.4 Sample Fabrication

The system simulated above was fabricated using sputtering, e-beam lithography, and dry etching techniques. The details of the fabrication process are outlined below.

9.6.4.1 Sputtering

The exchange-coupled multilayer structure is a CoFe (bottom layer)/Ru/CoFe (top layer) multilayer. The layers were sputtered on a silicon substrate that was thoroughly cleaned in acetone and isopropanol (IPA), and baked on a hot plate at 90°C for 10 min. before deposition. After being cleaned, the sample was inserted into an ultra-high-vacuum (UHV) magnetron sputtering system. Since the stack contains only metallic layers, DC sputtering was performed in the presence of Ar at room temperature. The sputtering chamber had a nominal base pressure of $\sim 5 \times 10^{-8}$ mbar, and the Ar pressure was nominally 10–20 μ bar, with a flow rate of 4 sccm. A 10 nm Ta layer was sputtered under the bottom CoFe layer as a seed layer, directly on the Si substrate. A 5 nm Ta capping layer was sputtered as a protection layer on top of the top CoFe layer in order to prevent oxidation or any unwanted layer formation [77]. The substrate was rotated (at 12 rpm) in order to ensure uniform film deposition [78].

A systematic characterization of the multilayers was done by sputtering the stacks with Ru of various thicknesses (0.4 nm, 0.6 nm, 0.8 nm, 1.0 nm, and 1.2 nm) and measuring the hysteresis loops using a vibrating sample magnetometer (VSM) by Microsense. The strongest AFM EC was obtained for the stack with 0.8 nm Ru [43], and from the hysteresis loop, a value of $J = -10^{-3}$ J/m^2 was extracted, which was used in the simulations presented above.

9.6.4.2 Lithography and deposition of the hard mask

Figure 9.36 shows the fabrication steps for the structure shown in Fig. 9.33. The structures in the left column correspond to the patterned nanomagnets, and the structures in the right column correspond to an unpatterned control sample, with both sputtered in the same batch at the same time. The top layer needs to be patterned into nanomagnets keeping the bottom layer intact. Patterning the top nanomagnets involves a very precise etch stop detection. We used the control sample in order to determine completion of etching by magnetic characterization (to be shown later). The top layer was patterned into an array of nanomagnet pairs, where the nanomagnets were separated by approximately 15–20 nm from each other in each pair, and the pairs themselves were placed approximately 200 nm apart. The footprint of each nanomagnet was chosen to be 200×100 nm^2. This larger footprint ensures a strong magnetic signal from the nanomagnets, which can be detected by MFM. The thicknesses of the layers were kept the same as those used in the simulations. However, for this geometry, simulations show that the interlayer EC energy is larger than the dipole coupling energy by a factor of 10.

The nanomagnet pairs were exposed by a 100 kV Vistec (now Raith) EBPG 5200/5000 electron beam lithography system on a 100-nm-thick PMMA C7 resist. The resist was spin-coated on the sample, followed by hot-plate baking at 180°C for 3 min. The e-beam exposure was done using a 200 pA beam current with an exposure dose of 540–560 μC/cm^2. After the exposure, resist development

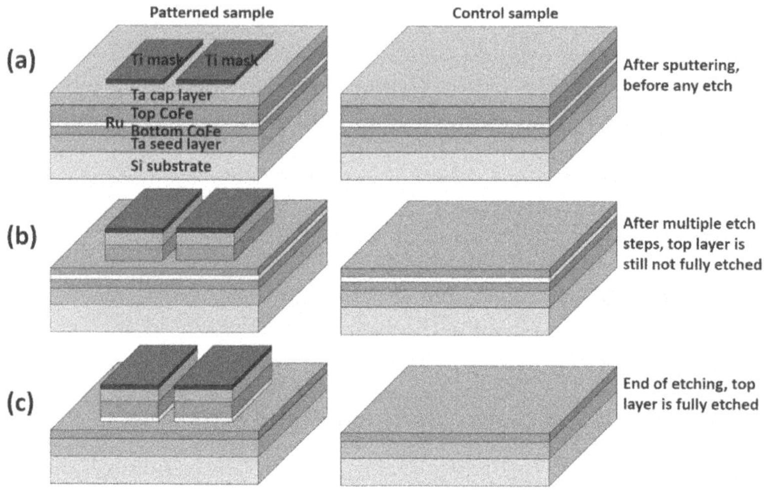

Figure 9.36 Process flow for sample fabrication; (a) after patterning and lifting off Ti, (b) after several etch steps, and (c) end of etching.

was performed in a solution of methyl-isobutyl ketone (MIBK), IPA, and methyl-ethyl ketone (MEK) for 32 s at room temperature [79].

Using electron-beam evaporation, an approximately 12 nm Ti layer was evaporated on the sample at a deposition rate of 1–2 Å/s. Then the deposited Ti layer was lifted off in dichloromethane. Ultrasonic agitation was used for about 10 min. in order to dislodge residual metal flakes, protrusions, and metal particles stuck between the patterns. Figure 9.36a depicts the schematic representation of the sample at this stage.

9.6.4.3 Etching in argon

Both control sample and patterned sample were etched in argon at the same time in several steps at 10 sccm flow rate, 6.5×10^{-3} mbar pressure, and 330 W plasma power (570 $\approx$ 580 V DC bias voltage). In the patterned sample, Ti protected the nanomagnets underneath, while the unprotected regions of the top layer were etched away. Each etch step took approximately 1–2 minutes. Figures 9.36b and 9.36c are schematics of the patterned sample and control sample after several etch steps and at the end of etching,

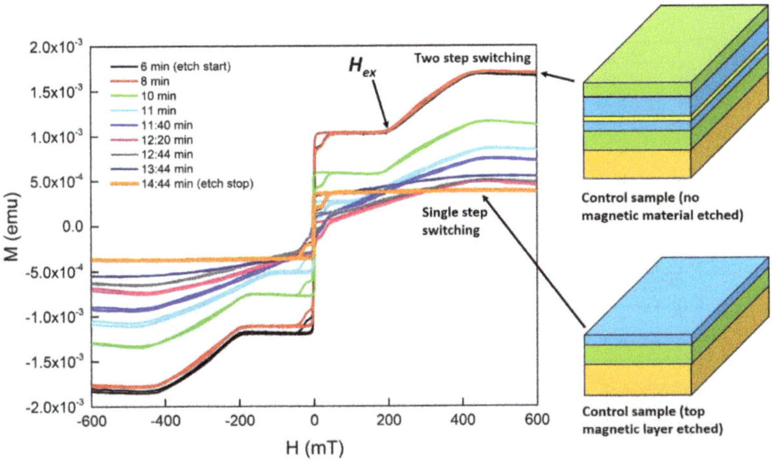

Figure 9.37 SEM micrograph of a section of the dot array. Four pairs are schematically shown, indicating nanomagnet dimensions, interpair separation, and separation between the magnets in a pair (image not to scale). Presence of the bottom film is illustrated by the light-blue region under the pairs.

respectively. A scanning electron micrograph (SEM) of an array of etched nanomagnets is shown in Fig. 9.37 (corresponding to the step in Fig. 9.36c). A cartoon illustration of four nanomagnet-pairs shows the dimensions of the nanomagnets, separation between the pairs, and spacing between nanomagnets within a pair (image not to scale).

To avoid overetching of the bottom layer, after every etch step, the magnetic moment of the control sample was measured by a VSM under an in-plane magnetic field. A detailed description of this VSM system can be found in Ref. [80].

Figure 9.38 is a collection of hysteresis loops obtained from the control sample after every etch step. The black (at 6 min.) and red (at 8 min.) curves correspond to the hysteresis loops after the first two etch steps, when the nonmagnetic Ta protecting layer was being etched but no magnetic layer was etched yet. From these hysteresis loops, the exchange field $\mathbf{H}_{ex}$ was found to be around 200 mT (indicated by a black arrow) and the saturation magnetization of CoFe, $\mathbf{M}_s = 14 \times 10^5$ A/m. These values were used to estimate

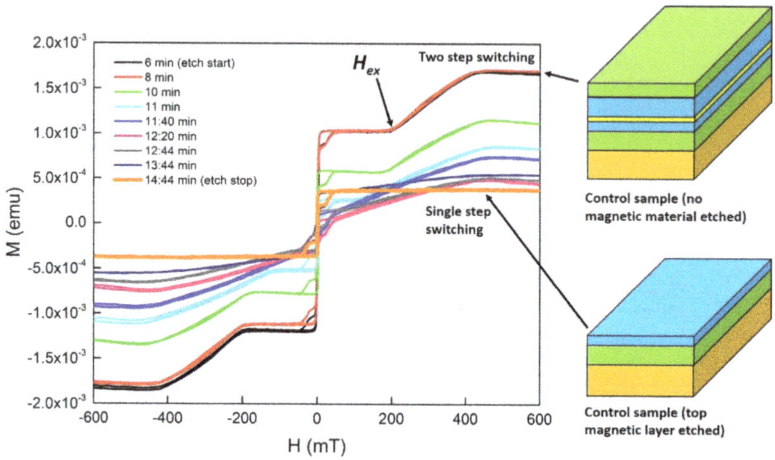

Figure 9.38 **M-H** loops of the control sample after each etch step; etching is stopped when the top layer is completely gone in the control sample (orange curve at 14:44 min. in the plot). Schematics of the control sample before etching and at the end of etching are also shown in inset.

the AFM EC strength as $J \approx \mathbf{M}_S t\mathbf{H}_{ex} \approx -10^{-3}$ J/m^2, as used in the simulations discussed above. Here, t is the thickness of the bottom layer. The exchange field is roughly the field at which the layers are perfectly antiparallel to each other and the interlayer EC completely overcomes the applied field [81]. The extracted value of the EC constant (J) is comparable to the published values for similar systems [77, 78, 81–84]. This value was used in the micromagnetic simulations shown in Section 9.6.3.

The two-step switching in the **M-H** loops clearly show the signature of strong AFM EC between the CoFe layers (shown in schematic). At each switching step, the two CoFe layers are oriented antiparallel to each other, exhibiting a reduction in the net magnetic moment of the whole stack. After etching for more than 10 min., as the top CoFe layer starts to get etched away, a gradual reduction in the saturation magnetic moment in the successive etch steps can be seen. Finally, when the top layer is completely removed, and only the bottom CoFe layer remains, a FM **M-H** loop with only a single switching step can be seen (orange curve, at 14:44 min.). This is expected because the remaining bottom layer behaves as a single FM

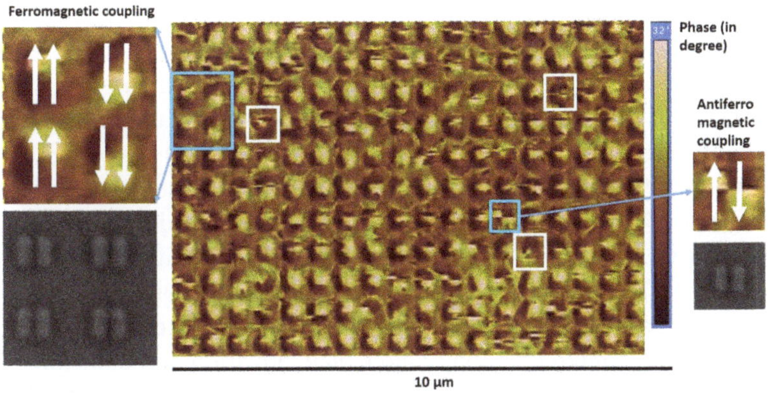

Figure 9.39 MFM image from a representative area of the sample, used for collecting switching statistics. Ferromagnetic and antiferromagnetic states are magnified in the insets. Pairs within the white boxes could not be resolved. The great majority of the magnets are ferromagnetically coupled.

film without having any coupling with any other layer (shown in the schematic). Therefore, at the end of etching, in the patterned sample, the top layer outside the Ti mask was gone, but the magnets under the mask and the bottom layer outside the mask remained intact. Once the sample was fabricated, the next step was to characterize it using the MFM.

9.6.5 MFM Measurements

Before the MFM measurement, the sample was magnetized and demagnetized in a rotating magnetic field by being mechanically rotated at 1800 rpm (30 Hz) within an in-plane magnetic field. The magnetic field was provided by a VSM electromagnet that was ramped up from 0 to 500 mT and then down to 0 mT over a period of 1 min. This rotating magnetic field essentially allows the magnets to settle in their ground state after the magnetization and demagnetization cycles [55]. Next, the magnetic states of the nanomagnets were imaged using a Veeco (now Bruker) Nanoscope V MFM. The MFM data from a representative area of the sample is shown in Fig. 9.39.

Magnetic force microscopy relies on the stray field emanating from a magnetic sample in order to correctly determine the magnetic polarity of the domains [8]. Typically, it is difficult to obtain a strong MFM signal from thin single-domain nanomagnets [19]. To obtain a strong MFM signal from the thin (~10 nm) nanomagnets used in the experiment, the MFM scans were performed very slowly, at a rate of 0.5 Hz (about 2 s per line) using a standard Bruker MESP probe [85]. A detailed description of this MFM system will be found in Ref. [19]. Switching statistics of the nanomagnet ensemble were collected from the scans. The MFM data in Fig. 9.39 show the ground state of the nanomagnet pairs after magnetization and demagnetization. Most of the nanomagnet pairs are ordered ferromagnetically, with a few ordered antiferromagnetically and some that could not be resolved. Within the blue boxes shown in the figure, some examples of ferromagnetically and antiferromagnetically ordered pairs are given. Snapshots from SEM micrographs are also shown. The white arrows are used to indicate the magnetization directions of the nanomagnets. Antiferromagnetically ordered magnets are randomly scattered on the sample. Examples of some unresolved nanomagnet pairs are shown within the white boxes in the figure. Possible origins of antiferromagnetically ordered pairs and unresolved pairs will be discussed in more detail in Section 9.6.6.

The switching statistics are tabulated in Table 9.1. There are 198 pairs of nanomagnets in total in Fig. 9.39. Of these, 161 pairs point along the same direction (both up or both down, shown in the zoomed in region of Fig. 9.39). Only 12 pairs point along antiparallel directions. A usably clean signal was not detected from the rest of the 25 pairs (three of such pairs are shown within the white boxes in Fig. 9.39). Excluding these 25 pairs from the total 198 pairs, almost 93% of the pairs are ferromagnetically coupled and only 7% are antiferromagnetically coupled. Including these

Table 9.1 Switching statistics collected from the MFM image in Fig. 9.39

	Number of pairs	Percentage of pairs (%)
Parallel state	161	81
Antiparallel state	12	6
Unresolved state	25	13

25 nonidentifiable pairs in the total count, 81% of the pairs are ferromagnetically coupled, 6% are antiferromagnetically coupled, and 13% are nonidentifiable (shown in the table).

9.6.6 Interpretation of the Antiparallel States and Undefined States in the MFM Data

Although the majority of the pairs were found in the parallel state in the MFM image, we try to understand the origin of the unidentifiable states and the antiparallel states. Process variations contribute to some of the unresolved states in the MFM image. During the lift-off and etch steps, some of the fabricated nanomagnets come out differently from the intended shape laid out in the design (e.g., rounded corners, bulging, missing corners, and bridging between magnets) [25–27, 42]. All these effects could lead to states that cannot be identified or resolved by MFM. For example, nanomagnets with weak shape anisotropy (e.g., caused by bulging in the middle) will end up in a vortex state and show no MFM contrast [27].

The AFM states, on the other hand, could be caused by IEC strength variation across the multilayers. Local nonuniformities in the EC strength can be introduced during the layer deposition or successive process steps. Especially defects, surface roughness, etc., could cause localized variation of interlayer coupling strength on the film [86, 87]. Since the films were deposited on a large area, there could also be local thickness variation of the layers. In particular, thickness variation of the spacer layer could contribute to a reduction in the IEC strength in certain regions of the film, because as discussed earlier, the coupling strength is a strong function of the spacer thickness [43]. In this way, the vertical EC between the layers could be lost under a few nanomagnet pairs and their ground state could be dictated by the lateral dipole coupling alone.

To investigate such possibilities, additional OOMMF simulations were performed for the nanomagnets with the geometry used in the experiment (200×100 nm^2) in the presence of random thermal fluctuations varying the IEC strength between the layers [15, 26]. Since the experiments were performed under a rotating magnetic field, we incorporated the real, oscillating demagnetizing field in the simulations. Figure 9.40 shows the time dependence of both

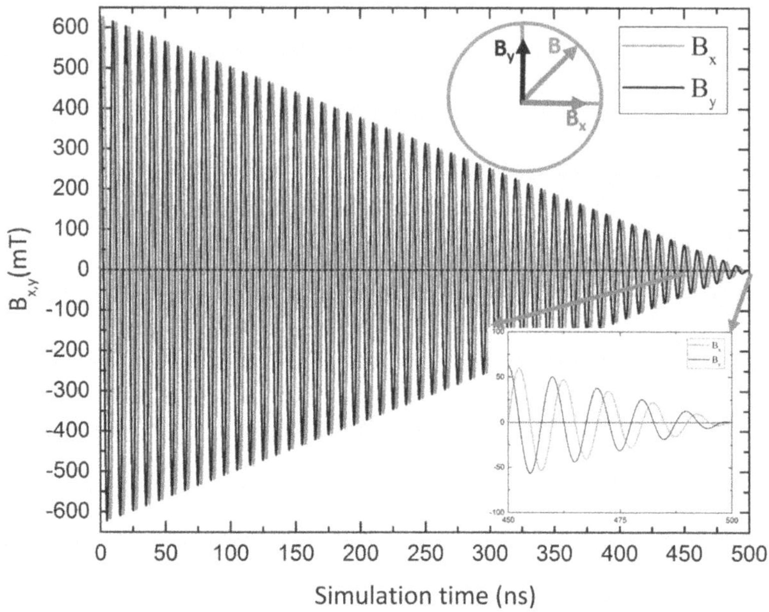

Figure 9.40 Time evolution of the applied rotating field; X and Y components of the applied field ($\mathbf{B}_x$, $\mathbf{B}_y$) reduce from a large amplitude to zero. A plot of the rotating field along with its components are shown in the inset.

components ($\mathbf{B}_x$ and $\mathbf{B}_y$) of the oscillating field with a 500-ns-long demagnetizing process and a 10 ns time period. To perform these computationally heavy simulations within a realistic time frame the simulated time scales are much faster than the experimental ones.

In these simulations, we systematically investigated the effect of EC strength on the ground state of the system in the presence of thermal fluctuations. Four different cases were chosen: (i) $J = -10^{-3}$ J/m^2, (ii) $J = -5 \times 10^{-4}$ J/m^2, (iii) $J = -10^{-4}$ J/m^2, and (iv) $J = 0$. $J = -10^{-3}$ J/m^2 and $J = -5 \times 10^{-4}$ J/m^2 correspond to strong EC. Weak EC is represented by $J = -10^{-4}$ J/m^2, and $J = 0$ corresponds to no EC. A large number of simulations were run using different seeds for the random thermal fields [88]. Figure 9.41 shows the time evolution of the system for different EC strengths. Three coupling strengths were chosen for this plot: strong coupling ($J = -10^{-3}$ J/m^2) in Fig. 9.41a, weak coupling ($J = -10^{-4}$ J/m^2)

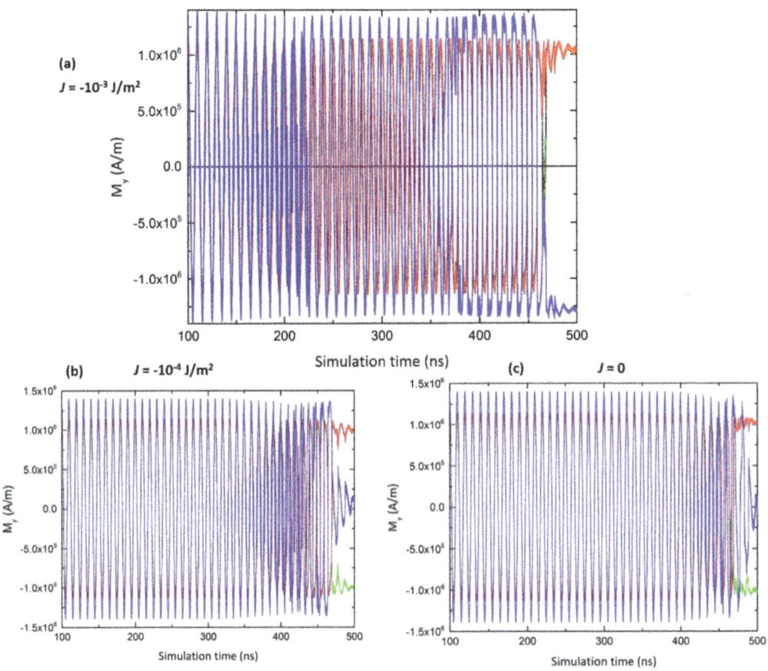

Figure 9.41 Time evolution of the Y component of magnetization ($\mathbf{M}_y$) of the bottom layer (blue curve) and nanomagnets (red and green curves) (a) strong exchange coupling ($J = -10^{-3}$ J/m^2). At the end of simulation, red and green curves overlap, showing the nanomagnets are settled in parallel direction, (b) weak exchange coupling ($J = -10^{-4}$ J/m^2) and (c) no exchange coupling ($J = 0$). In both (b) and (c), the nanomagnets end up pointing in antiparallel directions, shown by the red and green curves.

in Fig. 9.41b, and no coupling ($J = 0$) in Fig. 9.41c. Time evolution of the nanomagnets are shown by the red and green curves, and the blue curves represent time evolution of the bottom layer from one simulation run.

In Fig. 9.41a, initially, the magnetizations of the nanomagnets and bottom layer rotate in phase with each other following **B**. This is indicated by a complete overlap of the curves until ~200 ns. After that, the phase difference between magnetization of the bottom layer and nanomagnets slowly starts increasing. With the external field, **B**, decreasing, the effect of EC starts to become

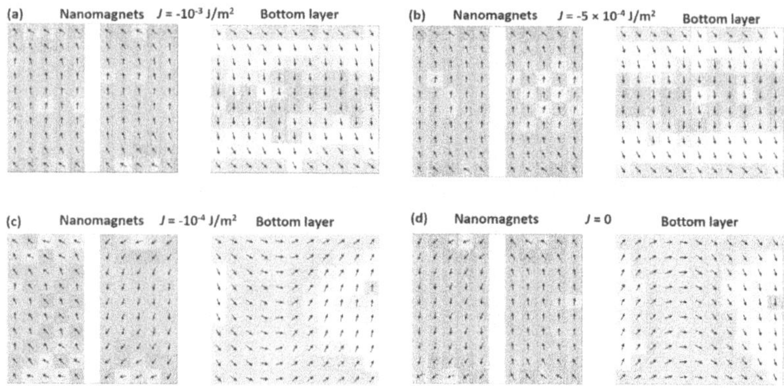

Figure 9.42 Domain configurations of the two nanomagnets (side by side) and the bottom layer: (a) $J = -10^{-3}$ J/m^2, (b) $J = -5\times 10^{-4}$ J/m^2, (c) $J = -10^{-4}$ J/m^2, and (d) $J = 0$ J/m^2. (a) and (b) correspond to strong exchange coupling, (c) corresponds to weak coupling, and (d) corresponds to no exchange coupling.

stronger, slowly driving the magnetizations toward their ground state. As the simulation time increases and the applied field further decreases, magnetizations of the bottom layer and nanomagnets rotate completely out of, phase. Eventually, at the end of the simulation ($\mathbf{B} = 0$ at 500 ns), the two nanomagnets find their ground state pointing to same direction, opposite to the direction of the bottom layer. Similar behavior is observed for $J = -5 \times 10^{-4}$ J/m^2, when EC strength is reduced by a factor of 2. In this case, the EC is still strong enough to drive the system to the same ground state (not shown in the figure). Figure 9.41b depicts the situation when the EC is much weaker (reduced by a factor of 10, from $J = -10^{-3}$ J/m^2 to $J = -10^{-4}$ J/m^2). In this situation, the nanomagnets may end up randomly in parallel or antiparallel configurations. An example of antiparallel configuration is shown in the figure. Unlike Fig. 9.41a, the magnetizations follow $\mathbf{B}$ for a much longer time (up to more than ~350 ns), eventually ending up in an antiparallel state as the EC is too weak to have any noticeable effect. If there is no EC between the layers ($J = 0$), the nanomagnets always end up in antiparallel configuration, as shown in Fig. 9.41c.

Some of the simulated domain patterns in the ground state are shown in Fig. 9.42 for strong, weak, and no AFM EC. Figures 9.42a and 9.42b correspond to the cases when EC is strong ($J = -10^{-3}$ J/m^2 and $J = -5 \times 10^{-4}$ J/m^2). In such cases, the nanomagnets always prefer parallel alignment between them. The bottom layer is settled perfectly along the opposite direction. For weak EC ($J = -10^{-4}$ J/m^2) (Fig. 9.42c), the magnets may prefer FM or AFM alignment. An example of antiparallel domain configuration is shown. If there is no EC between the layers ($J = 0$), and only lateral dipole coupling exists between the nanomagnets, the nanomagnets end up in an antiparallel state dictated by the dipole coupling alone, as shown in Fig. 9.42d. For weak or no EC, the bottom layer finds a ground state with complex domain patterns that can be predicted by the micromagnetic simulations.

9.6.7 Simulation of a Majority Gate Structure Using the Underlayer-Mediated Exchange Coupling Scheme

A three-input majority gate structure has been simulated in OOMMF, extending the underlayer-mediated EC scheme in Fig. 9.33. In the structure, the nanomagnets are on top of the magnetic underlayer separated by the spacer. The footprint of each nanomagnet was chosen to be $120 \times 60 \times 5$ nm^3. The intermagnet distance was chosen to be 20 nm. The EC strength was set to be $J = -10^{-3}$ J/m^2, which is the same as it was in demonstrated coupling scheme. All other parameters were kept the same as those used in Section 9.6.3 ($T = 0$ K and $\alpha = 0.5$).

The input nanomagnets are designated as A, B, and C. The nanomagnet in the middle and the output nanomagnet to the right of it are denoted as M and Out, respectively. The magnetizations of the bistable inputs were set in OOMMF in order to code them with 1 or 0. The nanomagnets M and Out were clocked by applying a strong magnetic field of 1 T along their hard axes, and the field was applied to the bottom layer as well. Then the field was gradually reduced to zero so that the clocked nanomagnets and the underlayer could settle into their ground states.

Figure 9.43 (left and right) shows the magnetization states of the majority gate after relaxation for four input combinations [89]. In the left column of Fig. 9.43, logical AND operation has been shown. In this case, input A was set to 0 and the other two inputs were

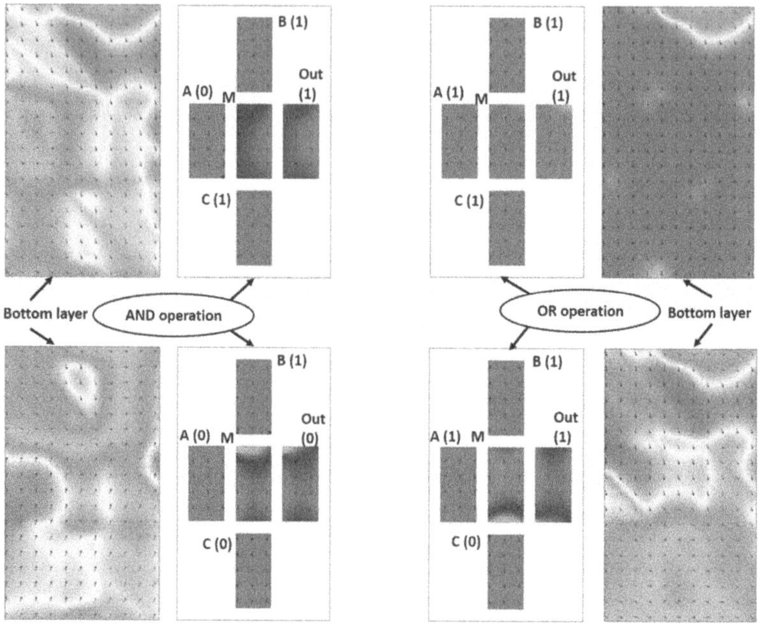

Figure 9.43 A majority gate simulated using the exchange-coupled multi-layers: (left) AND operation and (right) OR operation [89].

set to two different combinations (B = 1, C = 1 and B = 1, C = 0). As the nanomagnets relax to their ground state, the magnetizations of M and Out end up at 1 and 0, respectively, which are the logical outputs of an AND operation between B and C, given the input A is set at 0. In the right column of Fig. 9.43, an OR operation between B and C is shown, where the input A is now set at 1, keeping the other input combinations the same as in the AND operation. Since M and Out both are exchange-coupled to the bottom layer, upon relaxation, the magnetization state of M is essentially buffered to the output magnet. This behavior is different from a dipole-field-coupled NML majority gate, such as the SNML gates discussed in this chapter, in the sense that in a dipole-coupled majority gate, the magnetization of the output magnet will be in an inverted state to that of the nanomagnet in the middle (M). This is due to the dominance of EC over the dipole coupling between them. The domain patterns in the bottom layer for all the input combinations are also shown in the figure.

These preliminary simulation results clearly indicate that complex NML computations can be performed using exchange-coupled nanomagnets.

Next, we discuss two important implications of the proposed coupling scheme, (i) scaling and (ii) errors in a line of nanomagnets. Nanomagnets become more unstable due to thermal fluctuations as they are made smaller. EC offers much stronger interactions between nanomagnets, which keeps them stable even at a much smaller size scale. Energy calculations have already been shown for 120×60 nm^2 nanomagnets. In this section, we show some simulations in which the nanomagnet footprint has been reduced while keeping the aspect ratio, thickness of the magnets, and magnet-to-magnet separation the same. The EC constant J was set to -10^{-3} J/m^2 with a 1 nm spacer. Nanomagnets with four different footprints are simulated: 80×40 nm^2, 60×30 nm^2, 40×20 nm^2, and 30×15 nm^2. Figure 9.44 shows the simulated magnetization evolution and energy landscape for these cases [89].

For all the simulations, the two nanomagnets eventually end up pointing in the same direction upon relaxation. Moreover, the exchange-coupling energy is at least twice as strong as the dipole coupling energy in all the cases. For example, in Fig. 9.44d, for sub-30-nm nanomagnets (footprint: 30×15 nm^2), the change in exchange energy is $\sim$14 eV, which is twice the change in demagnetization energy ($\sim$7 eV). The simulations establish the fact that it is possible to have strong magnetic interaction even if the nanomagnet size is pushed close to the superparamagnetic limit.

Another interesting aspect of this system is that demagnetization energy scales with the volume of the magnets, while EC energy depends on the area of the interacting surfaces. Therefore, it is possible to further scale down the magnets vertically without compromising the exchange energy, whereas the strength of dipole interaction will continue to decrease.

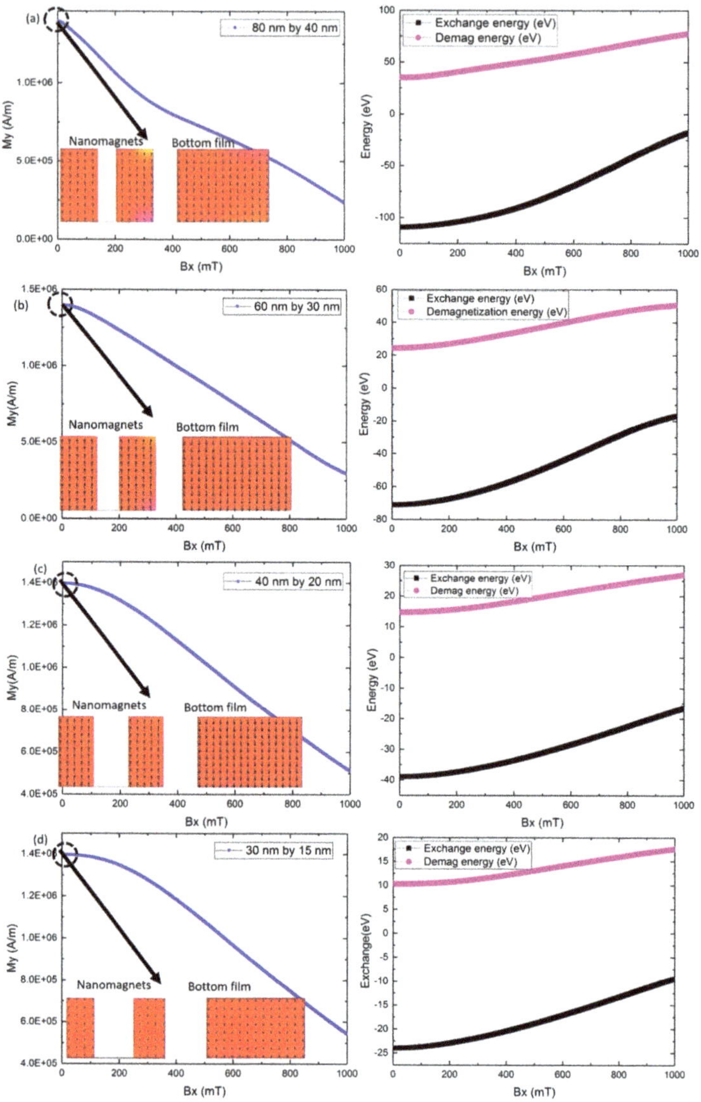

Figure 9.44 Magnetization evolution and energy landscape of the nanomagnets with different footprints: (a) 80 × 40 nm^2, (b) 60 × 30 nm^2, (c) 40 × 20 nm^2, and (d) 30 × 15 nm^2. Magnetizations of the nanomagnets and the bottom layers at their ground state are also shown. The pink curve and the black curve correspond to change in demagnetization and exchange energy, respectively [89].

9.6.8 Effect of Exchange Coupling on a Line of Nanomagnets

In conventional NML, correct magnetic ordering strongly depends on the interaction energy or coupling energy between neighboring nanomagnets [87]. From the simulations in Section 9.6.3, it can be seen that EC results in at least four times larger coupling energy than dipole coupling energy. When one fixed-magnetization magnet switches a single neighboring nanomagnet, this leads to a huge reduction of thermally induced errors. For example, in a dipole-coupled case, a 10 kT energy barrier yields an error rate of $\sim e^{-10} = 5 \times 10^{-5}$, while in the exchange-coupled case, the energy barrier is four times higher, leading to an error rate of $\sim e^{-40} = 4 \times 10^{-18}$. This is an extremely small error rate compared to the dipole-coupled case.

In an NML magnet chain, long wires residing in a single clocking zone are most susceptible to thermally induced errors, as magnets far away from the input or driver magnet feel less influence from the driver. Calculations [15] show that a 10 kT barrier would allow correct ordering for a chain of four to five nanomagnets only. Although it is hard to predict without actual simulations if stronger coupling will cure the architectural weakness of long NML chains, reliable ordering for 8–10 nanomagnets should be possible.

9.7 Summary

In this chapter, we discussed how exchange interactions can be used within and between nanomagnets to increase the robustness of NML. First, a novel coupling scheme called "synchronous nanomagnet logic" was described, wherein alternating magnets were biased using an exchange-coupled layer in each magnet so that clocking in a particular direction nulled only every second magnet. Combined with shape-induced asymmetry, a two-magnet pipeline is created, and only nearest neighbors couple.

Next, we introduced how exchange interactions between two laterally adjacent nanomagnets could also increase NML reliability. Micromagnetic energy calculations show that it is possible to achieve significantly stronger magnetic coupling between two nanomagnets, if they both are antiferromagnetically exchange-coupled to a common bottom layer.

In this chapter, we discussed two approaches for improving NML technology, but we did not discuss a comparison of NML with other potential beyond-Moore technologies since such comparisons have been performed before and the work presented here does not fundamentally change the basic NML approach. Specifically, the Semiconductor Research Corporation (SRC)-funded Nanoelectronics Research Initiative (NRI) sponsored several rounds of benchmarking of emerging device and circuit technologies, and the first such attempt was led by Kerry Bernstein from IBM [2], with subsequent efforts by Dmitri Nikonov and Ian Young from Intel [90, 91]. Basically, this benchmarking showed that NML, together with other magnetoelectric emerging device technologies, showed promise due to very low power (nonvolatility) but faced challenges due to a relatively slow speed (magnetic resonance) when compared with conventional CMOS technology. These findings are in agreement with our own work, which showed that the main advantage of NML is in its nonvolatility and performance gains can be derived from architectures using parallel processing, such as systolic arrays [92].

Both of these new variations in NML improve the promise of NML as a potential beyond-Moore technology that could lower power consumption, increase computational speed through pipelining, and increase density through the fabrication of multilayer magnetic systems. Further investigation is required to understand the switching behavior and energetics of these coupling schemes and offer the possibility of continued size scaling of nanomagnets. Nevertheless, room-temperature, low-energy computation through nanomagnetics may become an important alternative to CMOS.

References

1. Moore, G. E., and Fellow, L. (1998). Cramming more components onto integrated circuits, *Proc. IEEE*, **86**, pp. 82–85.
2. Bernstein, K., Cavin, R. K., Porod, W., Seabaugh, A., and Welser, J. (2010). Device and architecture outlook for beyond CMOS switches, *Proc. IEEE*, **98**, pp. 2169–2184.
3. Lent, C. S., Tougaw, P. D., Porod, W., and Bernstein, G. H. (1993). Quantum cellular automata, *Nanotechnology*, **4**, pp. 49–57.
4. Lent, C. S., and Tougaw, P. D. (1993). Lines of interacting quantum-dot cells: a binary wire, *J. Appl. Phys.*, **74**, p. 6227.
5. Lent, C. S., Tougaw, P. D., and Porod, W. (1993). Bistable saturation in coupled quantum dots for quantum cellular automata, *Appl. Phys. Lett.*, **62**, p. 714.
6. Tougaw, P. D., and Lent, C. S. (1994). Logical devices implemented using quantum cellular automata, *J. Appl. Phys.*, **75**, p. 1818.
7. Orlov, A., Snider, G., Lent, C., Bernstein, G. H., and Amlani, I. (1998). Demonstration of functional quantum-dot cellular automata cell, *J. Vac. Sci. Technol. B*, **16**, p. 3795.
8. Imre, A. (2005). Experimental study of nanomagnets for magnetic quantum-dot cellular automata (MQCA) logic applications, PhD Dissertation, University of Notre Dame.
9. Orlov, A. O., Alamani, I., Bernstein, G. H., Lent, C. S., and Snider, G. L. (1997). Realization of a functional cell for quantum-dot cellular automata, *Science*, **277**, pp. 928–930.
10. Gardelis, S., Smith, C., Cooper, J., Ritchie, D., Linfield, E., and Jin, Y. (2003). Evidence for transfer of polarization in a quantum dot cellular automata cell consisting of semiconductor quantum dots, *Phys. Rev. B*, **67**, p. 33302.
11. Bernstein, G. H., Bazan, G., Chen, M., Lent, C. S., Merz, J., Orlov, A. O., Porod, W., Snider, G. L., and Tougaw, P. D. (1996). Practical issues in the realization of quantum-dot cellular automata, *Superlattices Microstruct.*, **20**, pp. 447–459.
12. Bazaìn, G. (1996). Charge detector realization for AlGaAs/GaAs quantum-dot cellular automata, *J. Vac. Sci. Technol. B*, **14**, p. 4046.
13. Cowburn, R. P., Koltsov, D. K., Adeyeye, A. O., and Welland, M. E. (1999). Single-domain circular nanomagnets, *Phys. Rev. Lett.*, **83**, pp. 1042–1045.

14. Csaba, G. (2003). Computing with field coupled nanomagnets, PhD Dissertation, University of Notre Dame.
15. Csaba, G., and Porod, W. (2010). Behavior of nanomagnet logic in the presence of thermal noise, in *2010 14th International Workshop on Computational Electronics*, pp. 1–4.
16. Wu, A. Q., Kubota, Y., Klemmer, T., Rausch, T., Peng, C., Peng, Y., Karns, D., Zhu, X., Ding, Y., Chang, E., Zhao, Y., Zhou, H., Gao, K., Thiele, J., Seigler, M., Ju, G., and Gage, E. (2013). HAMR areal density demonstration of 1+ tbpsi on spinstand, *IEEE Trans. Magn.*, **49**, pp. 779–782.
17. Orlov, A., Imre, A., Csaba, G., Ji, L., Porod, W., and Bernstein, G. H. (2008). Magnetic quantum-dot cellular automata: recent developments and prospects, *J. Nanoelectron. Optoelectron.*, **3**, pp. 55–68.
18. Siddiq, M. (2014). Clock line and field coupled input for nanomagnet logic, PhD Dissertation, University of Notre Dame.
19. Varga, E. (2013). Experimental study of novel nanomagnet logic devices, PhD Dissertation, University of Notre Dame.
20. Imre, A., Csaba, G., Ji, L., Orlov, A., Bernstein, G. H., and Porod, W. (2006). Majority logic gate for magnetic quantum-dot cellular automata, *Science*, **80**, pp. 1–4.
21. Varga, E., Orlov, A., Niemier, M., Hu, X., Bernstein, G. H., and Porod, W. (2010). Experimental demonstartion of fanout for nanomagnet logic, *IEEE Trans. Nanotechnol.*, **9**, pp. 668–670.
22. Niemier, M., Varga, E., Bernstein, G. H., Porod, W., Alam, M. T., Dingler, A., Orlov, A., and Hu, X. (2010). Boolean logic through shape-engineered magnetic dots with slanted edges, *IEEE Trans. Nanotechnol.*, pp. 1–10.
23. Varga, E., Niemier, M., Csaba, G., Bernstein, G. H., and Porod, W. (2013). Experimental realization of a nanomagnet full adder using slanted-edge magnets, *IEEE Trans. Magn.*, **49**, pp. 4452–4455.
24. Varga, E., Niemier, M., Bernstein, G. H., Porod, W., and Hu, X. S. (2010). Programmable nanomagnet-logic majority gate, in *Device Research Conference*, pp. 85–86.
25. Li, P., Csaba, G., Sankar, V. K., Ju, X., Lugli, P., Hu, X., Niemier, M., Porod, W., and Bernstein, G. H. (2012). Switching behavior of lithographically fabricated nanomagnets for logic applications, *J. Appl. Phys.*, **111**, pp. 07B911-1–07B911-3.
26. Dey, H., Csaba, G., Hu, X., Niemier, M., Bernstein, G. H., and Porod, W. (2013). Switching behavior of sharply pointed nanomagnets for logic applications, *IEEE Trans. Magn.*, **49**, pp. 3549–3552.

27. Niemier, M. T., Bernstein, G. H., Csaba, G., Dingler, A., Hu, X., Kurtz, S., Liu, S., Nahas, J., Porod, W., Siddiq, M., and Varga, E. (2011). Nanomagnet logic: progress toward system-level integration, *J. Phys. Condens. Matter*, **23**, p. 493202.
28. Carlton, D. B., Emley, N. C., Tuchfeld, E., and Bokor, J. (2008). Simulation studies of nanomagnet-based logic architecture, *Nano Lett.*, **8**, pp. 4173–4178.
29. Sharrock, M. P. (1994). Time dependence of switching fields in magnetic recording media (invited), *J. Appl. Phys.*, **76**, p. 6413.
30. Niemier, M., Alam, M. T., Hu, X., Bernstein, G. H., Porod, W., Putney, M., and DeAngelis, J. (2007). Clocking structures and power analysis for nanomagnet-based logic devices, in *Proceedings of the 2007 International Symposium on Low Power Electronics and Design (ISLPED)*, pp. 26–31.
31. Carlton, D., Lambson, B., Scholl, A., Young, A., Ashby, P., Dhuey, S., and Bokor, J. (2012). Investigation of defects and errors in nanomagnetic logic circuits, *IEEE Trans. Nanotechnol.*, **11**, pp. 760–762.
32. Lambson, B., Gu, Z., Carlton, D., Dhuey, S., Scholl, A., Doran, A., Young, A., and Bokor, J. (2012). Cascade-like signal propagation in chains of concave nanomagnets, *Appl. Phys. Lett.*, **100**, p. 152406.
33. Kurtz, S., Varga, E, Siddiq, M., Niemier, M., Porod, W, Hu, X., and Bernstein, G. H. (2011). Non-majority magnetic logic gates: a review of experiments and future prospects for 'shape-based' logic, *J. Phys. Condens. Matter*, **23**, p. 53202.
34. Alam, M. T., Siddiq, M., Bernstein, G. H., Niemier, M., Porod, W., and Hu, X. (2010). On-chip clocking for nanomagnet logic devices, *IEEE Trans. Nanotechnol.*, **9**(3), pp. 348–351.
35. Alam, M. T., Kurtz, S., Siddiq, M., Niemier, M., Bernstein, G. H., Hu, X., and Porod, W. (2012). On-chip clocking of nanomagnet logic, *IEEE Trans. Nanotechnol.*, **11**, pp. 273–286.
36. Csaba, G., Lugli, P., Csurgay, A., and Porod, W. (2005). Simulation of power gain and dissipation in field-coupled nanomagnets, *J. Comput. Electron.*, **4**, pp. 105–110.
37. Alam, M. T. (2010). Design, fabrication and modeling of clocked nanomagnet logic circuit elements, PhD Dissertation, University of Notre Dame.
38. Dingler, A., Garrison, M., Hu, X. S., Niemier, M., and Alam, M. T. (2009). System-level energy and performance projections for nanomagnet-

based logic, in *IEEE/ACM International Symposium on Nanoscale Architectures (NANOARCH)*, pp. 21–26.

39. Lambson, B., Gu, Z., Bokor, J., Carlton, D., Dhuey, S., and Scholl, A. (2012). Error immunity techniques for nanomagnetic logic, in *International Electron Devices Meeting*, pp. 267–270.
40. Butler, K. C., Bernstein, G. H., Csaba, G., Porod, W., Hu, X., and Niemier, M. (2014). Contiguous clock lines for pipelined nanomagnet logic, *J. Comput. Electron.*, **13**, pp. 763–768.
41. Gu, Z., Nowakowski, M., Carlton, D., Storz, R., Hong, J., Lambson, B., Bennett, P., Alam, M. T., Marcus, M., Young, A., Scholl, A., and Bokor, J. (2014). Speed and reliability of nanomagnetic logic technology, *Condens. Matter Mesoscale Nanoscale Phys.*, arXiv:1403.6490 [cond-mat.mes-hall].
42. Shah, F. (2015). Exchange-biased magnetic multilayers for nanomagnet logic applications, PhD Dissertation, University of Notre Dame.
43. Parkin, S. (1991). Systematic variation of the strength and oscillation period of indirect magnetic exchange coupling through the 3d, 4d, and 5d transition metals, *Phys. Rev. Lett.*, **67**, pp. 3598–3601.
44. Dey, H., Csaba, G., Bernstein, G. H., and Porod, W. (2016). Exchange coupling between laterally adjacent nanomagnets, *Nanotechnology*, **27**, p. 395202.
45. Nogués, J., and Schuller, I. K. (1999). Exchange bias, *J. Magn. Magn. Mater.*, **192**, pp. 203–232.
46. Prejbeanu, I. L., Bandiera, S., Alvarez-Herault, J., Sousa, R. C., Dieny, B., and Nozieres, J. P. (2013). Thermally assisted MRAMs: ultimate scalability and logic functionalities, *J. Phys. D: Appl. Phys.*, **46**, p. 74002.
47. Guo, Z., Wu, Y., Qiu, J., Zong, B., and Han, G. (2008). Exchange bias and magnetotransport properties in IrMn/NiFe/FeMn structures, *Phys. Rev. B*, **78**, p. 184413.
48. Shah, F., Csaba, G., Butler, K., and Bernstein, G. H. (2013). Closely spaced nanomagnets by dual e-beam exposure for low-energy nanomagnet logic, *J. Appl. Phys.*, **113**, p. 17B904.
49. Beaujour, J. M. L., Chen, W., Kent, A. D., and Sun, J. Z. (2006). Ferromagnetic resonance study of polycrystalline cobalt ultrathin films, *J. Appl. Phys.*, **99**, p. 08N503.
50. OOMMF software package, Available at: http://math.nist.gov/oommf/.
51. Spedalieri, F. M., Jacob, A. P., Nikonov, D. E., Member, S., and Roychowdhury, V. P. (2011). Performance of magnetic quantum cellular automata

and limitations due to thermal noise, *IEEE Trans. Nanotechnol.*, **10**, pp. 537–546.

52. Becherer, M., Csaba, G., Porod, W., Emling, R., Lugli, P., and Schmitt-Landsiedel, D. (2008). Magnetic ordering of focused-ion-beam structured cobalt-platinum dots for field-coupled computing, *IEEE Trans. Nanotechnol.*, **7**, pp. 316–320.
53. Ju, X., Wartenburg, S., Rezgani, J., Becherer, Kiermaier, J., Breitkreutz, S., Schmitt-Landsiedel, D., Porod, W., and Lugli, P. (2012). Nanomagnet logic from partially irradiated Co / Pt nanomagnets, **11**, pp. 97–104.
54. Breitkreutz, S., Kiermaier, J., Eichwald, I., Ju, X., Csaba, G., Schmitt-Landsiedel, D., and Becherer, M. (2012). Majority gate for nanomagnetic logic with perpendicular magnetic anisotropy, *IEEE Trans. Magn.*, **48**, pp. 4336–4339.
55. Bernstein, G. H., Imre, A., Metlushko, V., Orlov, A., Zhou, L., Ji, L., Csaba, G., and Porod, W. (2005). Magnetic QCA systems, *Microelectron. J.*, **36**, pp. 619–624.
56. Skumryev, V., Stoyanov, S., Zhang, Y., Hadjipanayis, G., Givord, D., and Noguyes, J. (2003). Beating the superparamagnetic limit with exchange bias, *Nature*, **423**, pp. 2–6.
57. Lavrijsen, R., Lee, J., Fernandez-Pacheco, A., Petit, D., Mansell, R., and Cowburn, R. P. (2013). Magnetic ratchet for three-dimensional spintronic memory and logic, *Nature*, **493**, pp. 647–650.
58. Parkin, S., Bhadra, R., and Roche, K. (1991). Oscillatory magnetic exchange coupling through thin copper layers, *Phys. Rev. Lett.*, **66**, pp. 2152–2155.
59. Das, J., Scott, K., Rajaram, S., Burgett, D., and Bhanja, S. (2015). MRAM PUF: a novel geometry based magnetic PUF with integrated CMOS, *IEEE Trans. Nanotechnol.*, **14**, pp. 436–443.
60. Parkin, S., Jiang, X., Kaiser, C., Panchula, A., Roche, K., and Samant, M. (2003). Magnetically engineered spintronic sensors and memory, *Proc. IEEE*, **91**, pp. 661–679.
61. Pham, H., Cimpoesu, D., Plamadă, A.-V., Stancu, A., and Spinu, L. (2009). Dynamic critical curve of a synthetic antiferromagnet, *Appl. Phys. Lett.*, **95**, p. 222513.
62. Piramanayagam, S. N. Perpendicular recording media for hard disk drives, *J. Appl. Phys.*, **102**, pp. 1–22.
63. Bhanja, S., Karunaratne, D. K., Panchumarthy, R., Rajaram, S., and Sarkar, S. (2016). Non-Boolean computing with nanomagnets for computer vision applications, *Nat. Nanotechnol.*, **11**, pp. 177–183.

64. Nikonov, D. E., Member, S., Bourianoff, G. I., and Ghani, T. (2011). Proposal of a spin torque majority gate logic, *IEEE Electron Device Lett.*, **32**, pp. 1128–1130.
65. Mathon, J. (1991). Theory of magnetic multilayers. Exchange interactions and transport properties, *J. Magn. Magn. Mater.*, **100**, pp. 527–543.
66. Dennis, C. L., Borges, R. P., Buda, L. D., Ebels, U., Gregg, J. F., Hehn, M., Jouguelet, E., Ounadjela, K., Petej, I., Prejbeanu, I. L., and Thornton, M. J. (2002). The defining length scales of mesomagnetism: a review, *J. Phys. Condens. Matter*, **14**, pp. R1175–R1262.
67. Ruderman, M. A., and Kittel, C. (1954). Indirect exchange coupling of nuclear magnetic moments by conduction electrons, *Phys. Rev.*, **96**, pp. 72–75.
68. Kasuya, T. (1956). Electrical resistance of ferromagnetic metals, *Prog. Theor. Phys.*, **16**, pp. 58–63.
69. Yosida, K. (1957). Magnetic properties of Cu-Mn alloys, *Phys. Rev.* **106**, pp. 893–898.
70. Coey, J. M. D. (2010). *Magnetism and Magnetic Materials*, Cambridge University Press, Cambridge.
71. Bruno, P. (2002). In *Magnetism: Molecules to Materials III* (Miller, J. S., and Drillon, M., eds.), Wiley-VCH Verlag GmbH, Weinheim, pp. 329–353.
72. Majkrzak, C., Cable, J. W., Kwo, J., Hong, M., McWhan, D. B., Yafet, Y., and Waszczak, J. V. (1986). Observation of a magnetic antiphase domain structure with long-range order in a synthetic Gd-Y superlattice, *Phys. Rev. Lett.*, **56**, pp. 2700–2703.
73. Grunberg, P. (1986). Layered magnetic structures: evidence for antiferromagnetic coupling of Fe layers across Cr interlayers, *Phys. Rev. Lett.* **57**, pp. 2442–2445.
74. Heinrich, B. (1990). Ferromagnetic and antiferromagnetic exchange coupling in bcc epitaxial ultrathin Fe (001)/Cu (001)/ Fe (001) trilayers, *Phys. Rev. Lett.*, **64**, pp. 673–677.
75. Parkin, S., More, N., and Roche, K. (1990). Oscillations in exchange coupling and magnetoresistance in metallic superlattice structures: Co/Ru, Co/Cr, and Fe/Cr, *Phys. Rev. Lett.*, **64**, pp. 2304–2307.
76. Dey, H., Csaba, G., Bernstein, G. H., and Porod, W. (2017). Study of switching behavior of exchange-coupled nanomagnets using transverse magnetization metrology, *AIP Adv.*, **7**, p. 056321.
77. Sbiaa, R., Piramanayagam, S. N., and Law, R. (2009). Magnetization reorientation in antiferromagnetically coupled Co films and (Co/Pd) multilayers, *Appl. Phys. Lett.*, **95**, p. 242502.

78. Kirk, T., Hellwig, O., and Fullerton, E. (2002). Coercivity mechanisms in positive exchange-biased Co films and Co/Pt multilayers, *Phys. Rev. B*, **65**, p. 224426.
79. Bernstein, G. H., and Hill, D. A. (1992). On the attainment of optimum developer parameters for PMMA resist, *Superlattices Microstruct.*, **11**(2), pp. 237–240.
80. Li, P. (2014). Enhanced permeability dielectrics in nanomagnetic logic, PhD Dissertation, University of Notre Dame.
81. Margulies, D. T., Schabes, M. E., McChesney, W., and Fullerton, E. E. (2002). Interlayer coupling and magnetic reversal of antiferromagnetically coupled media, *Appl. Phys. Lett.*, **80**, p. 91.
82. Fullerton, E. E., Margulies, D. T., Schabes, M. E., Carey, M., Gurney, B., Moser, A., Best, M., Zeltzer, G., Rubin, K., Rosen, H., and Doerner, M. (2000). Antiferromagnetically coupled magnetic media layers for thermally stable high-density recording, *Appl. Phys. Lett.*, **77**, p. 3806.
83. Kurebayashi, H., Sinova, J., Fang, D., Irvine, A. C., Skinner, T. D., Wunderlich, J., Novak, V., Campion, R. V., Gallagher, B. L., Vehstedt, E. K., Zarbo, L. P., Vyborny, K., Ferguson, A. J., and Jungwirth, T. (2014). An antidamping spin-orbit torque originating from the Berry curvature, *Nat. Nanotechnol.*, **9**, pp. 211–217.
84. Hellwig, O., Berger, A., Kortright, J. B., and Fullerton, E. E. (2007). Domain structure and magnetization reversal of antiferromagnetically coupled perpendicular anisotropy films, *J. Magn. Magn. Mater.*, **319**, pp. 13–55.
85. Bruker (2016). MESP magnetic force microscopy tips by Bruker, Available at: http://www.brukerafmprobes.com/p-3309-mesp.aspx. (Accessed: 1st January 2016).
86. Zoll, S., Dinia, A., Jay, J., Meny, C., Pan, G., Michel, A., El Chahal, L., Pierron-Bohnes, V., Panissod, P., and Van der Berg, H. (1998). Influence of the growth technique on the coupling and magnetoresistance of Co/Ru sandwiches, *Phys. Rev. B*, **57**, pp. 4842–4848.
87. Jung, J. H., Lim, S. H., and Lee, S. R. (2012). Interlayer exchange coupling between [Pd/Co] multilayers and CoFeB/MgO layers with perpendicular magnetic anisotropy, *Appl. Phys. Lett.*, **101**, p. 242403.
88. Dey, H., Csaba, G., Shah, F., Bernstein, G. H., and Porod, W. (2016). Shape-dependent switching behavior of exchange-coupled nanomagnets, **52**, pp. 4–8.
89. Dey, H. (2017). Experimental study of exchange coupled nanomagnets, PhD Dissertation, University of Notre Dame.

90. Nikonov, D. E., and Young, I. A. (2013). Overview of beyond-CMOS devices and a uniform methodology for their benchmarking, *Proc. IEEE*, **101**, pp. 2498–2533.

91. Nikonov, D. E., and Young, I. A. (2015). Benchmarking of beyond-CMOS exploratory devices for logic integrated circuits. *IEEE J. Explor. Solid-State Computat. Devices Circuits*, **1**, pp. 3–11.

92. Ju, X., Niemier, M., Becherar, M., Porod, W., Lugli, P., and Csaba, G. (2013). Systolic pattern matching hardware with out-of-plane nanomagnet logic devices, *IEEE Trans. Nanotechnol.*, **12**, pp. 399–407.

Index

For Product Safety Concerns and Information please contact our EU
representative GPSR@taylorandfrancis.com
Taylor & Francis Verlag GmbH, Kaufingerstraße 24, 80331 München, Germany

www.ingramcontent.com/pod-product-compliance
Ingram Content Group UK Ltd.
Pitfield, Milton Keynes, MK11 3LW, UK
UKHW022159190926
475702UK00003B/116
9789814800112